Future of Education in Asia

Ambigapathy Pandian · Vighnarajah ·
Weng Marc Lim · Huey Fen Cheong
Editors

Future of Education in Asia

Sustainability, Technology and Internationalisation

Springer

Editors
Ambigapathy Pandian
Centre of Knowledge Advancement
and Languages (CeKAL)
i-CATS
Kuching, Sarawak, Malaysia

Weng Marc Lim
Sunway University
Subang Jaya, Selangor, Malaysia

Vighnarajah
Open University Malaysia
Petaling Jaya, Selangor, Malaysia

Huey Fen Cheong
Universiti Malaya
Kuala Lumpur, Malaysia

ISBN 978-3-032-07284-9 ISBN 978-3-032-07285-6 (eBook)
https://doi.org/10.1007/978-3-032-07285-6

This Springer imprint is published by the registered company Springer Nature Switzerland AG
The registered company address is: Gewerbestrasse 11, 6330 Cham, Switzerland

Foreword

In an era marked by rapid technological advancements, global interconnectedness, and pressing environmental challenges, the future of education in Asia is undergoing a profound transformation. The education systems across the region must not only adapt to these shifts but also lead the way in fostering sustainable, inclusive, and future-ready learning environments. *Future of Education in Asia: Sustainability, Technology, and Internationalisation* is a timely exploration of these critical themes, offering insights and strategies for shaping the educational landscape of tomorrow.

This volume is thoughtfully structured into three thematic sections, each addressing a pivotal aspect that will influence the trajectory of education in Asia for years to come. The first section, *Fostering Sustainable Education*, emphasises the importance of creating educational systems that are both resilient, equitable and environmentally conscious. From the role of digital and hybrid mobility programmes in promoting global competence to inclusive practices across diverse classrooms, the chapters in this section examine how sustainability can be integrated into educational strategies to benefit all students.

The second section, *Technological Trends in Education*, explores the transformative role of emerging technologies in reshaping teaching and learning. As artificial intelligence (AI) and other digital innovations continue to evolve, these chapters offer valuable perspectives on how technology can enhance educational outcomes, improve accessibility, and equip students with the skills necessary to thrive in a rapidly changing world. The focus is on how educational institutions can leverage technological advances to meet the demands of the twenty-first century.

The third section, *Internationalisation of Education*, highlights the importance of cross-border collaboration and regional cooperation. This section addresses the current state of internationalisation in higher education, focusing on how Asia's educational systems can become more globally competitive while fostering stronger academic exchanges and international partnerships. Through an in-depth exploration of various internationalisation models, this section provides valuable insights into how the region can build a more integrated and dynamic educational ecosystem.

Throughout the chapters, the research is organised into thematic clusters that align with the current and future challenges facing education in Asia. These thematic clusters drawing on the works of Ferhnani (2019), such as *Past & Futures*, *Humanity at the Limen*, *Environmental Futures*, *Post-Normality & Complexity*, and *Technological Trends*, serve as the foundation for exploring the evolving nature of education and its intersection with sustainability, technology, and internationalisation. The thoughtful analyses in these chapters contribute to the ongoing conversation about how Asia can build educational systems that are both innovative and inclusive, addressing the needs of a rapidly changing world.

This collection serves as an essential resource for educators, policymakers, and stakeholders invested in advancing education across Asia. It provides a comprehensive framework for advancing education in the region, ensuring that future generations are equipped with the knowledge, skills, and values necessary to thrive in a globalised, technology-driven world. *Future of Education in Asia: Sustainability, Technology, and Internationalisation* is not just a reflection of the present but a roadmap for the future, guiding us towards a more inclusive, innovative, and sustainable future in education.

Professor Datuk Dr. Mohd Tajudin Md Ninggal

Titles on the *Futures of Asia*

Future of Business in Asia: Exploring Pathways to Tomorrow's Business Landscape
Future of Cities in Asia: Future Direction of Urban Planning, Technology and Environmental Sustainability
Future of Economy in Asia: Geopolitical Diplomacy, Technological Revolutions and Sustainable Pathways
Future of Education in Asia: Sustainability, Technology and Internationalisation
Future of Healthcare in Asia: Advancing Equity, Ethics and Technology in Asian Healthcare
Future of Media in Asia: Artificial Intelligence, Digital Technology and Media Practice
Future of Work in Asia: Transforming Work Models, Workforce Dynamics and Workplace Technology

Acknowledgements

The completion of a book project such as this would not have been possible without the vision and support of many people. Our sincere gratitude goes to the Springer editorial and production team, Ms. Doris Bleier, Ms. Saranya Kalidoss, Ms. Catalina Sava, and Mr. Madanagopal Deenadayalan, for their professionalism, guidance, and encouragement throughout the critical stages of the publication process. Their support has been invaluable in bringing this volume to completion.

We also wish to thank the reviewers for their constructive feedback, as well as the contributing authors for their dedication, scholarship, and patience during the review and production stages. The support of families, colleagues, and friends is likewise gratefully acknowledged.

Special acknowledgment is made to Mrs. Selvarani for her unwavering encouragement, determination, and resilience, which continue to serve as a source of inspiration. Appreciation is also extended to Ms. Amutha from Reach Advertising for her valuable assistance in editing and managing the chapters, which greatly facilitated the final preparation of this book.

List of Reviewers

1. Assoc Prof Dr Jamilah binti Ahmad, Faculty of Educational Sciences and Technology, Universiti Teknologi Malaysia, Johor Bahru, Malaysia.
2. Dr Anita Jimmie, School of Pre-U and Continuing Education, Curtin University, Sarawak Campus, Malaysia.
3. Mr Amin Sarman, Faculty of Education, UiTM Puncak Alam, Malaysia.
4. Dr Esther Tong, College of Professional and Continuing Education, The Hong Kong Polytechnic University, Hong Kong.
5. Ms Adibah binti Azman, Faculty of Social Sciences and Humanities, Open University Malaysia, Malaysia.
6. Dr Haida Umiera Binti Hashim, Academy of Language Studies, UiTM Shah Alam, Malaysia.

7. Associate Professor Dr Lara Promnitz-Hayashi, Faculty of International Studies, Keiai University, Japan.
8. Prof. Dr. Vishalache A/P Balakrishnan, Faculty of Education, Universiti Malaya.
9. Ms Sheela Faizura Nik Fauzi, Academy of Language Studies, UiTM Samarahan, Sarawak, Malaysia.
10. Dr. Noel Jimbai Anak Balang, Selangau District Education Office, Sarawak, Malaysia.
11. Ms Wong Woei Ling, Kuching District Education Office, Sarawak, Malaysia.
12. Assistant Professor Dr Ku Faridah Binti Ku Ibrahim, Faculty of Social Sciences and Liberal Arts, UCSI University, Malaysia.
13. Ms Sujatha Krishnan, School of Pre-U and Continuing Education, Curtin University, Sarawak Campus, Malaysia.
14. Professor Dr Neeta Inamdar, Faculty of Humanities, Liberal Arts and Social Sciences, Manipal Academy of Higher Education (MAHE), India.

Contents

Editors' Note

The clusters of future studies in each chapter are based on Fergnani (2019, 2020), who identified six (6) key research/thematic clusters from reviews of future studies since 1968: "corporate foresight; past and futures; humanity at the limen; environmental futures; post-normality and complexity, and technological trends". While the chapter topics present different aspects of work in Asia, the clusters of future studies show different dimensions of futures and future studies.

References

Fergnani, A. (2019). Mapping futures studies scholarship from 1968 to present: A bibliometric review of thematic clusters, research trends, and research gaps. *Futures, 105*: 104–23.

Fergnani, A. (2020). Mapping fifty years of futures studies scholarship (1968–2017). In R. Slaughter & A. Hines (Eds.), *The Knowledge Base of Futures Studies 2020* (pp. 24–47). US: Association of Professional Futurists and Foresight International.

Chapter 1
Introduction: Unveiling the Future Landscape of Education in Asia

Vighnarajah and Ambigapathy Pandian

Abstract This chapter serves as an introduction to the book, laying the foundation for understanding how education in Asia is being shaped by key drivers such as sustainability, technology and internationalisation. This chapter emphasises the urgency of adapting educational systems to address the evolving demands of a rapidly changing global landscape. Drawing on Fergnani(Futures 105: 104–123, 2019; Mapping fifty years of futures studies scholarship (1968–2017). Association of Professional Futurists and Foresight International, US, pp. 24–47, 2020) research clusters, the chapter addresses the increasing importance of sustainability and internationalisation in helping to shape policies that foster global collaboration and knowledge exchange. Further, the chapter outlines the future directions for educational systems in Asia, presenting actionable guidelines for policymakers to create inclusive, equitable and sustainable educational frameworks. The chapter sets the stage for the ensuing discussions throughout the book, offering valuable insights for educators, students and stakeholders on how to prepare for the future of education while navigating the complexities of technological, environmental and international challenges.

Keywords Future of education · Asia · Sustainability · Technology · Internationalisation

1.1 Introduction: About the Book

This book features contributions from educators and researchers who bring deeply rooted, region-specific perspectives to the evolving field of education in Asia. Framing the future through an Asian lens reflects the region's growing influence

Vighnarajah (✉)
Faculty of Education, Open University Malaysia, Petaling Jaya, Malaysia
e-mail: vighnarajah@oum.edu.my

A. Pandian
i-CATS University College, Kuching, Malaysia

A. Pandian et al. (eds.), *Future of Education in Asia*,
https://doi.org/10.1007/978-3-032-07285-6_1

and the widespread view that Asia will be central to shaping global learning landscapes. This outlook is evident not only in scholarly forums like the 25th ASAA Biennial Conference 2024 (*Asia Futures: Studies of, in and with Asia*), but also in policy-level conversations such as the Asia–Pacific Regional Education Minister's Conference (APREMCII) in June 2022. There, ASEAN education ministers began work on the ASEAN Education 2050 Roadmap as part of the region's broader Post-2025 Vision. These developments underscore the rising significance of Asia in both academic and policy dialogues. They also point to increasing interest from institutions and individuals looking to embed Asian perspectives into teaching, research and curriculum design. As such, this book comprises chapters reflecting the diversity of educational thought, practice and innovation across the wider Asian region.

The chapters in this book are grounded in the lived experiences of educators and researchers who are directly engaged with the evolving challenges and possibilities of education across Asia. Their work reflects the region's educational diversity, spanning across rural classrooms, digital innovation hubs, multilingual environments and cross-border academic networks. Asia's learning systems are far from uniform; they are shaped by intersecting histories, cultural traditions and localised policy priorities. The chapters included here are uniquely positioned to examine how their practices are redefining what learning means in their respective contexts and underscoring how educators across Asia are not just adapting to change but actively shaping education's future from within.

Grounded in the perspective that futures studies focus not on predicting what will happen, but on exploring a range of plausible and desirable outcomes, the book draws on Dator's (2019) framework, which highlights the importance of envisioning "alternative futures" alongside the ongoing pursuit of 'preferred futures'. From here, the chapters are organised around six key research clusters identified by Fergnani (2019, 2020): *corporate foresight*, *past and futures*, *humanity at the limen*, *environmental futures*, *post-normality and complexity* and *technological trends*. Each contribution is linked to one or more of these clusters, providing a structured approach to understanding the complex and evolving educational landscape in Asia.

Looking ahead, this book seeks to provide educators, policymakers and stakeholders with the insights and tools necessary to navigate the fast-evolving educational landscape and to foster a more equitable, adaptable and sustainable future for learning across Asia.

1.2 Understanding the Interplay of Sustainability, Technology and Internationalisation

To grasp the concept of interplay in education, we must first understand what interplay entails and how it shapes systemic change. Interplay refers to the mutual influence and dynamic interaction between multiple elements, where each force does not operate in isolation, but continuously reshapes and is reshaped by the others. In the context of this book, we examine the interplay among sustainability, technology and internationalisation, three powerful and intersecting forces that are fundamentally altering educational environments in Asia. While each of these dimensions has been studied extensively on its own, their combined effects present more complex and layered implications for how education is conceived, structured and experienced.

Drawing on this understanding, we explore interplay not simply as co-occurrence, but as a process of interaction that produces new educational possibilities, tensions and trade-offs (Savelyeva, 2022). For example, digital technologies may enable sustainable practices by reducing physical infrastructure needs, yet they also raise concerns about energy consumption and digital inequality. Similarly, internationalisation may foster knowledge exchange, yet its influence can disrupt local cultural and pedagogical practices (Suguku, 2023).

This book approaches these dynamics by treating interplay as both context-dependent and future-oriented, shaped by historical legacies and institutional frameworks, but also by emerging trends and collective aspirations.

In framing our analysis, we align with futures thinking, particularly through the lens of Fergnani's (2019, 2020) six research clusters, which help situate the interplay within broader conceptual landscapes: from technological trends and environmental futures to post-normality and complexity. Each cluster offers a lens through which to understand the layered impacts of this interplay. These clusters allow us to explore how the interaction between sustainability, technology and internationalisation not only reflects but also contributes to evolving notions of educational purpose, equity and innovation. Thus, the book views interplay not as a static theme, but as a methodological and conceptual tool to investigate the future of education in Asia, where change is continuous, forces are interwoven and futures are co-constructed.

The COVID-19 pandemic intensified the urgency of rethinking education in Asia. It disrupted established systems, exposed long-standing inequities and accelerated the adoption of digital learning technologies. As schools shifted online, many learners faced barriers related to digital access, technological infrastructure and socio-economic disparities, issues closely tied to both sustainability and internationalisation. These disruptions highlighted the need for resilient and adaptive educational models that can withstand future crises while remaining inclusive and equitable. At the same time, the pandemic showcased the transformative potential of technology and international collaboration in mitigating learning loss and fostering innovation.

Going forward, the future of education in Asia will depend on how effectively these interwoven forces are balanced, leveraging technology for good, embedding sustainability into policy and practice and fostering global engagement without undermining local relevance. Policymakers must therefore develop strategies that balance these dimensions, ensuring education systems remain inclusive, equitable and resilient amid shifting environmental, social and economic landscapes. For educators, understanding this interplay is critical to adopting new pedagogical approaches that integrate digital tools and foster global competencies, while nurturing awareness of sustainability challenges. Meanwhile, students must be prepared with the skills, adaptability and values necessary to navigate an increasingly complex and interconnected world where sustainability and technology are central to personal and professional success.

1.3 Book Structure and Overview

1.3.1 Section a: Fostering Sustainable Education

The concept of sustainable education has gained significant traction across Asia, aligning with global efforts such as the UN's Sustainable Development Goals (SDGs). Broadly defined, sustainable education aims to equip learners with the knowledge, skills, values and attitudes necessary to contribute to a more just, inclusive and environmentally responsible world. In practice, this includes integrating sustainability themes into curricula, promoting environmental stewardship through institutional practices and cultivating critical thinking and ethical decision-making. Across the region, schools and universities are experimenting with approaches such as green campus initiatives, climate literacy programmes and community-based sustainability projects. However, these efforts vary widely in scale and depth, shaped by local economic resources, governance structures and educational philosophies.

Despite encouraging momentum, the adoption of sustainable education is uneven across Asia. While some countries have embedded sustainability deeply into their national education strategies, others remain at exploratory stages. Factors such as funding limitations, teacher preparedness and competing policy priorities often constrain implementation. Furthermore, cultural perceptions of sustainability, ranging from spiritual stewardship in parts of Southeast Asia to innovation-driven models in East Asia, mean that sustainability is interpreted and enacted in diverse ways. These localised understandings are critical in determining how sustainable education is accepted, practised and sustained. For instance, incorporating sustainability into early childhood education or technical and vocational training may require entirely different approaches than doing so at the tertiary level. A one-size-fits-all model is neither feasible nor effective.

The chapters in this section reflect this complexity, offering research-driven insights into how sustainability intersects with issues such as inclusivity, mobility, early learning and intercultural competence. Collectively, they contribute to a more nuanced understanding of how educational practices in Asia are evolving, not through a single framework, but through interconnected efforts that are sensitive to both global imperatives and local realities.

Chapter 2 explores the potential of international mobility programmes (IMPs) in fostering global competence, while highlighting their limitations due to cost and accessibility. In response, the chapter introduces virtual mobility, emerging during the COVID-19 pandemic, as a digital alternative that expands access through cost-effective means. Since 2023, these have evolved into hybrid models, blending virtual and in-person experiences. Grounded in constructivist learning theory and aligned with the OECD's Global Competence framework, the chapter presents a more inclusive and scalable approach to global learning.

Chapter 3 examines inclusive education practices through a comparative lens, anchored by a case study from Bali, Indonesia, where a school has implemented inclusive strategies since 2012. Drawing on additional examples from China, Thailand and Korea, the chapter identifies key methods such as differentiated instruction, the whole-person approach and constructivist teaching. It evaluates their impact on literacy and social-emotional learning (SEL), offering evidence of meaningful progress despite systemic challenges. The chapter provides actionable insights for educators working towards more inclusive and equitable classroom environments.

Chapter 4 focuses on early childhood education as a foundation for sustainability, with case-based insights from Kagoshima Prefecture, Japan. It explores how elements of Education for Sustainable Development (ESD) are embedded in kindergarten routines and curricula, promoting sustainability-related knowledge, attitudes and behaviours from a young age. Emphasising culturally rooted practices, the chapter highlights their relevance and adaptability across Asian contexts, reinforcing the critical role of early learning in driving long-term societal change.

Chapter 5 investigates how short-term exchange programmes shape intercultural awareness, based on a study of 104 Vietnamese students at Universiti Malaysia Sarawak. The chapter shows that even limited exposure abroad can significantly shift cultural perceptions and foster ethnorelativism, despite challenges like adapting to unfamiliar environments. It positions experiential learning as a catalyst for developing intercultural sensitivity and calls for more structured approaches to intercultural education in a globally connected Asia.

1.3.2 Section B: Technological Trends in Education

The COVID-19 pandemic has accelerated the adoption of digital technologies in education across Asia, prompting a rapid shift from traditional classroom settings to virtual, hybrid and technology-enhanced learning environments. This shift has highlighted both the potential and challenges of integrating emerging technologies such as Artificial Intelligence (AI), Virtual Reality (VR) and online platforms into diverse educational contexts. Post-pandemic, educational institutions are increasingly exploring how these technologies can support more flexible, personalised and inclusive learning experiences while addressing persistent issues related to digital equity and infrastructure (Santos, 2024).

This section examines the evolving role of technology in transforming teaching, learning and assessment, with a focus on how Asian countries are navigating this complex terrain. Drawing on case studies and research from China, Malaysia, South Korea, Thailand, Vietnam and the Philippines, the chapters explore how innovations like AI and generative technologies are reshaping academic practices and learner engagement. The section also investigates the implications of these trends for cultural values, academic integrity and pedagogical strategies. Together, these contributions offer a nuanced understanding of technological integration in education, emphasising the need for context-sensitive approaches that balance innovation with equity and ethical considerations in a post-COVID era.

Chapter 6 introduces the evolving role of Artificial Intelligence (AI) in higher education and academic publishing across China, South Korea and Singapore. It explores how AI is reshaping student engagement, teaching practices and research workflows while highlighting the delicate balance between innovation and academic integrity. By examining regional case studies, the chapter invites readers to reconsider how AI might enhance learning and scholarship without compromising the rich academic traditions of Asian institutions.

Chapter 7 examines the growing influence of Generative AI technologies on education in Malaysia, especially in the post-COVID-19 context. The chapter highlights how AI tools are transforming teaching methodologies, student participation and assessment practices. It addresses the ethical challenges and equity issues arising from AI adoption, offering practical strategies for educators and policymakers to foster inclusive and responsible use of AI in Malaysian classrooms, positioning the country as a regional leader in educational innovation.

Chapter 8 explores the integration of AI, Virtual Reality (VR) and Geographic Information Systems (GIS) within the built environment industry's continuing education across Malaysia, Thailand and Singapore. The chapter demonstrates how these technologies support digital literacy and innovation, while helping balance efficiency, cultural heritage, sustainability and urban development. It offers insights into practical applications that prepare professionals for future challenges in education and city planning within the Asian context.

Chapter 9 investigates translanguaging practices in online mathematics classrooms in the Philippines, focusing on how multiple languages and semiotic resources facilitate or constrain communication and learning in virtual environments. Using interviews, video recordings and student focus groups, the study reveals both the affordances and limitations of technology in supporting collaborative problem-solving and academic engagement. The chapter concludes with recommendations to improve online multilingual learning, particularly in mathematics education.

Chapter 10 evaluates the impact of hybrid learning models in Vietnamese higher education, assessing how the blend of digital and face-to-face instruction affects student learning and autonomy. Drawing on interviews and focus groups with students and faculty, the chapter identifies the benefits of increased flexibility and digital skill development, alongside challenges such as infrastructure gaps and uneven digital literacy. The chapter offers recommendations to scale hybrid learning effectively, emphasising equitable access and curriculum adaptation for resilient education systems across Asia.

1.3.3 Section C: Internationalisation of Education

Internationalisation in education refers to the deliberate integration of international, intercultural and global dimensions into the purposes, functions and delivery of educational systems. It encompasses activities such as student and staff mobility, cross-border academic collaboration and the incorporation of global perspectives into curricula and research (Chasi & Heleta, 2022). Within the Asian context, internationalisation involves balancing global standards with local cultural values, fostering regional cooperation, especially among ASEAN nations and addressing disparities in access and educational quality. This multifaceted process aims to prepare learners to engage effectively in an increasingly interconnected world while promoting academic excellence and mutual understanding.

Post-COVID-19 dynamics and evolving geopolitical considerations further complicate the internationalisation agenda, prompting institutions and governments to rethink strategies for sustainable and equitable international engagement (Mbah et al., 2024). The insights presented here offer a valuable contribution to understanding how Asian higher education systems are navigating this evolving landscape and aligning with broader regional goals such as ASEAN integration.

In the light of this, the section explores how internationalisation is shaping higher education across Asia by examining the diverse approaches, policies and challenges experienced by countries in the region. The chapters highlight the complex interplay between national education policies, global academic rankings, research collaboration and funding mechanisms. They also reveal the varying stages of internationalisation, from emerging initiatives to well-established frameworks, reflecting each country's unique socio-cultural and economic landscape.

Chapter 11 investigates the internationalisation status of Philippine higher education using World System Theory and bibliometric data across six dimensions: Quality Assurance, Research Output and Citation, Collaboration, Funding and University Rankings. The chapter reveals the Philippines' borderline position between the semi-periphery and periphery, highlighting gaps and opportunities to strengthen its international academic presence. The findings have broader implications for ASEAN's regionalisation goals, calling for a reassessment of networks like the ASEAN University Network to better facilitate integration and academic cooperation.

Chapter 12 examines India's approach to internationalisation of higher education through its 2020 National Education Policy (NEP), emphasising the socio-cultural grounding of the policy and its systemic scope. Using qualitative and comparative analysis, the chapter highlights how India's IoHE policies align with, yet differ from, those of China, Japan, Malaysia and Singapore. It unpacks the nuanced interplay between policy intentions, regulatory frameworks and on-the-ground practice, providing a layered understanding of how Asian nations navigate international collaboration and educational reforms within their unique contexts.

1.4 The Way Forward

The future of education in Asia is rapidly evolving, influenced by technological advancements, socio-cultural dynamics and global interconnectivity. Educational institutions, policymakers and stakeholders must foster adaptable and inclusive learning environments that respond to these complex changes. At the same time, learners and educators alike need to develop the competencies and mindsets necessary to thrive in diverse, digitally enhanced and globally connected contexts. Building resilience, cultural sensitivity and digital literacy will be critical as education systems strive to meet the needs of future generations.

The chapters in this volume collectively underscore the importance of innovation, collaboration and equity in shaping the future educational landscape across Asia. Emphasising regional perspectives alongside global trends, these studies highlight both challenges and opportunities that call for responsive policies and practices. Sustainable education, internationalisation and technology integration must be pursued with intentionality and a commitment to inclusivity to ensure that no learner is left behind.

The insights presented here highlight emerging challenges and opportunities within Asian education systems. Collectively, they serve as a valuable resource for educators, policymakers and researchers aiming to foster innovation and resilience. Readers are encouraged to engage with these perspectives to inform practice and inspire ongoing dialogue towards building a more inclusive and adaptable educational future.

Highlights

- The book applies Fergnani's (2019, 2020) Six Research Clusters to explore how sustainability, technology and internationalisation are transforming education in Asia, providing a holistic view of future educational trends.
- Sustainability is reshaping education in Asia, with institutions integrating sustainable development goals (SDGs) and green initiatives into curricula, preparing students for climate change challenges.
- Technology, including AI and online learning platforms, is revolutionising education by enhancing accessibility, flexibility and personalised learning experiences.
- Internationalisation drives cross-border partnerships and research initiatives, expanding global perspectives in curricula and providing students with international mobility opportunities.
- Policy-making in Asia is prioritising the integration of sustainability, technology and internationalisation to adapt educational systems to future global demands, ensuring access and equity for all students.

References

Chasi, S., & Heleta, S. (2022). Towards more sustainable, equitable and just internationalisation practices: The Case of internationalisation conferences. *Journal of Studies in International Education, 27*(4), 603–620. https://doi.org/10.1177/10283153221139924

Dator, J. A. (2019). Jim Dator: A Noticer in Time – Selected Works 1967–2018. Cham: Springer.

Fergnani, A. (2019). Mapping futures studies scholarship from 1968 to present: A bibliometric review of thematic clusters, research trends, and research gaps. *Futures, 105*, 104–123.

Fergnani, A. (2020). Mapping fifty years of futures studies scholarship (1968–2017). In R. Slaughter & A. Hines (Eds.), *The Knowledge Base of Futures Studies 2020* (pp. 24–47). US: Association of Professional Futurists and Foresight International.

Mbah, M. F., Clifton, N., & Kushnir, I. (2024). Internationalising higher education curricula for sustainable development: Considerations for indigeneity and (inter)culturality. *Discover Sustainability, 5*, 307. https://doi.org/10.1007/s43621-024-00520-y

Santos, L. F. (2024). Internationalisation in the digital transformation: A scoping review. *Higher Education Quarterly, 78*(3), 807–824.

Savelyeva, T. (2022). An introduction to the sustainability of tertiary education in Asia: collaborations, transformations, global–local tensions, and future developments. In T. Savelyeva & G. Fang (Eds.), *Sustainable Tertiary Education in Asia.* Springer, Singapore. https://doi.org/10.1007/978-981-19-5104-6_1

Suguku, D. (2023). Pillar of internationalization in higher education: The contribution of international collaborations and online delivery approaches to internationalization in HEIS. In *SHS Web of Conferences* (Vol. 156, p. 05004). EDP Sciences.

Vighnarajah is a Senior Lecturer and currently, the Director of the Master of Education programme at Open University Malaysia. With over 16 years of experience, he has held various roles, including school teacher, university lecturer, teacher trainer and consultant, working with students and educators across all educational levels, from primary to postgraduate. His leadership experience extends to significant management positions, such as Acting Dean, Head of School, Head of Department, Chair of the University Teaching and Learning Committee and Executive Editor for academic journals. Dr. Vighnarajah possesses a Ph.D. in Pedagogy, and his areas of expertise include pedagogy, curriculum instruction and instructional technology. He has also been honoured with multiple teaching awards, Best Paper awards and sponsorships from the Commonwealth of Learning (COL) and CEMCA.

Ambigapathy Pandian is a Professor and currently, the Director of the Centre of Knowledge Advancement and Languages (CeKAL) at i-CATS university college. As a scholar, Professor Ambigapathy's field of research interest includes Language, Literacy education and Teaching English to Speakers of Other Languages. Ambigapathy has conducted workshops for educators in Asia, such as professional development and teaching practices. He is the author and editor of more than 60 Books published by Scholar's Press, Cambridge Scholar Publishing, Oxford Fajar Press, Lang Publishers (New York) and Common Ground Publications (Australia) and many articles featured in prestigious International journals.

References

Chapter 2
Future of Sustainable Education in Asia: Equitable Learning Through Digital and Hybrid Mobility Programmes

Hiroyuki Ishizaki, Maria Anityasari, and Isma Rosila Ismail

Abstract In this borderless age, one mission for educators is to develop students as global talent who can thrive in multicultural settings. International mobility programmes (IMPs) can be an effective way to facilitate the acquisition of Global Competence (GC) through practical experiences. Nowadays, most educational institutions worldwide are encouraging their students to attend these IMPs. In reality, however, such chances tend to be only for a limited number of students, mainly due to budget limitations. Know-how regarding programme design and implementation is another issue for academic and administrative staff. Thus, the challenge for educators is providing equitable learning opportunities by overcoming financial and resource constraints. In 2021, the virtual mobility programme was started as a countermeasure against the COVID-19 pandemic travel restrictions by taking advantage of Digital Transformation (DX). Because of its huge benefits, the programme has been upscaled to a hybrid one since 2023. This chapter elaborates on the details of this innovative programme series, programme design methodologies based on constructivist learning theory, and programme evaluation in conjunction with the OECD Global Competence skillsets and Knowledge, Skills, Attitudes and Values (KASV) framework. We describe insights to solve constraints and share the details to replicate this methodology. The goals here are to share the method for offering IMPs as fair and equal learning opportunities utilising DX and to reveal the findings of outcomes from an interdisciplinary IMP with various out-of-the-classroom learning activities.

H. Ishizaki (✉)
Shibaura Institute of Technology (SIT), Tokyo, Japan
e-mail: ishizaki.hiroyuki.n3@sic.shibaura-it.ac.jp

M. Anityasari
Institut Teknologi Sepuluh Nopember (ITS), Surabaya, Indonesia
e-mail: maria@its.ac.id

I. R. Ismail
Department of Language and Communication, Centre for Fundamental and Continuing Education, Universiti Malaysia Terengganu, Kuala Terengganu, Malaysia
e-mail: ismarosila@umt.edu.my

A. Pandian et al. (eds.), *Future of Education in Asia*,
https://doi.org/10.1007/978-3-032-07285-6_2

Keywords Equitable learning · Collaborative online international learning (COIL) · Interdisciplinary education · International mobility · Global Competence · Engineering education

2.1 Introduction

For many years, Asian nations have endeavoured to enhance quality and accessible education for all. In this globalisation era, where countries become borderless and skilled workers can move anywhere, higher education should better equip university students to face the global job market, fierce competition, and solve unknown future problems. It is not easy to address those pressures, as there is a large discrepancy in higher education quality across Asian nations, and even across cities in individual nations, especially among developing countries. Equity in accessing good quality education is a serious concern. This became more critical when the COVID-19 pandemic hampered the whole world, with severe impacts on education.

This chapter aims to introduce a new international mobility programme (IMP) to provide all Asian students with fair, effective and equal learning opportunities regardless of locational and/or financial issues. The main goal of IMPs is to develop global talent who can actively demonstrate their capabilities in the borderless world. Meanwhile, the problem statement is how to provide young students with their first international experience, especially during the pandemic. Then, the objective is to ignite their curiosity and kickstart their educational journeys by taking advantage of Digital Transformation (DX). One such programme, Virtual Asia Exploration (VAx), was started in 2021 during the pandemic. Led by Shibaura Institute of Technology (SIT) Japan as the main coordinator and by Institut Teknologi Sepuluh Nopember (ITS) Indonesia as the co-chair, this interdisciplinary and multi-ethnic programme is organised by 6 universities from 5 countries (Indonesia, Japan, Malaysia, Thailand and the Philippines). It has been facilitating mutual understanding and long-lasting collaboration not only for participating students but also for academic staff.

In 2023, a hybrid programme (HAx) was conducted. The onsite programme was hosted by ITS in Surabaya, Indonesia for seven (7) days. This programme can cultivate global skills through equitable access to all Asian university students. As a result of four (4) cohort batches, a total of 292 students from more than 10 countries have successfully received a certificate. Figure 2.1 shows an overview of HAx 2023 programme contents.

This chapter shares the concepts, the operational aspects of the programme, the implementation and the participants' learning outcomes as well as the involved lecturers and staff. The authors hope that this programme can be modified and replicated in other contexts. Further analysis related to Knowledge, Skills, Attitudes and Values (KSAV) of the participants has been conducted. Moreover, utilising the Global Competencee (GC) Framework, we try to reveal the effectiveness of a multicultural group learning environment and interaction with local communities in participants' open mindset and intercultural communication.

Fig. 2.1 HAx 2023 programme poster

Because of its favourable responses from participants and their home universities, HAx 2024 was conducted and concluded with another successful results. Thus, ITS is planning to host HAx 2025, too.

2.2 Literature Review

The current era witnessed that the political, economic and social development of a country is significantly influenced by the internationalisation of higher education (Atherton et al., 2020). At the heart of the global landscape of international education and student mobility is ASEAN because it is home to 20 million students worldwide. With increased regional cooperation, ASEAN member states appear to be on the way towards implementing international education and the establishment of a knowledge-based society that is driven by knowledge creation in higher education institutions and student mobility and collaboration with the (Yodpet et al., 2023). Therefore, IMPs are actively implemented around the world and Asia with various expected learning outcomes such as academic enrichment, soft skills practice, personal development and global citizenship (Teichler, 2017).

From a pedagogical perspective, an IMP emphasises active and experiential learning, cultural competence, interdisciplinary approaches and the development of skills essential for global citizenship and success in a diverse and interconnected world (Robertson-Geibel, 2020). As Pacheko (2020) notes, 'Effective education for Global Competence allows students to mobilise and use their knowledge, attitudes, skills and values together while exchanging ideas on a global issue in and outside of school or interacting with people from different cultural backgrounds (for example, engaging in a debate, questioning viewpoints, asking for explanations or identifying directions for deeper exploration and action)'. Thus, many educational institutes and associations are working hard to provide students with more IMP opportunities to expand their future career horizons.

In Asian regions, most of the governments have some policies and initiatives for promoting IMPs (See Table 2.1). In terms of practice, however, it is subject to the capability and resources of each university. Needless to mention, big universities in the urban area tend to have a bigger budget, human resources and locational advantages as compared with the schools located in the local or rural areas.

Intraregional student mobility can strengthen the ASEAN identity and nurture future ASEAN citizens, entrepreneurs and leaders of the ASEAN community (Chao Jr., 2017). In order to increase IMP opportunities for Asian students as equitable education, some institutes are taking their own initiatives. The Asian International Mobility for Students (AIMS) Programme is an initiative to support the IMPs of students and enhance cooperation in higher education among Asian countries. The Asia University Network (AUN), an ASEAN intergovernmental higher education network established in 1995, is also conducting several official initiatives including academic mobility and exchange programmes. Asia Technological University Network (ATU-Net), a consortium of engineering institutes throughout Asia, is actively coordinating collaborative programmes among nearly 40-member institutions (MIs).

Intra-ASEAN inbound/outbound mobility has increased over the 2013 to 2018 period. Still, in 2018 less than 10% of all inbound/outbound mobility was between ASEAN Member States. Inbound students are distributed unevenly across ASEAN.

Table 2.1 Comparison of IMP situations in major Asian countries

Country	Demand	Governmental	Key Initiatives	Challenges
Thailand	High demand due to ASEAN integration and national	Education Reform Policies promoting student exchanges	Exchange programs under ASEAN frameworks (e.g.,	Language barriers, funding limitations, and administrative
Malaysia	Significant demand, supported by national policies like the Education Blueprint to	Malaysia Education Blueprint 2015-2025 enhancing internationalization in higher education,	Outbound programs supported by Malaysia Education Blueprint and scholarships.	Financial constraints, visa issues, and ensuring credit transfer recognition.
Indonesia	Growing demand as students seek global exposure to enhance employability.	DIKTI (Directorate General of Higher Education) promoting partnerships with	LPDP scholarships by the covernment, partnerships with global institutions.	Funding limitations, language barriers, and inconsistent credit recognition.
The Philippines	High demand, especially in healthcare, IT, and business fields for	CHED (Commission on Higher Education) supporting regional mobility through the	The Commission on Higher Education (CHED) support, dual-degree programs,	Limited funding, administrative capacity, and rural student
Vietnam	Increasing interest due to modernization and exposure to	Higher Education Reform Agenda (HERA) increasing participation in	Collaborations with Japan, Europe, and the U.S.; emphasis on STEM and	Language proficiency issues, credit transfer barriers, and funding
Singapore	Strong demand, leveraging Singapore's global education hub	Global Schoolhouse Initiative Positions Singapore as an education hub and	Global research collaborations, high-quality student exchange programs.	Maintaining quality amidst high demand; ensuring equitable access.
Japan	Stable demand, focused on high-quality research and outbound mobility.	Top Global University Project Enhancing international partnerships and promotes study abroad opportunities.	Outbound mobility supported by scholarships and bilateral agreements.	Language barriers and attracting inbound students beyond traditional markets.
Korea	Increasing demand driven by competitive job markets and government initiatives.	Global Korea Scholarship (GKS) Providing Korean students studying abroad and international students in Korea.	Global Korea Scholarship (GKS), joint degree programs with global institutions.	Funding, limited awareness, and adapting programs to international standards.
China	High demand, emphasizing both outbound mobility and inbound attraction programs.	Belt and Road Education Initiative promoting student and faculty exchanges with countries along the Belt and Road route.	Belt and Road Education Initiative, transnational education partnerships.	Visa policies, credit transfer mechanisms, and balancing inbound-outbound flows.

More than 80% study in Thailand, Vietnam and Malaysia, with over 50% in Malaysia alone. Of outbound students leaving ASEAN, over 80% go to Australia, the US, UK and Japan (Atherton et al., 2020). This is an initial objective of organising this HAx programme as a vehicle of facilitating to build confidence of Asian youngsters through the richness of various culture and histories through the understanding of their origin and neighbouring countries.

As a countermeasure for IMP suspensions caused by the COVID-19 pandemic, six (6) universities among the ATU-Net Mobility Special Interest Group (SIG) (see Table 2.2) launched an online programme named "Virtual Asia Exploration (VAx)" in early 2021. Figure 2.2 illustrates the geographic distances among the host institutions.

At that time, MIs were desperate to provide IMPs in order to avoid sacrificing their study opportunities. The content was carefully designed considering programme outcomes and student benefit. For creating a diversity-driven environment in the

Table 2.2 List of six (6) host member institutions (MIs)

Country	Institution name	Acronym
Indonesia	Institut Teknologi Sepuluh Nopember	ITS
Japan	Shibaura Institute of Technology	SIT
Malaysia	Universiti Malaysia Terengganu	UMT
Malaysia	Universiti Malaysia Perlis	UniMAP
Thailand	King Mongkut's University of Technology Thonburi	KMUTT
The Philippines	Cebu Technological University	CTU

Fig. 2.2 Geographic location of 6 host MIs

virtual settings, it was determined that both lecturers and participants should be multi-national. The following features were then embedded:

- An interdisciplinary curriculum to combine engineering and social science delivered by multi-national lecturers' team
- Mingling participants from various countries to create multi-ethnic circumstances
- Many group discussions and presentations for facilitating mutual understanding beyond nationalities
- 2 courses: a Certificate of Completion Course (CCC) equivalent to 2 credits upon completion of all assignments, final presentation and more than 80% attendance, and Certificate of Attendance Course (CAC) with more than 60% attendance and no assignment submissions.

VAx was eventually a perfect match with the Collaborative Online International Learning (COIL) category (Appiah-Kubi & Annan, 2020). In short, an inevitable countermeasure for the pandemic consequentially created a new methodology, COIL, which became a powerful option for equitable education. After the lifting of travel restrictions, VAx was upgraded to Hybrid Asia Exploration (HAx) so that students can choose virtual or onsite options based on their preference and affordability (Bates & Vancouver, 2019).

2.2.1 New Approach from Interdisciplinary Perspectives

The VAx and HAx IMPs are designed with a total of 14 classes; six (6) engineering lectures, another six (6) lectures about the culture and history of five (5) Asian countries and lastly group final presentations. The three (3) VAx cohort batches in 2021 and 2022 were online. Regarding the HAx one (1) in 2023, engineering lectures in June and July were online, and culture and history lectures were in hybrid mode, with onsite programmes at ITS in Surabaya, the second biggest city in Indonesia.

These courses are designed to provide participants with integrated learning experiences with interdisciplinary perspectives (Van den Beemt et al., 2020). The contents mainly consist of three (3) components. The first part is technological subjects since ATU-Net MIs are mostly engineering schools. Six (6) professors from various fields lectured on their technologies and applications to the social issues of their countries.

The second part is social science aspects. This programme intentionally addressed this issue by seeking to persuade more understanding of their own background and surrounding Asian neighbours. The National Education Association (NEA), the labour union representing current teachers, teachers in training and faculty and staffers at colleges and universities in the United States, describes that GC has already become an imperative skill for twenty-first century education (National Education Association, 2010). Table 2.3 demonstrates the VAx/HAx interdisciplinary content structure.

Table 2.3 VAx/HAx content structure

No.	Technological lectures	Culture and history
1	Evolution of robotics	Malaysia northern region
2	Point-of-care medical tests	Thailand
3	Innovative technologies for sustainability	The Philippines
4	Electric power generation	Malaysia eastern region
5	Smart farming	Indonesia
6	P2P network and electronic information systems	Japan

The third part is group work such as discussions, a final presentation for both programmes and extracurricular activities during the HAx onsite component. Face-to-face communication facilitates mutual understanding, so that Asian youth will lead in an inclusive and sustainable world (Ramos & Schleicher, 2018). While the virtual approach was a good start, our evidence suggests that in-person was far more attractive and meaningful for participants. Therefore, HAx incorporated visits to historic museums, cultural exposure and interaction with local communities.

Before the beginning of this series, this interdisciplinary approach was experimental (Debnath, 2005). After completing two batches, the favourable responses made organisers confident about the effectiveness.

2.2.2 Programme Philosophy, Content Planning and Course Design

(i) ***Confidence as Asians Through a Better Understanding of Rich Culture and History***

'Building confidence as Asians' is embedded in the design philosophy of the VAx/HAx series. It is a holistic appreciation to nurture a deep understanding of the rich culture and history of Asian countries. By embracing their roots, individuals can cultivate a strong sense of identity, pride and purpose, empowering them to contribute positively to the global community (Rychen & Salganik, 2003). This programme promotes understanding and appreciation of Asian cultures beyond stereotypes by fostering a more nuanced global perspective.

This programme also fosters cultural exchange and dialogue. This will help participants to realise and recognise their interconnectedness with Asian neighbours, and also the importance of collaboration in addressing shared challenges. Practice of cross-cultural competencies is also expected to nurture the minds of future leaders (Caligiuri & Tarique, 2012).

(b) ***Practical Opportunities to Acquire Soft Skills in a Multi-National Learning Environment***

'The Future of Education and Skills—Education 2030', published by the OECD, clearly articulated that 'Future-ready students need to exercise agency, in their education and throughout life. Agency implies a sense of responsibility to participate in the world and, in so doing, to influence people, events and circumstances for the better. "Agency" requires the ability to frame a guiding purpose and identify actions to achieve a goal'. The OECD Learning Framework 2030 asserts that students who are best prepared for the future are change agents. They should acquire competencies that consist of Knowledge, Skills, Attitudes and Values (Skovsgaard, 2018).

Our arising problem statement, then, is 'How can educators offer chances for developing this agency to students regardless of their circumstances, especially soft skills trainable through practical experiences?' Soft skills are essential for effective communication, collaboration and overall success in diverse and dynamic settings. Intercultural communication could be effectively acquired through group activities that simulate real-world scenarios as well as during their in-person programme, allowing them to practice effective communication with people from various cultural backgrounds (Bennett & Hammer, 2017).

As English is not a mother tongue for many Asian countries, proficiency can occasionally be a barrier, but participants are still able to show respect and build mutual trust by embracing differences or even deficiencies. They also learn about the cultural sensitivity and awareness for developing understanding of cultural nuances, norms and values to enhance their ability to navigate multicultural interactions. They are also expected to acquire teamwork skills for achieving common goals, learning to navigate cultural differences and appreciating diverse perspectives. After this programme, the participants are expected to become confident with intercultural skills applicable to their future workplace (The British Council, 2013).

(iii) ***Building Long-Lasting Friendships Beyond Ethnic Differences***

Building friendships requires open-mindedness, empathy and a genuine appreciation for diversity. To foster such friendships, the participants start by sharing their experiences and common interests. They are advised to look into their similarities instead of emphasising differences, concentrating on shared and common interests. These settings provide a neutral ground for individuals from various ethnic backgrounds to connect and bond over personal stories and shared passions.

The participants also face each other's cultural, traditional, or religious celebrations. For example, the course schedule reflects holidays for the Islamic Hari Raya celebration and the Indian Deepavali festival. This mutual sharing creates a sense of inclusivity and strengthens the bond by acknowledging and respecting each other's cultural backgrounds. This cross-cultural competency would be applicable all around the world (Pierre et al., 2021).

(iv) ***Content Planning as an Interdisciplinary Approach to Engineering and Social Science***

Content planning with an interdisciplinary approach to engineering and social science requires thoughtful consideration to bridge the gap between the technical and societal aspects of the curriculum. Participants were encouraged to analyse their societal issues that can be solved by the technologies covered in lectures (see Table 2.3) and recognise the overlaps between engineering and society. For instance, consider problems related to sustainable development, technology adoption, or the impact of engineering solutions on communities (Hmelo-Silver, 2004). The participants need to define the goals from both perspectives. They need to clearly outline objectives that address technical challenges while considering social implications. These ideas should be included in their final group presentation.

In the case of HAx onsite component, many extracurricular activities and societal interactions are incorporated under a 'Get out of the classroom, jump into the real-world' concept. For example, participants visit vocational schools for cultural exchange with local students, who usually have little interaction with foreigners. Through the presentation of their own countries, participants recognise the features and beauties of their own nations. They also realise the meaning of social contribution by stimulating learning in younger generations (Taylor, 2014). Figure 2.3 outlines the components of an interdisciplinary structure.

(v) ***Underlying Framework and Theory for Developing Global Competence***

This programme is also designed for developing Global Competence, or GC (Organisation for Economic Co-operation and Development (OECD), 2005). Further, the US-based National Education Association (NEA) has declared, 'Our increasingly interconnected and interdependent global society mandates that American students be educated to develop habits of the mind that embrace tolerance, a commitment

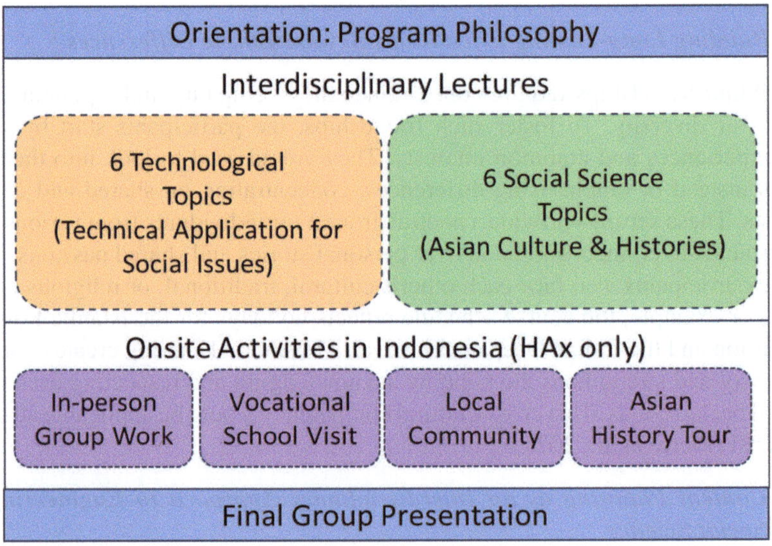

Fig. 2.3 VAx/HAx content components

Fig. 2.4 OECD PISA GC KSAV framework

to cooperation, an appreciation of our common humanity, and a sense of responsibility—key elements of Global Competence' (National Education Association (NEA), 2010).

NEA's GC definition contains four (4) basic elements: international awareness, appreciation of cultural diversity, proficiency in foreign languages and competitive skills. The Programme for International Student Assessment (PISA) GC definition, meanwhile, also consists of 4 dimensions: Knowledge, Skills, Attitudes + Values (KSAV) (see Fig. 2.4).

Hence, HAx content was designed in response to cognitive assessment of the programme, while the background questionnaire addressed seven (7) educational policy questions. See Table 2.4 for a comparison of PISA Policy and HAx content design.

From a theoretical perspective, Constructivism Learning Theory (CLT) is highly applicable to IMPs (Givi Efgivia et al., 2021). The central idea of constructivism is that human learning constructs new knowledge upon the foundation of previous learning (Applefield et al., 2000). Accordingly, it is important to accumulate a wide range of learning experiences so that this construction will be built upon robust structures. Honebein summarised seven pedagogical goals of a constructivist learning environment, which include the knowledge construction process, experiences in and appreciation for multiple perspectives, learning in realistic contexts, student-centred learning, social experiences, multiple modes of representation, and awareness of the knowledge construction process with reflection (Honebein, 1996). Then, the question is 'When and where can students find the first opportunity to build the initial structure for starting GC KSAV acquisition?' A basic foundation is essential

Table 2.4 Comparison of PISA policy and HAx content design

No.	PISA Educational Policy Questions	HAx content design
1	To what degree are students able to critically examine contemporary issues of local, global, and intercultural significance?	Experiential learning through face-to-face interactions with multi-ethnic participants
2	To what degree are students able to understand and appreciate multiple cultural perspectives (including their own) and manage differences and conflicts?	Developing multiple cultural perspectives by sharing social issues of own countries through group discussions and presentations
3	To what degree are students prepared to interact respectfully across cultural differences?	Appreciation of other cultures and history through the lectures by multi-national lecturers
4	To what degree do students care about the world and take action to make a positive difference in other peoples' lives and to safeguard the environment?	Visiting a new country with foreign teammates and interacting with Indonesian local schools and communities
5	What inequalities exist in access to education for global competence between and within countries?	Sharing each country's educational issues through group discussions
6	What approaches to multicultural, intercultural and global education are most commonly used in school systems around the world?	Creating multicultural and intercultural learning circumstances that could be extended to long-lasting friendships
7	How are teachers being prepared to develop students' global competence?	Designing an interdisciplinary curriculum, which integrates technology, social science, and out-of-the-classroom practical activities

to keep constructing these goals. Therefore, this programme aims to provide the first chance to trigger a long-lasting educational journey.

Basic characteristics of CLT and its learning environment are also reflected in the content design: knowledge sharing between teachers and students, sharing of authority between teachers and students, the teacher's role as a facilitator or a guide, and small-sized heterogeneous learning groups (Tam, 2000). Thus, the roles of both lecturers and supporting staff are important for designing insightful content to stimulate participants' curiosity and provide interactive activities to construct solid frames.

(vi) ***Affordable Programme for Equitable Opportunities***

Financial factors are always a significant barrier for any IMPs. In the case of VAx, the first three (3) cohort batches were totally virtual and free of charge. Regarding the HAx onsite component, while it is a 7-day course spread across the ITS campus participants, it is still highly affordable because ITS generously bears the whole cost of accommodation at on-campus hostels, ground transportation in Surabaya, and some meals, with only a USD60 programme fee.

The participants can obtain real international experiences by paying for their international flight tickets to/from Surabaya, domestic transportation, travel insurance and miscellaneous. This has achieved a dramatically low price with no involvement of commercial servicers. As the round-trip ticket between Surabaya and major ASEAN capitals costs around USD100–300 (see Table 2.5), even the students in the rural areas

Table 2.5 Cost reference for in-person attendance

Route from major capitals	Round trip ticket	Trip duration (1 way)
KUL - Surabaya	$80	>3 hours
Bangkok - Surabaya	$190	9 hours (1 transit)
Manila - Surabaya	$220	8-10 hours (1 transit)
Hanoi - Surabaya	$290	8-9 hours (1 transit)
Ho Chi Min	$160	7-8 hours (1 transit)
Tokyo - Surabaya	$500	11-12 hours (1 transit)

of ASEAN countries can enjoy real IMP experiences with a less than USD500-600 budget.

(vii) ***Ethical Considerations and Sensitivities***

Greater diversity brings increased complexity in ethical considerations. With participating students and lecturers representing a wide range of ethnicities and nationalities, individual perspectives on religion, culture, history and political systems may vary significantly. Participants come from diverse religious backgrounds, including Islam, Buddhism, Hinduism, Taoism, Christianity, Shintoism and others. Organisers must take into account various factors, such as dietary restrictions, prayer schedules and cultural norms regarding behaviour.

In the Asian region, political systems also vary, with some countries being monarchies and others governed by single-party systems. Understandings of political correctness can differ across countries, adding another layer of complexity. Modern history poses an additional challenge. Discussions of each country's history inevitably involve topics such as colonisation, war and occupation, especially when addressing the struggles for independence. Although these events are historical and occurred in the time of participants' ancestors, they may evoke strong emotions.

Some participants might feel offended or uncomfortable. However, this also presents a valuable opportunity to foster mutual trust and understanding. By learning about and discussing these shared histories together, participants can develop deeper insights into one another's experiences. This challenge has the potential to benefit both students and lecturers, enriching their perspectives and promoting cross-cultural empathy.

2.3 Methodology

2.3.1 VAx: All Online, More Participants, Free from Transportation Time and Costs

This programme was primarily designed as a 14-class course, which consists of an orientation on programme philosophy, 6 technological topics and another 6 classes for Asian culture and history (see Fig. 2.3), and the final group presentation. All

Table 2.6 VAx/Hax number of registrants and completion ratio

Programme Name	VAx 1	VAx 2	VAx 3	HAx 1
Programme period	May-Aug 2021	Sep-Dec 2021	Sep-Dec 2022	Jun-Sep 2023
No. of registrants	276	187	75	89
No. of CCC completions	35	38	20	41
No. of CAC completions	53	64	29	12
Completion ratio	31.90%	54.50%	65.30%	59.60%

participants were assigned to a group with a minimum of three (3) nationalities to create multi-ethnic teams. At the end of every class, a group discussion was conducted to share the lessons learned through each lecture.

These were during the COVID-19 pandemic, and therefore all online. The best advantages of this approach for students and institutions included almost no cost or time for transportation, and expansion beyond select students, thanks to the elimination of multiple constraints such as budget, operational know-how and the capacity of international office staff. However, this 'easy to register, free-of-charge (FOC)' condition also caused an 'easy to withdraw' issue.

At the beginning of the first cohort batch, out of nearly 300 registrants, nearly half disappeared or didn't show up at all (see Table 2.6). As a personal rule of thumb among authors, the completion ratio of FOC courses is generally around 30% or less. Therefore, the quota is set as a maximum of 20 registrations per institute. From cohort batch three (3), a recommendation letter from the home university was set as a requirement. Consequently, the completion rate improved, while the number of registrants decreased.

The Certificate of Completion course (CCC) requires submission of all assignments and a group presentation with more than 80% attendance for the certificate and transcript. The Certificate of Attendance course (CAC) requires more than 60% attendance without assignment submission for the certificate issuance.

2.3.2 *HAx: A Combination of Virtual and Onsite, Close Communications Among Participants, Interactions with the Local Community*

In response to the shift to endemic status for COVID-19, the 2023 programme was held in hybrid mode by leveraging the platform developed through the VAx series and generous support from the ITS Global Engagement Office. The programme affordability described in Table 2.5 resulted in 44 participants flying to Surabaya. Those who did not travel to Indonesia could still join online for lectures and the final group presentation.

While not quantitatively measurable, the perceived benefits from onsite activities among participants were notable. The participants enjoyed various multicultural activities that lasted beyond the total 12 h of six (6) lectures, especially interactions

Table 2.7 HAx onsite programme agenda

Day 0	-	Arrival at Surabaya
Day 1	AM	Opening ceremony/Lecture: "Asian Past, Present, & Future"/Cultural Dance
	PM	Campus tour/Team building/Indonesian traditional games
Day 2	AM	Surabaya city tour 1: City Hall & Library, 2: Monument & Museum
	PM	Visiting local community village 1
Day 3	-	Choice of mountain tour/free time to hang out with new friends
Day 4	AM	Lecture about the Culture & History of Northern Malaysia and Thailand
	PM	Lecture about the Philippines, group discussions, Indonesian martial arts
Day 5	AM	Lecture about Indonesia and western Malaysia
	PM	Lecture about Japan/group discussions/fun Indonesian language
Day 6	AM	Visit & presentation to local vocational schools
	PM	Visiting local community village 2&3
Day 7	AM	Group work for preparing the final presentation
	PM	Final presentation & feedback/learning reflection/closing ceremony
Day 8	-	Departure for home country

with local communities and vocational students. These experiences are expected to develop participants' GC based on the KSAV framework. Table 2.7 shows the overview of the onsite programme agenda.

2.4 Findings

Major expectations for the onsite programme among 44 participants surveyed were "Making friends" (69.0%) and 'Learning other cultures' (52.4%) (see Fig. 2.5). The participants were mostly engineering students and showed a strong interest in interacting with various people regardless of their background. Expectations about 'Technical knowledge' and 'English practice' were both 35.7%. As English is a foreign language for most students, they have a desire to improve their proficiency. These are quite insightful findings and will be reflected in the next programme.

This analysis is focused on HAx 2023. Out of the initial eighty-nine (89) registrants, 53 participants completed the programme (the completion ratio: 59.6%). The nationality composition is 40% Malaysian, 35% Filipino, 11% Indonesian, 8% Japanese and 6% Vietnamese, respectively (Fig. 2.6). Each team had a minimum of three (3) nationalities and, in some cases, more ethnic diversity.

Analysis of learning outcomes identified positive insights. A post-programme questionnaire conducted on a five (5)-point scale ('5' being 'strongly agree', '1' being 'strongly disagree') revealed that learning experiences for participants were more related to attitudinal factors such as having an 'Open mind to make new friends' (4.71), 'Forward-looking mindset to create a better future' (4.55) and 'Comfort to work together beyond ethnic differences' (4.38), respectively, (see Fig. 2.7). In terms of Values, 'Learning opportunities outside the campus' (4.48) and 'Awareness about

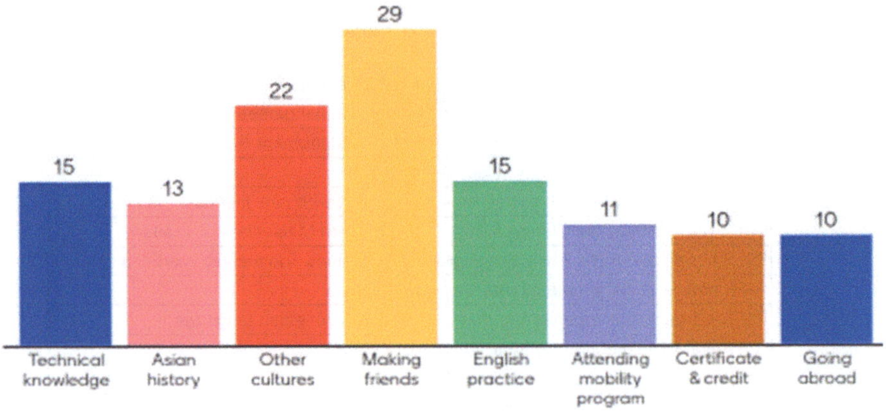

Fig. 2.5 Participants' objectives and expectations

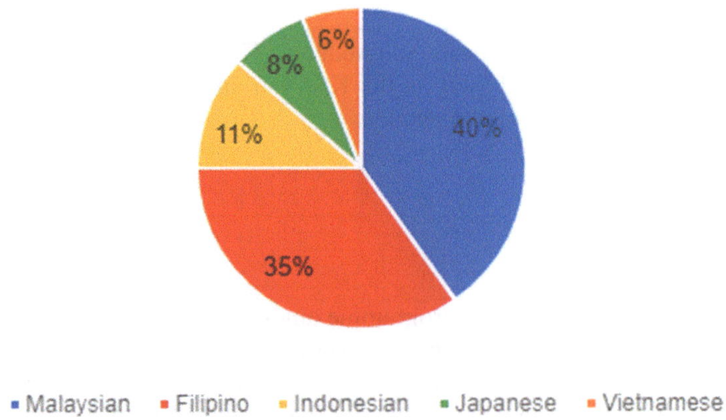

Fig. 2.6 Nationality of HAx participants

representing your country' (4.36) were highly recognised. Skill factors were similarly high, with 'Intercultural communication skill' and 'English conversation skill as a foreign language' scoring 4.43 and 4.38, respectively.

This result is unsurprising since these factors were correlated with course objectives and/or participants' expectations. Nevertheless, it is reasonable to conclude that the programme contents were effective for enhancing participant attitudes and skills, and eventually improving their GC. Regarding the question 'Which field studies gave you effective learning experiences beyond the lectures?' the most named responses were 'Day 7: Group work & final presentation', 'Day 6: Kampong (local community) 3' and 'Others: Informal interaction with professors and ITS committee', all at 4.48, followed by 'Day 6: Local Vocational School Visit' (4.45) and 'Day 6: Kampong (local community)' (4.43).

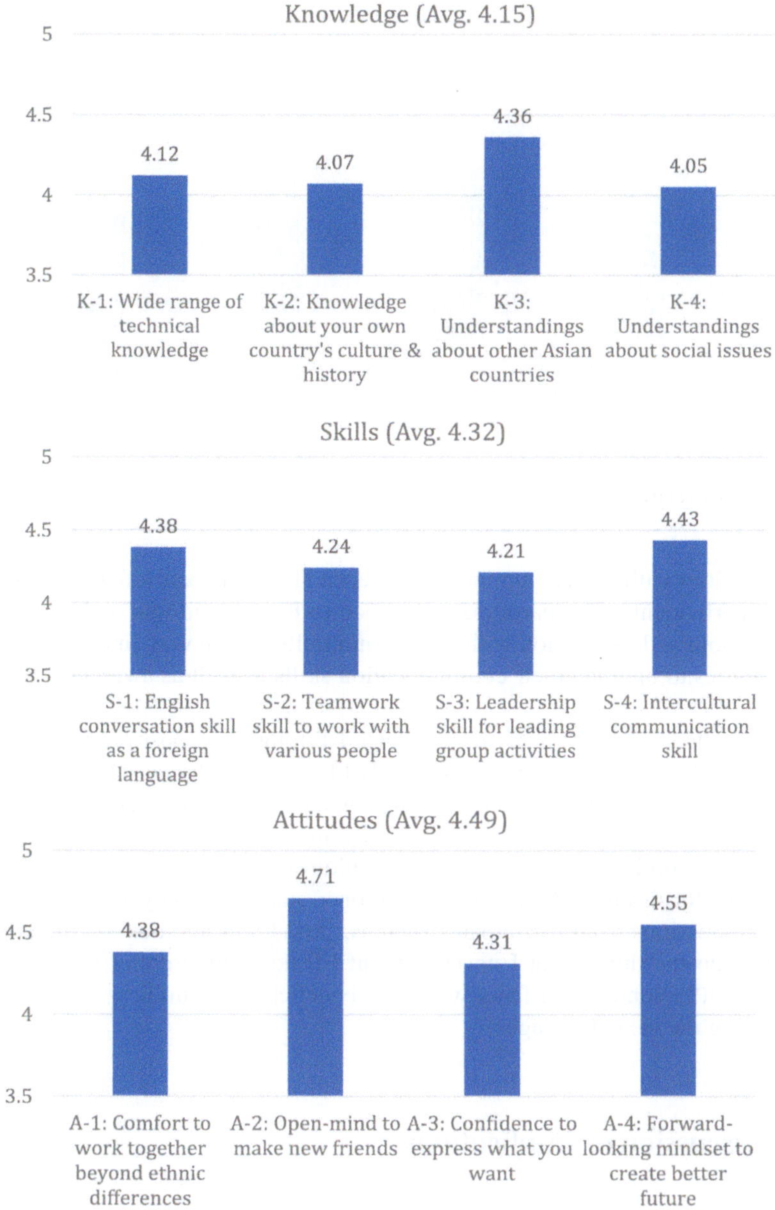

Fig. 2.7 KSAV Self-evaluation

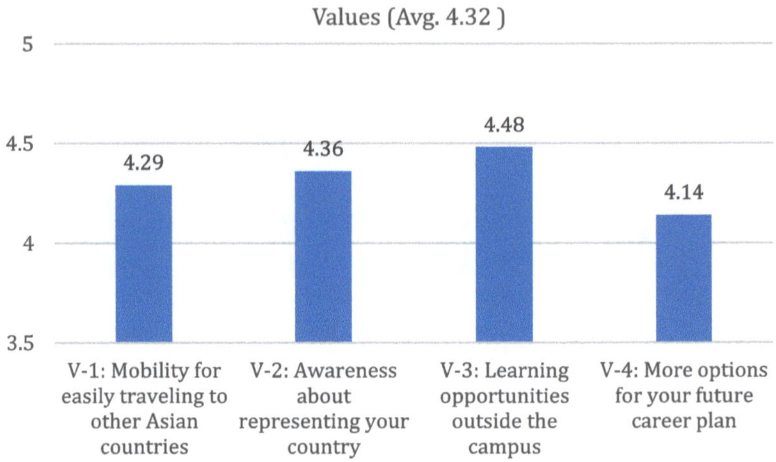

Fig. 2.7 (continued)

Interactions with local communities and even ITS volunteers left a strong impression on participants. The participants seemed to have found that activities outside the classroom and interaction with local communities were very important learning opportunities to upgrade their communication skills as well as awareness as global citizens.

Free-writing feedback was also collected through the post-programme questionnaire. While many wrote simple words like 'fun' or 'great experience', there were several insightful comments for improving the next programme. Frequently mentioned terms were 'communication skill', 'interaction with other Asian people', 'challenge', 'confidence', 'learning other cultures' and 'making foreign friends'. These were consistent with the above-mentioned statistical analysis. It is interesting to find confessions from participants such as, 'first I was nervous' or 'I challenged myself to do the things that I never thought I'll do'. This implies that a sense of achievement beyond initial fears would be beneficial for building confidence and overcoming the next challenges.

2.5 Discussion of Implications

There is a huge demand for IMPs, which will effectively develop GC by enhancing the knowledge, skills and attitudes necessary for exploring future careers. Asian youth appear eager to build friendships by embracing differences and facilitating mutual understanding. Group learning environments will create a positive spiral for improving English communication, teamwork skills and an open and forward-looking mindset. Going abroad and interacting with foreign and/or local people raises

Fig. 2.8 Photo of local school visit and presentations

awareness about the attractiveness of intercultural communication. In particular, post-programme interviews with several students identified that interaction with local vocational school teenagers made a strong impression (see Fig. 2.8).

The participants were requested to briefly introduce their country's outlook and attractiveness, and therefore needed to familiarise themselves with their origins. Generally speaking, such ambassadorial roles are rare opportunities for engineering students. The local students and their teachers also appreciated this session as they could interact with several foreign students at one time to experience Asian diversity. This also inspired some local Indonesian students to consider their potential to explore overseas just like these visitors. For example, a couple of students asked about the possibility of studying in Japan in the future, and received advice for obtaining scholarships. This was a good chain reaction. Thus, HAx could create an affordable programme option for educators to provide students with such learning opportunities.

The benefit of IMPs is to experience differences in culture, religion and even sexual orientation. In this case, about 50% of onsite participants were non-Muslims coming to Indonesia, the most populous Muslim country in the world. They found Muslim people were very friendly and welcoming. In addition, there were a couple of LGBT students from The Philippines. In the post-programme survey, they both wrote about their relief, as they were fairly treated. Meanwhile, some other students personally explained to me that it was their first time teaming up with LGBT individuals, and this turned out to be a good chance to facilitate their understanding of what they have known only through the media. This highlights a real advantage of IMPs to enhance fairness and equality across multiple dimensions.

Another interesting finding is the demand for this type of interdisciplinary content amongst engineering students. While most participants were majoring in engineering, their primary objectives were making friends and interests in other cultures rather than subject knowledge. Consequently, they adhered to more learning outcomes in Attitudes and Values. Hence, it can be concluded that an interdisciplinary IMP is an effective approach for developing well-balanced engineers.

There were some shortcomings to be improved in the next programme. For instance, the concepts of GC and KSAV were barely mentioned at the orientation. If we had explained more details and facilitated participants' understanding at the beginning, they may have paid more attention to these concepts throughout the programme and gained more competencies. Another incident was a health issue. Some students encountered stomach pain or even diarrhoea due to the dietary and hygiene differences from their home countries. Preventive measures should be more clearly elaborated at the pre-departure online orientation.

Based on this success, HAx 2024 will accommodate a maximum of 150 students. This upscaling strategy is connected to publications and presentations at international education conferences such as the Asia–Pacific Association for International Education (APAIE) and the American Society for Engineering Education (ASEE). These are productive cycles to attract more participants from new institutes regardless of the location and time zone thanks to the advantage of hybrid mode.

This upscaling will contribute to more equitable education and enhance the sense of pride of Asian students. The contents will be fine-tuned by adding more on/off-campus group activities, including more interactions with vocational school students, the lecturers and organising staff. From a research perspective, the learning outcomes from these interactions will be articulated by assessing the effects on KSAV in conjunction with leadership awareness. Additionally, a new analytical method for GC and KSAV is slated for development. There is almost no standard assessment methodology for IMPs since most IMPs are custom-made. Currently, the data mining tool for analysing correlations between the learning outcomes and their causes is under development. It will be applied to the next programme.

2.6 Future Directions

Diversity is integrated into all aspects of the programme. Instructors provided preparatory guidance, such as cultural norms around physical contact. For example, in Malaysia, men typically avoid shaking hands with women, while in Indonesia, some Muslim women may initiate a handshake, which men can accept. In other cultures, even hugs between men and women are acceptable. Participants are encouraged to adapt their behaviour to each context. Historical discussions included topics like World War II and Japan's occupation of Southeast Asia. Lecturers from Malaysia, Indonesia and Japan highlighted the war's dual legacy: a tragic past but also a catalyst for independence and enduring peace in Asia.

Students were reminded that sustainable peace requires ongoing efforts to foster mutual understanding. During the programme in Surabaya, participants visited the Heroes Museum, learning about Indonesia's independence movement and the significance of November 10th, commemorated as Heroes Day. They also discovered the name of Universiti Teknologi Sepuluh Nopember is related to the Battle of Surabaya on 10th November, 1945.

Gender diversity naturally emerged in the programme, with LGBTQ + participants contributing unique perspectives. A Filipino student reflected in their final report, saying, 'Being transgender, or being part of a minority (LGBTQ +) makes me quite afraid of being in danger, judged, or misunderstood by others. However, this programme has opened my eyes and shown me that people from other cultures can be accepting and even share similarities with me'. This highlighted the growing importance of inclusivity regarding gender orientation in modern Asia.

Looking to the future, educational programmes in Asia must further integrate digital platforms to facilitate intercultural learning, particularly as hybrid learning continues to expand across the region. The use of virtual exchange programmes, alongside traditional face-to-face mobility, will enable a broader exchange of ideas, perspectives and experiences. These digital tools could allow students to engage in real-time discussions with peers from different countries, thus fostering greater mutual understanding and respect. Moving forward, the inclusion of interactive online platforms could help bridge geographical gaps, ensuring that students who are unable to participate in physical mobility programmes still gain access to the rich cultural and educational experiences that such programmes offer.

Moreover, future programmes must place an even stronger emphasis on sustainability and inclusivity, particularly in the context of gender and sexual orientation. As students continue to come from increasingly diverse backgrounds, the curriculum must evolve to provide more comprehensive support for minority groups, such as LGBTQ+ students. The future direction of these programmes should involve developing mentorship networks and support structures that empower students to navigate and share their diverse identities in safe and inclusive environments. By strengthening the focus on inclusivity and providing tailored support for students from all walks of life, educational institutions can ensure that all learners feel represented, valued and capable of thriving in a globalised world.

2.7 Conclusions

An IMP-based interdisciplinary curriculum can provide unique and valuable learning opportunities that standard lectures cannot deliver, such as multi-ethnic group work, direct interaction with the local community and discussion about social issues and solutions (Rychen & Salganik, 2003). GC is a very useful skill set both for individual education and societal benefits. KSAV is effectively enhanced through IMP learning experiences, especially Attitudes and Values through out-of-the-classroom activities.

Digital Transformation (DX) forced by the pandemic has effectively enabled everyone to attend virtual mobility programmes regardless of location and financial limitation. In addition, the HAx featured in this chapter offers an onsite programme with very reasonable costs, especially for ASEAN students. This practical experience is effective for nurturing open-minded attitudes and communication/teamwork skills, together with knowledge about technology, history and culture and Asian values and pride.

This VAx/HAx series has established new IMP methodologies of equitable learning opportunities for all Asian students. The organiser team intends to continue this programme to develop future global talent with practical GC. It is our strong belief that it is the educators' duty to create equitable educational opportunities where

Fig. 2.9 Group photo

more and more students will explore the world by overcoming mental and financial hurdles. The next programme will be upscaled to 150 participants in 2024, and open to participants from all Asian countries. Additional perspectives from European individuals will also be added to provide a balanced understanding for participants about the strengths of Asian populations and room for improvement (Fig. 2.9).

Highlights

- Virtual and hybrid mobility programmes (IMPs) are breaking down financial and geographical barriers, enabling broader access to global learning opportunities.
- Interdisciplinary IMP curricula foster Global Competence by providing students with hands-on experiences like multi-ethnic group work and community engagement.
- Cultural and gender diversity, including LGBTQ + inclusion, is integrated into the programme, encouraging students to embrace diverse perspectives and cultural norms.
- Future programmes must prioritise hybrid learning, sustainability and inclusivity to ensure equitable access and support for all students, especially those from minority backgrounds.

Acknowledgements The authors gratefully wish to thank all of the professors and academic staff who contributed to this educational journey. We also appreciate all participants in this ambitious programme. Last but not least, we thank Dr. Rongal Watson, Associate Professor of Beloit College, for adding insights for fine-tuning this article.

References

Appiah-Kubi, P., & Annan, E. (2020). A review of a collaborative online international learning. *Engineering Management and Systems Faculty Publications*. https://ecommons.udayton.edu/enm_fac_pub

Applefield, J. M., Huber, R., & Moallem, M. (2000). Constructivism in theory and practice: Toward a better understanding. *The High School Journal, 84*(2), 35–53.

Atherton, G., Crosling, G., Shuib, M., & Azizan, S. N. (2020). International student mobility in ASEAN. In *International student mobility in ASEAN* (pp. 18–34). https://doi.org/10.4018/978-1-7998-4489-1.ch002

Bates, T., & Vancouver, L. (2019). *Teaching in a digital age—Second edition: Guidelines for designing teaching and learning* (2nd ed.). https://pressbooks.bccampus.ca/teachinginadigitalagev2/

Bennett, M. J., & Hammer, M. (2017). A developmental model of intercultural sensitivity. *The International Encyclopedia of Intercultural Communication, 1*(10). www.idrinstitute.org

Caligiuri, P., & Tarique, I. (2012). Dynamic cross-cultural competencies and global leadership effectiveness. *Journal of World Business, 47*(4), 612–622. https://doi.org/10.1016/j.jwb.2012.01.014

Chao Jr., R. Y. (2017). Mobility, mutual recognition and ASEAN community building: The road to sustainable ASEAN integration. *Journal of International and Comparative Education, 6*(2), 105–121. https://doi.org/10.14425/jice.2017.6.2.105

Debnath, S. C. (2005). College student motivation: An interdisciplinary approach to an integrated learning systems model. *The Journal of Behavioral and Applied Management, 6*(3), 168–188.

Givi Efgivia, M., Rinanda, A. R., Hidayat, A., Maulana, I., & Budiarjo, A. (2021). Analysis of constructivism learning theory. http://repo.iain-tulungagung.ac.id

Hmelo-Silver, C. E. (2004). Problem-based learning: What and how do students learn? *Educational Psychology Review, 16*(3).

Honebein, P. C. (1996). Seven goals for the design of constructivist learning environments.

National Education Association (NEA). (2010). Global competence is a 21st century imperative.

Organisation for Economic Co-operation and Development (OECD). (2005). *The definition and selection of key competencies: Executive summary.*

Pacheco, Emily-Marie. (2020). Culture learning theory and globalization: Reconceptualizing culture shock for modern cross-cultural sojourners. *New Ideas in Psychology, 58*, 100801. https://doi.org/10.1016/j.newideapsych.2020.100801

Pierre, Y., Rathee, N. K., & Rathee, V. S. (2021). Developing cross-cultural competency through multicultural perspective: An exploratory inquiry. *European Scientific Journal, 17*(27), 324. https://doi.org/10.19044/esj.2021.v17n27p324

Ramos, G., & Schleicher, A. (2018). Preparing our youth for an inclusive and sustainable world. *The OECD PISA Global Competence Framework*, 1–63. https://www.oecd.org/education/Global-competency-for-an-inclusive-world.pdf

Robertson Geibel, W. (2020). A pedagogy of student mobility: Facilitating humanistic outcomes in internationalization and student mobility. *Journal of Comparative and International Higher Education,* Rychen, D. S., & Salganik, L. H. (2003). *Key competencies for a successful life and a well-functioning society*. Hogrefe & Huber Publishers.

Rychen, D. S., & Salganik, L. H. (Eds.). (2003). Key competencies for a successful life and a well-functioning society. Hogrefe & Huber Publishers.
Skovsgaard, J. (2018). The future of education and skills: Education 2030. *OECD Education Working Papers*. http://www.oecd.org/education/2030/E2030%0APosition%0APaper%0A(05.04.2018).pdf
Tam, M. (2000). Constructivism, instructional design, and technology: Implications for transforming distance learning. *Educational Technology & Society, 3*(2).
Taylor, A. (2014). Community service-learning and cultural-historical activity theory. *Canadian Journal of Higher Education, 44*(1), 95–107. https://doi.org/10.47678/cjhe.v44i1.183605
Teichler, U. (2017). Internationalisation trends in higher education and the changing role of international student mobility. *Journal of International Mobility, N° 5*(1), 177–216. https://doi.org/10.3917/jim.005.0179
The British Council. (2013). *Culture at work—The value of intercultural skills in the workplace*. www.britishcouncil.org
Van den Beemt, A., MacLeod, M., Van der Veen, J., Van de Ven, A., van Baalen, S., Klaassen, R., & Boon, M. (2020). Interdisciplinary engineering education: A review of vision, teaching, and support. *Journal of Engineering Education, 109*(3), 508–555. https://doi.org/10.1002/jee.20347
Yodpet, W., Salvador Quetzal, A., Siek, N., Vebrina Sihite, F., Alegado, P. J. E., Balakrishnan, V., Green, B., & Hollings, S. (2023). International education within ASEAN and the rise of the Asian century. *Educational Philosophy and Theory, 55*(1), 21–34. https://doi.org/10.1080/00131857.2022.2080545

Hiroyuki Ishizaki is a visiting professor at the Shibaura Institute of Technology (SIT), one of Japan's leading technological universities. Since 2014, he has been serving as the director of SIT's Malaysia Representative Office, where he designs international mobility programs and Collaborative Online International Learning (COIL) initiatives in partnership with universities across Southeast Asia. His responsibilities also include coordinating and implementing industry-university collaborations, leveraging his practical experience as an entrepreneur. Additionally, he is actively engaged in interdisciplinary education research, collaborating with international co-authors to publish his findings. Committed to fostering industry-ready global engineers, he continuously promotes a lifelong learning mindset.

Prof. Ir. Maria Anityasari S.T., ME, Ph.D., IPU, ASEAN Eng., is an accomplished academic and professional with extensive experience in industrial engineering and international academic collaboration. Currently, serving as a professor at the Department of Industrial and Systems Engineering at Institut Teknologi Sepuluh Nopember (ITS) Surabaya, Indonesia, she has been contributing to the department since 1996. She also served as the director of ITS Global Engagement from 2012 to 2024, significantly enhancing the institute's international presence. In addition to her academic roles, she has been actively engaged as a consultant for the Surabaya City Government since 2013.

Isma Rosila Ismail (Ph.D.) is a senior lecturer at the Department of Language and Communication, Universiti Malaysia Terengganu. With a strong background in communication studies, her research interests include human communication, intercultural communication, environmental communication, and culture and heritage. She is dedicated to fostering student learning and promoting effective communication practices. She has contributed to teaching and curriculum development and mentored students in academic and professional growth. She also served as the deputy director of International Centre, Universiti Malaysia Terengganu from 2016 to 2019 and from 2022 to January 2024 significantly enhancing the university mobility program and international presence.

Chapter 3
Teaching Strategies for Promoting Sustainable Inclusive Practices in Heterogeneous Classrooms: A Comparative Study of Indonesia, China, Thailand and Korea

Ni Nyoman Padmadewi, Luh Putu Artini, and Ni Made Ratminingsih

Abstract Inclusive education is essential for achieving quality education, which is one of the Sustainable Development Goals (SDGs) reflecting a collective commitment by the worldwide community. Despite the global focus on inclusive education, numerous nations still encounter challenges in its execution, and research on its practical implementation is of utmost importance. This chapter is written based on the study that analyses the strategies implemented by a school in Bali, Indonesia, which has been applying best practices of inclusive education since 2012. It additionally offers empirical facts regarding its implementations in China, Thailand and Korea. Data were collected over three years at a school in Bali through intensive observations, interviews and document analysis. The analyses were accompanied by empirical reviews of studies conducted in China, Thailand and Korea. The data were analysed both qualitatively an quantitatively. The study in Bali reveals three primary approaches: differentiated instruction, the whole-person approach and constructivism, all of which employ numerous strategies. The results indicate significant progress in the students' basic literacy, despite numerous hurdles encountered. The students' social-emotional learning (SEL) must be adequately prepared for maximising an effective classroom. In comparison with China, Thailand and Korea, differentiated instruction, individualised teaching and collaboration are also employed. Although all schools acknowledge that teachers encounter challenges in instructing children with special needs in inclusive settings, it is suggested that the strategies employed remain pertinent for future consideration.

N. N. Padmadewi (✉) · L. P. Artini · N. M. Ratminingsih
Universitas Pendidikan Ganesha, Kabupaten Buleleng, Bali, Indonesia
e-mail: nym.padmadewi@undiksha.ac.id

L. P. Artini
e-mail: putu.artini@undiksha.ac.id

N. M. Ratminingsih
e-mail: made.ratminingsih@undiksha.ac.id

A. Pandian et al. (eds.), *Future of Education in Asia*,
https://doi.org/10.1007/978-3-032-07285-6_3

Keywords Inclusive education · Sustainable · Teachers' strategies · Differentiated instruction · Basic literacy · Social-emotional learning · Special needs

3.1 Introduction

Inclusive education is an essential foundation for realising the rights of every child, especially those with special needs, to gain access to quality education (Bryant et al., 2008; Kusimo et al., 2019). These children, with all the limitations they have, require a special pedagogical approach so that teachers can provide appropriate learning and support their development. In a global context, the commitment to inclusive education is reflected in Sustainable Development Goal (SDG) number 4 from UNESCO, which emphasises the importance of providing equitable education for all individuals (Rambla & Langthaler, 2016; Tonegawa, 2023).

However, the importance of inclusive education is not only related to children with special needs. Differences in social backgrounds, abilities, learning styles, cultures and other factors make inclusive education relevant for all students, and multiculturalism therefore becomes prominent. An inclusive multicultural environment creates a space where each student is recognised for their uniqueness and value, awakening each individual's potential (Amthor & Roxas, 2016; Jansen et al., 2015).

The COVID-19 pandemic demands rapid and creative adaptations in the education system, especially for children with special needs. In this context, the importance of social-emotional learning (SEL) in student well-being becomes increasingly apparent (MacCann et al., 2019; Schonert-Reichl, 2017), and learning practices in inclusive classrooms become increasingly demanding. Therefore, in-depth research is needed to understand how teachers adapt SEL practices to help students return to classroom learning offline after the pandemic in the context of inclusive education.

This research is crucial for understanding how the sudden shift during the pandemic impacts the choice of educational approaches, learning quality and student engagement in inclusive education. It aims to identify practical solutions and strategies to prioritise inclusive education during and post-pandemic. Recognising the significance of inclusive education implementation in improving educational quality and equity, additional research is needed to discover optimal strategies for advancing inclusive education. As a result, this study seeks to explore the strategies used by educators at an inclusive school in Bali that has been implementing inclusive education. Given that this has been a worldwide issue, it is imperative to compare the implementation of Inclusive education with that of China, Thailand and Korea. So, the research questions are formulated as follows:

(1) What are the appropriate approaches to teaching in an inclusive school?
(2) What are teachers' strategies for teaching students with special needs in an inclusive system?
(3) How is the achievement of special needs students after three years of attending the inclusive education programme?

(4) How are inclusive education implemented in China, Thailand and Korea?
(5) What future consideration can this research offer?

3.2 Literature Review

Inclusive education, an approach to ensuring the rights of every child, including those with disabilities, has become the focus of global attention to receive quality education in regular schools. The ultimate goal is to increase the capacity of schools to remove barriers to the attendance, achievement, access and engagement of all students (Kusimo et al., 2019). In Slee's (2018) view and Rapp and Corral-Granados (2021), this involves increasing school capacity to ensure all students, including those with special needs, have an optimal educational experience. This inclusivity is related to SDGs, especially SDG 4, which emphasises equality, inclusion and quality of education.

Inclusive education is a technique and ideas for teaching students with special needs in conventional schools (Bryant et al., 2008). This education focuses on the rights of each student, and its goal is to support the independence of students with disabilities and ensure equal access to educational materials, activities and practices (Anupriya & Salim, 2014). This view is consistent with SDG 4, which promotes access to quality education for all individuals, regardless of background or personal circumstances (Rambla & Langthaler, 2016).

Most countries, including Indonesia, have adopted inclusive education as part of their vision to achieve Sustainable Development Goals. In developing the SDGs in 2015, UNESCO recognised the importance of providing accessible and quality education for children with special needs. This global movement aligns with the principles of inclusivity promoted by inclusive education (Tonegawa, 2023). SDG 4 is not only about ensuring access to education for all but also about ensuring inclusive and equal quality education. Inclusive education is a solution to this goal and a critical element that contributes to achieving other SDGs by addressing social inequalities and promoting equality.

Apart from global goals, inclusive education also has local implications in Indonesia. In the Indonesian context, Sipayung and Hsu (2023) investigated how Indonesian teachers align their beliefs and competencies with curriculum reforms, especially the Merdeka Curriculum and SDG 4. This study shows that teachers see SDG 4 as a framework to guide their teaching practices, increasing overall understanding of the relevance of these objectives in the Indonesian educational context.

Teaching and learning strategies, which refer to various kinds of techniques used by teachers to promote efficient knowledge transfer, are crucial in determining educational outcomes (Freeman et al., 2014), including play-based learning (Vasoya & Vansdadiya, 2023). Vansdadiya et al., (2023) add that play-based learning is essential for literacy and numeracy needed by all students. Bryant et al., (2008) noted that teaching techniques and ideas for students with special needs can be considered inclusive education, where various strategies are used to support the success of all students. In the context of learning strategies, inclusive education applies a differentiation, comprehensive and constructivist approach. Through a differentiation approach, teachers can address individual student differences, including those with special needs, to ensure that each student can learn according to their level of understanding.

Inclusive education, integrating students with special needs, fosters a multicultural classroom environment, promoting interaction among individuals of diverse cultural backgrounds and abilities (Jansen et al., 2015). This approach goes beyond academics, incorporating social-emotional learning (SEL) perspective to enhance students' emotional and social well-being (Schonert-Reichl, 2017; Weissberg et al., 2015), including those with particular disabilities (Amthor & Roxas, 2016). Through this inclusive, multicultural and SEL-focused education, classrooms become spaces that respect diversity, emphasise inclusivity and recognise special needs as integral to the learning process.

Research reveals that teachers play an important role in adapting to children who face learning difficulties (Khayati et al., 2020). This adaptation includes adjustments to teaching methods, strategies, materials, learning tools and learning environments. Jansen et al., (2015) found the effectiveness of a comprehensive multicultural approach compared to a standard multicultural approach. This inclusive approach results in anticipated higher inclusion levels, especially for those needing acceptance. This study also highlights that explicitly diverse group involvement increases support for internationalisation as individuals feel a stronger sense of belonging to their group. In the Indonesian context, Sipayung and Hsu (2023) indicate that teachers view SDG 4 as a framework that guides their teaching practices and contributes to an increasingly comprehensive understanding of the relevance of objectives in Indonesian education. Understanding SDG 4 significantly improves teaching practices by equipping teachers with a strong foundation and comprehensive framework for delivering quality education. Despite their studies, they have yet to investigate how the teaching and learning process can be explored during and after the pandemic and how students' achievement develops during those periods.

3.3 Methodology

The research concentrated on a bilingual school in North Bali that adheres to the inclusive education philosophy. The school's selection is intentional and complies with several criteria, as follows:

i. A sustained commitment to inclusive education for over 10 years
ii. An established system for parent engagement
iii. Instruction of English as a foreign language to all students
iv. Recognition as a model school within the region.

Table 3.1 Time frame and mode of teaching during data collection

Time	Modes of teaching and learning
January–March 2020	Regular process
April 2020–June 2021	Online
July 2021–June 2022	Transition
July 2022–December 2023	Regular process

The study spanned three years at the school, allowing researchers to examine teachers' techniques regarding pupils' behavioural modifications and the growth of their communication skills. The timeline for data collecting and the instructional methods is summarised in Table 3.1.

Two regular mainstream classrooms, Grades 2 and 3, were involved, where three cases of students with autism spectrum disorders (ASD) were included. Each classroom had a primary teacher, an assistant and two teachers responsible for educating the four students with ASD. The data collection was conducted from December 2021 until December 2023, allowing the researchers to observe the period before and after the pandemic.

The data was collected by observing the process of teaching and learning, interviewing the teachers and the principal of the school, and analysing the individual education plans designed by the teachers, the assessment instruments and their results, and their task performance analysis. In an attempt to provide more thorough insights into teaching practices in inclusive schools, the research on inclusive education in China, Thailand and Korea was also reviewed.

3.4 Finding and Discussion

3.4.1 Approaches of Teaching in an Inclusive School in Indonesia

In Indonesia, inclusive education is governed by Permendiknas No. 70 of 2009. The regulation pertains to inclusive education for pupils with disabilities and those with exceptional intelligence and/or outstanding capabilities. It advocates for equal educational rights to ensure that all children, including those with special needs, have access to a quality education devoid of discrimination; it raises public awareness that every individual has the right to learn in the same environment without encountering stigma; it promotes the development of children's potential by supporting students with special needs and talents in achieving their fullest capabilities within a nurturing context; it encourages the establishment of inclusive schools to motivate traditional institutions to evolve into welcoming, inclusive environments for all students, thereby enhancing shared values—specifically, creating a learning atmosphere that fosters cooperation, tolerance and respect for diversity.

Data collection at a school reveals multiple approaches that inform the implementation of various strategies. Table 3.2 provides a summary of the current inclusive education implementation, allowing for the identification of effective practices. Long-term observations, interviews and document analysis at the school being studied showed that an inclusive policy was consistently put into place using three main methods: differentiated instruction, constructivism and the whole-person approach, which are explained in Table 3.2.

The school incorporates multicultural practices to empower inclusivity for mainstream heterogeneous students and students with ASD. The following table summarises the practices (Table 3.3).

Multicultural practices conducted reconfirm dimensions introduced by Banks (1995a). By embracing multiculturalism, the learning process builds bridges between differences and strengthens the value of inclusivity in education (Fig. 3.1).

The activities for multicultural inclusion are carried out through project-based learning under the constructivist philosophy. Through collaborative projects and group discussions, students learn to understand others' perspectives, develop communication skills and build knowledge together. This creates a more inclusive learning environment and prepares students to participate in broader society. Those techniques improve comprehension and retention and foster a better knowledge of the material (Freeman et al., 2014).

The whole-person approach considers various aspects of students' competencies and manifests in joyful learning activities. Along with a focus on academics, the whole-person approach recognises the importance of physical, emotional, social and spiritual development as an integral part of learning. The school combines learning in and out of the classroom in formal and informal situations for fun and cheerful learning.

Table 3.2 Approaches of teaching in inclusive classrooms

Differentiated instruction	Differentiated instruction is characterised by differences in content, procedure, product and environment based on the requirements of the students who are frequently heterogeneous in the classroom
Constructivism	Constructivism is not just about teaching facts or information but provides a foundation for developing critical thinking, creativity and deep understanding. By shifting students' role from passive recipients to active constructors, constructivism creates a dynamic learning environment and supports students' development as lifelong learners
The whole-person approach	The whole-person approach to learning emphasises the development of all individual potential, including physical, intellectual, emotional, social and spiritual aspects. This approach responds to students' comprehensive needs and seeks to create a learning environment that supports holistic growth

Table 3.3 Multicultural practices for empowering inclusivity

Dimension	Description
Content integration	• Multicultural topics in reading texts • English Novel study: Promote cultures from different parts of the world • Virtual International field trip: Introducing the culture of different countries virtually • Extensive reading programme—virtual library/library
The knowledge construction process	1. Inclusion of visiting international students (Japan, Italy, Hungary, Canada 2. Inclusion of visiting international educators (Australia, Hong Kong, Canada, France, Malaysia) 3. Parents should be encouraged to volunteer as ambassadors for their respective countries 4. Performing Art Day—integrating stories of different cultures 5. Students' geography fair introduces information about any country with their culture 6. Cultural Gallery 7. Cultural, seasonal events
Prejudice reduction	1. Forbidding any racial bullying, conversations, or jokes 2. Inserting values of cultural perceptions and understanding in any relevant instructions, talks and exposures
An equity pedagogy	Teachers modify their teaching to facilitate the academic achievement of students from diverse racial, cultural and social class groups by: 1. Providing additional programmes if needed 2. Promoting values of respecting differences
Empowering school culture and social culture	Integrating the culture in ways that enable students from diverse racial, ethnic and gender groups to experience equality and equal status

Fig. 3.1 Few examples of multicultural practices for empowering inclusivity

3.4.2 *Teachers' Strategies in Teaching Students with Special Needs in an Inclusive System*

The teaching strategies are classified into four stages: strategies of teaching before the pandemic, strategies during the COVID-19 pandemic, strategies during the transition time, and strategies after the pandemic. The following tables summarise the strategies implemented in those stages of the period (Table 3.4).

In addition to the school's three approaches, teachers employ different teaching strategies to adapt to students' needs and characteristics. Students with ASD are primarily visual learners, so most strategies are based on how the students can learn by imitating others, watching others, or manipulating instructional activities that resemble real-world life (Table 3.5).

Besides causing flexibility and resilience in education, COVID-19 requires innovation in teaching approaches. Implementing technology for online learning, using digital learning platforms and adapting curricula to meet the challenges of distance learning have become imperative. The school decided to employ the Schoology application as the central learning management system, combined with weekly worksheet

Table 3.4 Strategies for teaching students with special needs (ASD) before the COVID-19 pandemic

Period of teaching	Kinds of strategies utilised by the teachers	Description
Before COVID-19 pandemic	Various types of cooperative learning:	Cooperative learning fosters positive and social relationships in which they can imitate and learn by observing other students and feel well accepted, respected and equally treated. On the other hand, regular students can exercise leadership and be a model for the students with ASD
	Buddy programme	
	Older-younger cooperative learning	
	Multimodal teaching: Audio-visual teaching strategies	Every student has a different learning style. Multimodal teaching includes using various techniques to reach students with various learning styles, including visual, auditory and kinesthetics. This provides an opportunity to understand and internalise the learning material in a way that is most effective for them
	Craft/art	
	Mingled with regular main-stream students through PE, art and music	
	Group work through Task-based teaching and project-based learning	Task-based learning emphasises the applicability of exercises and activities to real-world circumstances, making it easier for students to connect what they learn in school to what happens in real life
	Literacy programme of reading every day at home	The school implemented a policy of reading at home daily using a system that involved students and parents in the activity

Table 3.5 Strategies of teaching during the COVID-19 pandemic

Period of teaching	Kinds of strategies utilised by the teachers	Students' activities	
		Mainstream students in regular classroom	Students with ASD
During COVID-19 pandemic	Change the process of teaching into remote teaching by incorporating technology	Using a learning management system (Schoology + Kahoot + Zoom and Flip grid application) to empower students' engagement and interactions Literacy programme by keep reading every day and empowered by Schoology detective	At-home learning + 15 15-min Zoom synchronous learning accompanied by parents
	Made a learning protocol		
	Parents' orientation		
	Individualised learning	A special package (students' worksheet) taken weekly from and to school	Students' work-sheets taken weekly from and to school
	Scaffolding provision	Individual and small group guidance outside the classroom schedule	Individualised instructional video
	Literacy programme	Read daily by choosing a story from the e-library provided by the school. Pick-up and drop-off of library books	Reading with parents Pick-up and drop-off library books (by parents)

exchanges and assessments. Learning protocols were created, and comprehensive orientations were provided to students and parents.

Educators must also be more sensitive to aspects of students' mental and emotional well-being, accommodating psychological challenges that may arise due to dramatic changes in how they learn. Therefore, additional applications, such as Kahoot and the Flip Grid, were applied to maintain students' engagement and interactions. The following are teachers' strategies during the transition from remote learning during COVID-19 to in-person learning at school (Table 3.6).

Figure 3.2 shows the students having a morning circle to start the day, practising kindness by distributing mattresses to those in need during the social care programme, building a vocabulary of emotion using an emotion card, having a tent in the classroom as a place to relax and calm-down emotion, and doing activities like drawing to encourage expressions. After the pandemic, teachers had more variations of strategies, as revealed in Table 3.7.

Table 3.7 reveals that the strategies applied by the teachers are the manifestations of the three approaches, which emphasise the need to tailor the strategies to the student's needs and preferences, prioritise the students' active learning and

Table 3.6 Strategies for teaching during the transition from remote learning to in-person learning at school

Period of teaching	Kinds of strategies utilised by the teachers	Students' activities	
		Mainstream students in a regular classroom	Students with ASD
Transition in-person learning at school	The gradual process of grouping students according to numbers and time	In conjunction with the use of LMS and packages, and minimising Zoom meetings	Individualised learning every day
	Preparing social emotional Learning (SEL) of the students	Re-orientation to students and parents about school life	Re-orientation and re-induction about the physical setting of the school environment
		Incorporate SEL in the classroom: 1. Start the day with students' routine 2. Nurture a culture of kindness and social interactions through Project-based learning and meaningful plays	Incorporate SEL in the classroom: 1. Start the day with students' routine and morning circle 2. Nurture a culture of kindness 3. Build social-emotional vocabulary 4. Incorporate a "calm-down corner' 5. Encourage expression through art 6. Providing sensory spaces

Note Technology is no longer the main medium for teaching regular students and students with ASD, except for the occasional use of Schoology for sharing documents and files with students and the occasional use of Edu-games to stimulate students' engagement. As a transitional period, the school applies several activities to incorporate SEL for students. The activities vary, but all are intended to make learning joyful as shown in Fig. 3.2

Fig. 3.2 Several pictures incorporating SEL in the teaching and learning process

Table 3.7 Strategies for teaching after the COVID-19 Pandemic

Period of teaching	Kinds of strategies utilised by the teachers	Students' activities	
		Mainstream students in regular classroom	Students with ASD
Transition in-person learning at school	Direct instruction	In a regular classroom with innovative strategies (in or outside the classroom)	Pull-out in a special pod; integrated in 3 courses (PE, Art and Music class)
	Various types of cooperative learning	Similar-grade group work Different-grade group work	Buddy programme Older-younger cooperative learning
	Task-based learning	Task-based learning emphasises the student's active learning to empower them as the agents of their learning	Task-based learning (individual/group work) based on the individual student's ability and needs
	Project-based learning	Weekly, annually, and based on certain events at the school	Project-based learning group with mainstream students
	More socialisation and interactions	Providing opportunities for mainstream students to have socialisation and social interaction with the students with ASD	Multimodal scaffolding by empowering social environment programme (morning circle activity, Cooperative Buddy programme, Sensory Garden programme, Modelling, go-play activity)
	Using media and scaffolding strategies	Using media and scaffolding strategies for effective learning and better understanding	Using multimodal scaffolding and various media for behaviour modifications and better understanding
	Literacy programme	Reading-at-home daily programme and club activities	At-home reading with parents
	Gardening Project	Thematic and integrated skills	Nature/garden-life skill project

Note Despite the transition period during which the school designed and applied SEL and realised the learning loss experienced by most students, the school focused on reviewing the previous materials and providing students with remedial treatment in the first three months.

have direct real-world experiences, and manage mindful learning that balances the dimensions of cognitive, affective and psychomotor aspects

The students with ASD, for example, are pulled out in a particular pod and taught based on the individual education plan. They are integrated when they learn in three courses to enhance social interactions and connections through several programmes (like literacy programmes and gardening projects). To maximise the guidance based on the student's ability level, multimodal scaffolding is provided until the students can be more independent.

3.4.3 The Achievements of the Students After Three Years

After experiencing the moments of remote learning, transition period and return to having offline sessions entirely at school, it is essential to acknowledge their achievement at the end of the observations and identify their communication ability development (Tables 3.8, 3.9 and 3.10).

The data exhibit a distinct and noteworthy trend based on a three-year observation period. During the pandemic, students who studied from home did not demonstrate any improvement and even experienced learning loss. The previously observed improvements, particularly in behaviour modification and social interaction, reverted to their initial state. To reacclimatise the students to routines and the art of social interaction, social environment programmes are created throughout the transition period, with SEL enabled and taken into consideration. Improvements are evident after three years of using multimodal teaching approaches and media and a teaching style tailored to the students' needs and characteristics. Mainstream children develop more assertive characters, are more accepting and sympathetic, and have better communication abilities, whereas students with ASD also show noticeable development.

Table 3.8 Achievement of mainstream students

Student	Achievement
Students grade 2, 3, where the students of special needs are included in three courses	Having better character: 1. More understanding 2. More acceptance 3. More leading 4. More responsible
	More inclusive
	1. Willingness to be inclusive with students of special needs 2. More social awareness
	Better communication skills
	1. Able to guide students with special needs to communicate with others

Table 3.9 Behavioural development of students with ASD before and after the COVID-19 pandemic

Students with ASD	Before pandemic (2021 up to pandemic)	After pandemic
Case 1 (APT)	Cannot communicate but mimicking simply repeating what others say	Can communicate (even though communication needs simple stimulation)
	Mostly silent	Can socialise; Can be involved in group work and be involved in simple conversation
	Almost no social interactions	Can respond to questions of others; Can express his basic needs
Case 1 (DV)	Frequent tantrum	Less tantrum
	Cannot socialise with other students	Can socialise with other students
	Cannot get much involved in group conversations	Can be involved in a group conversation
	Can say something but frequently not connected to the questions (random talks; cannot focus on any topics)	Can answer questions appropriately most of the time, even though not able to be involved in regular classroom interactions, despite still initiating random talks
	Frequently refuse to go to school	Willingness to come to school happily
Case 1 (SN)	Ability to say words related to self, family and hobby, but very low social awareness and social interaction; like to draw about specific topic/electric appliances	He could answer almost all the questions, interact socially, and be involved in conversation, but he still lacked eye contact

Figure 3.3 indicates how each student with ASD has improved at school. Nine improvement points were identified: social interaction, the ability to say words about oneself, read a sentence, understand instructions, respond verbally, make eye contact, initiate conversations, read a paragraph and write sentences.

The case of APT suggests that he made slow progress. He has been unable to read and write since the beginning of school, and he continues to be unable to read even after three years. However, after the pandemic, with multimodal media strategies and the empowerment of social environment programmes, he can improve in social interactions so much that he can say words about himself and his feelings, understand instructions and give short responses despite no improvement in reading or writing (3.4).

Regarding DV, it is also clear that the student has improved significantly in reading Bahasa Indonesia, the student's native language, and English. His ability to read the material taken from National Geography: Kids Readers. He can respond to the teacher's questions in English despite the teacher's need to give them extra time to process and repeat the same questions twice or three times. This clearly shows that he understands what he is reading well. However, the student has made less progress

Table 3.10 Communication development of the students with ASD before and after the COVID-19 pandemic

Students with ASD	Before the pandemic (2021 up to the pandemic)	After pandemic
Case 1 (APT)	Know and articulate very limited vocabulary (short responses of yes/no)	Can respond to questions but sometimes still mimics; Can ask basic questions with support during morning circle
Case 1 (DV)	Knows and can articulate words related to himself and random topics; cannot socialise properly or make social interactions Initiate Verbal random topics but not in conversational modes	Develop a lot of vocabulary related to himself, his family and the topics he likes Can make conversation and improve in sustaining specific topics of high interest in conversations
Case 1 (SN)	He knows and can articulate words related to himself, has a hobby of drawing electrical appliances, has low eye contact, has low social interactions/and cannot make reciprocal conversations	Can make basic conversations and sustain a conversation on the topics of interest

in writing, even though he can read. His writing skills are restricted to writing words or short phrases; he cannot compose sentences (Fig. 3.5).

Regarding SN, Fig. 3.6 illustrates a systematic pattern indicating that SN likewise suffered from learning loss in nearly every area during the pandemic. After the pandemic, when the process of learning returned to on-site sessions at school, there was a clear improvement. As seen in the accompanying Fig. 3.7, his language and communication skills have improved to the point where he can write two sentences in the style of a simple picture that depicts a social interaction.

Figure 3.8 indicates that SN correctly completed the worksheet. This requires translating the material and repeating the questions two or three times with the help of his teacher. Despite the constraints and difficulties encountered, it is clear that he can write the words in English.

Students' behaviour also reflects their progress, as demonstrated in Table 3.11.

Case 1: APT

See Table 3.11.

Case 2: DV

See Table 3.12.

Case 3: SN

See Table 3.13.

The data previously displayed in Tables 3.11 and 3.12 and Fig. 3.9 show that there has been a noticeable improvement in the student's behaviour, particularly after the pandemic period. The student's behaviour returned to its baseline during the transition period. After the students had on-site classroom sessions, it only significantly improved after implementing numerous socialisation programmes with various multimodal media and teaching strategies (Table 3.13).

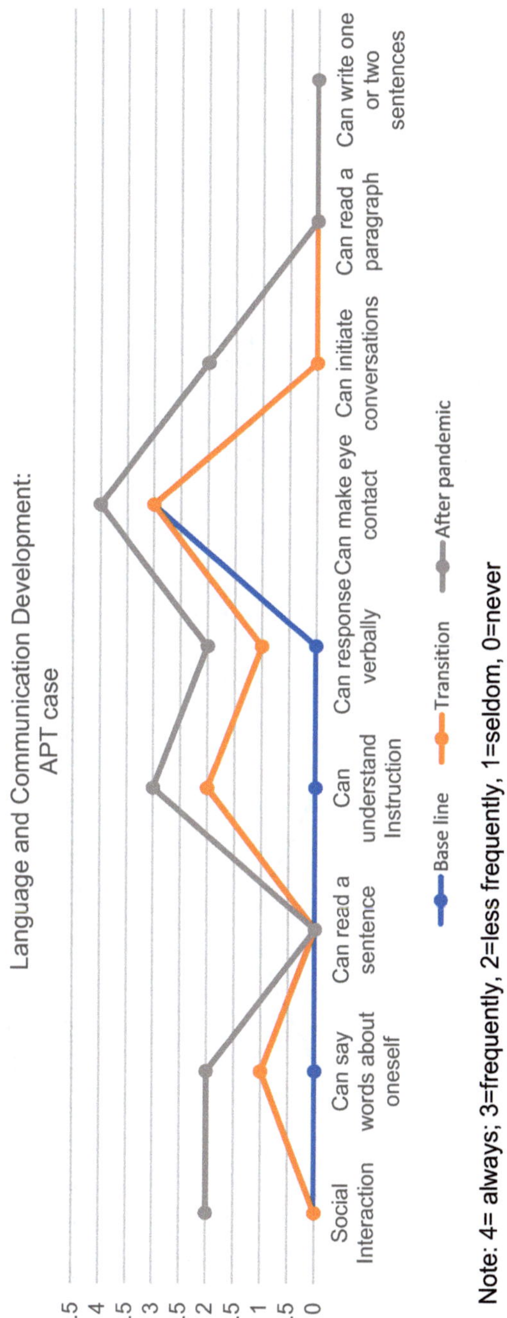

Fig. 3.3 Results of observation about APT Case

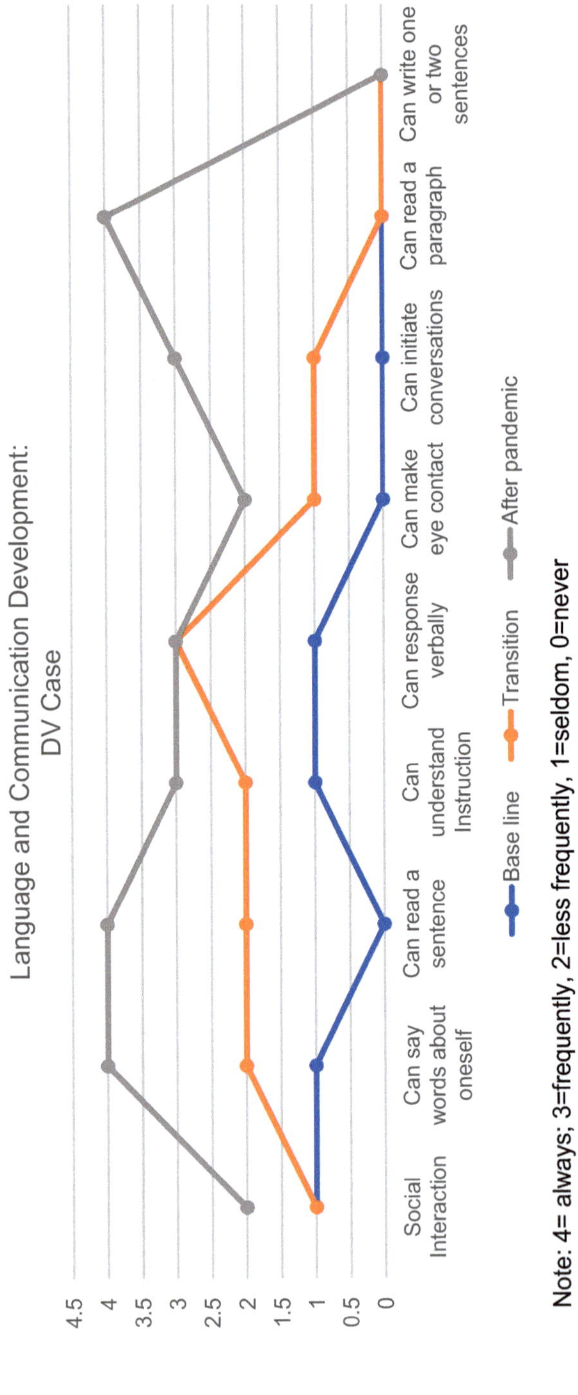

Fig. 3.4 Results of observation about DV

Fig. 3.5 Ability of DV to read and answer English text

Despite the implementation of various programmes aimed at enhancing equity and educational quality for all children, interviews with principals of other regular schools in the area reveal that they have not yet adopted inclusive education due to a deficiency in understanding the requisite knowledge and skills associated with it. They acknowledge that they infrequently encounter opportunities for training and professional development regarding the implementation of inclusive education.

3.5 The Implementation of Inclusive Education in China, Thailand and Korea

To acquire comprehensive insights into teachers' strategies for implementing inclusive education and to obtain an overarching understanding of inclusive education practices throughout Asia, several studies carried out in China, Thailand and Korea were reviewed.

3.5.1 Implementation of Inclusive Education in China

Inclusive education began in China in 1980, supported by legislation. The Chinese government promoted a distinctive inclusive approach to execute inclusive education that combined specialised educational institutions with ordinary schools (Zhang, 2010). China has sought to advance inclusive education through national government policies and regulations, extending to local implementations. The Learning in Regular Classroom (LRC) project, a localised strategy for instructing students with disabilities in mainstream classrooms of traditional schools, is socialised to implement inclusive education.

In 2019, China expanded the rationale for LRC to achieve equitable and high-quality education, designating inclusive education as a vital component of educational modernisation. In the same year, the Chinese government released a pivotal policy document, China's Education Modernisation 2035, which seeks to achieve

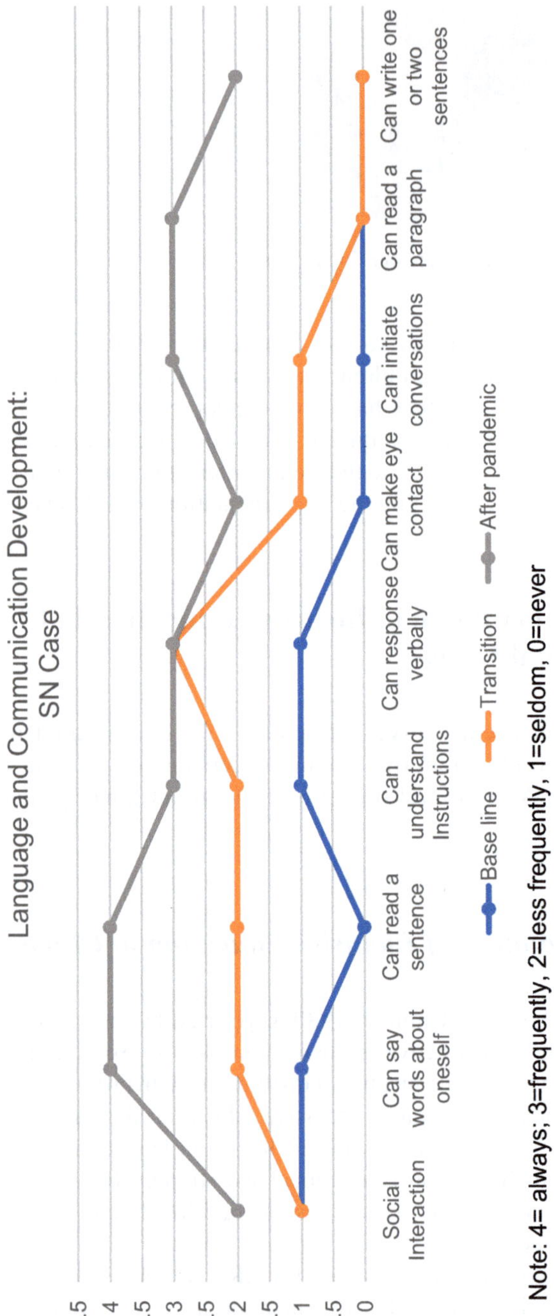

Fig. 3.6 Results of observation about SN

Fig. 3.7 Student (Case SN) can write a conversation in his native language taken on December 2023

Fig. 3.8 Worksheet in English done by SN

high-quality education, underscoring ‘inclusive development’ as a core educational principle and robustly promoting inclusive education as a vital objective for 2035 (The State Council of the People’s Republic of China, in (Wang et al., 2024). The inclusive education project aims to reduce discrimination, especially against students with disabilities. This technique has led to a 49.41% increase in the enrolment of students with disabilities in public schools (Ji, 2024).

LRC has encountered ongoing criticism for only ‘occupying standard classrooms,’ implying that students with disabilities are integrated only physically, rather than academically and socially (Zhao & Kokhan, 2022). Inclusive education encompasses more than just special education; it also transcends the integrative notion of incorporating children with disabilities into the regular educational framework (Zhang, 2010).

The rights to inclusive education in China are expanding; nonetheless, discrepancies persist between policy and implementation (McCabe & Jiang, 2024). Li et al., (2022) stated that there are several strategies utilised by teachers in implementing inclusive education, such as:

(1) Curriculum differentiation
(2) Universal design for learning
(3) Co-teaching

Table 3.11 Behaviour modification of APT after three years

Task performance assessment	Task performance analysis														
	Before COVID-19					Transition					After COVID-19 (November 2023)				
	1	2	3	4	5	1	2	3	4	5	1	2	3	4	5
Coming to the school gate	x						x								x
Washing hands before entering the school area	x						x								x
Greeting teachers at the yard	x					x								x	
Walking across the yards	x						x								x
Opening Shoes		x						x							x
Entering the classroom		x						x							x
Pre-activity before learning	x					x									x
Greetings: saying good morning		x						x						x	
Washing hand activity before snack		x						x							x
Having meal				x					x						x
Playing at a playground	x						x					x			
Exchanging uniform	x							x					x		
Toilette training	x					x					x				
Cleaning a desk/table	x						x								x
Greeting: saying goodbye	x						x								x

Note 1 = rejection; 2 = need verbal and physical guidance; 3 = need verbal but no physical guidance; 4 = responding/understanding instruction but need stimulation; 5 = independence

(4) Differentiated Instruction
(5) Collaborative learning
(6) Peer tutoring
(7) Individualised teaching
(8) Structured teaching
(9) Assessment strategies.

The implementation of these techniques demonstrates that educators working with children with special needs in inclusive classrooms must modify the curriculum to accommodate the students' needs. To recognise students from diverse backgrounds, teachers should understand Universal Design for Learning and integrate it into their teaching and learning process, as a one-size-fits-all approach proves insufficient.

Table 3.12 Behaviour modification of DV after three years

Task performance assessment	Task performance analysis														
	Before COVID-19					Transition					After COVID-19 (November 2023)				
	1	2	3	4	5	1	2	3	4	5	1	2	3	4	5
Coming to the school gate		x				x									x
Washing hands before entering the school area				x					x						x
Greeting teachers at the yard	x					x						x			
Walking across the yards	x					x									x
Opening Shoes		x					x								x
Entering the classroom		x					x								x
Pre-activity before learning	x					x									x
Greetings: saying good morning	x					x								x	
Washing hand activity before snack		x					x								x
Having meal			x					x						x	
Playing at a playground	x						x								x
Exchanging uniform	x							x							x
Toilette training		x					x								x
Cleaning a desk/table	x								x					x	
Greeting: saying goodbye		x					x								x

Note 1 = rejection; 2 = need verbal and physical guidance; 3 = need verbal but no physical guidance; 4 = responding/understanding instruction but need stimulation; 5 = independence

Teachers must tailor instruction to the students' qualities and adjust it to meet their needs.

Consequently, classroom management must align with the specified instruction, and coordination with stakeholders is of utmost importance. All of these tasks are complex and require specialised knowledge and skills. Regular teachers lacking sufficient experience in special education find it challenging to plan and develop an effective individualised curriculum (Li et al., 2022). Teachers with an understanding of inclusion are essential for effective inclusion (Thomas et al., 1998).

However, a Chinese study found that a significant number of educators lacked the pedagogical strategies or techniques required to meet the needs of children with disabilities (Li et al., 2022). Wang et al., (2024) assert that educators possess a favourable disposition towards individualised education programmes

Table 3.13 Behaviour modification of SN after three years

Task performance assessment	Task performance analysis														
	Before COVID-19					Transition					After COVID-19 (November 2023)				
	1	2	3	4	5	1	2	3	4	5	1	2	3	4	5
Coming to the school gate		x				x									x
Washing hands before entering the school area		x					x								x
Greeting teachers at the yard	x					x								x	
Walking across the yards	x					x									x
Opening Shoes		x					x								x
Entering the classroom		x					x								x
Pre-activity before learning		x					x								x
Greetings: saying good morning	x					x								x	
Washing hand activity before snack		x					x								x
Having meal		x					x								x
Playing at a playground	x					x						x			
Exchanging uniform	x					x									x
Toilette training	x					x					x				
Cleaning a desk/table	x					x								x	
Greeting: saying goodbye		x					x								x

Note 1 = rejection; 2 = need verbal and physical guidance; 3 = need verbal but no physical guidance; 4 = responding/understanding instruction but need stimulation; 5 = independence

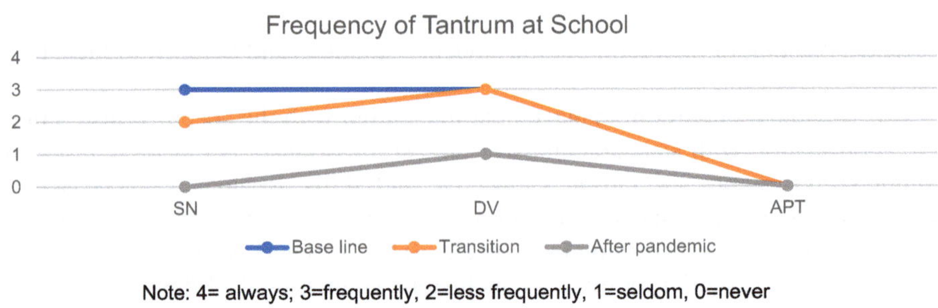

Fig. 3.9 Tantrum frequency of the three students at school

(IEPs) but encounter challenges in their implementation due to insufficient professional knowledge and abilities, limited resources and collaboration and inadequate teacher training. This necessitates specialised training on inclusive education and programmes for children with special needs. Implementing comprehensive teacher training that focuses on special education approaches can effectively bridge a considerable knowledge gap (Ji, 2024).

3.5.2 Inclusive Education in South Korea

In South Korea, the Constitutional Law that 'every person has the same right to receive an education commensurate with his ability', was launched in 1948. This was followed by the release of the national education regulations a year later. This regulation reaffirmed the policy of Inclusive education in the country (Kim, 2014). An Act of special education for persons with disabilities regulates the implementation of inclusive education from kindergarten to university, was launched in 1977. The 2021 statistics of school age children with special needs in South Korea can be categorised into four groups: (1) intellectual disability (n = 51,788, 52.8%), (2) autism (n = 15,215, 15.5%), (3) physical disability (n = 9,695; 9.9%) and (4) developmental delays (n = 9,367; 9.5%).

The emphasis of education in Korea is high academic achievement and performance, as indicated by the position of the country as the first highest performing country in the academic performance (OECD, 2018). In the highly competitive educational atmosphere in Korea, teachers and parents tend to prioritise high performing students and not giving equal public attention to diverse students' learning (Kang & Shin, 2022). The unequal attention hinders the implementation of effective inclusive education in the country. The country rates themselves as having lack of preparedness for working with students with disabilities (OECD, 2018).

Special education is mentioned in the general law for education but not regulated in a specific law until 1977 (Kim & Kim, 2020). Despite the massive improvement in inclusive education laws and policies nowadays, some challenges in the implementation of inclusive education are reported. For example, the teacher in regular schools have the general teaching qualification so that special training should be provided so that they can facilitate learning opportunities for special needs students. As a matter of fact, according to Seo (2020), there is a scarcity of specific guidelines on training preservice general education teachers to interact with students with disabilities. At present public disability awareness was reported as not adequate (Shin & Choi, 2021).

Similar to Indonesia, South Korea adheres a system of 6 years of elementary school, three year each in junior and senior high schools and 4 years in the university. Preschool facilities accepting children with disabilities under the supervision of the Ministry of Health and Welfare include (1) regular nursery schools, (2) nursery schools specialised for children with disabilities ('specialised nursery schools') and (3) integrated nursery schools for children with disabilities ('integrated nursery schools'). In particular, 'specialised nursery schools' and 'integrated nursery schools' are formally classified/specified as nursery schools providing special childcare for infants and young children with disabilities. More specifically, specialised nursery schools accept a minimum of 12 infants/children with disabilities and provide childcare focusing on medical treatments and support for these children. In addition, as specialised nursery schools provide inclusive education and childcare, they may also accept typically developing children up to 40 per cent of their capacity.

There is a strict screening of students in inclusive education. The 2022 Special Education Annual Report provided by the Ministry of Education mentions that approximately 73% of students eligible for special education received either part-time (56% of students) or full-time (17% of students) inclusive education. Special needs students are given the opportunities to be able to attend university. However, they often had difficulty taking classes even when they were in college. This is due to the fact that special schools in South Korea place a greater emphasis on a vocational curriculum, which does not cover the knowledge requirements required at the university level, namely Korean, English and Mathematics. Special needs children have the full support from the government from the basic education level up to college (Kim, 2013). However, the system makes students with special needs who want to continue their education to the university level should rely on extracurricular activities to learn the materials required at the university level.

Teachers in inclusive education must be qualified as "Formal Kindergarten teachers for Special Schools' ('Qualified SNE Teachers'), which is a 2nd-grade (or above) teaching certification issued by the Ministry of Education. To acquire this certification, teachers must major in special needs education at education/normal universities and complete at least 80 credits in special needs education and early childhood education (including practical training). In addition, national/public kindergartens set higher standards for Qualified SNE Teachers. Therefore, Qualified SNE Teachers must pass a recruitment examination to work at national/public kindergartens. If those with a formal teaching certification for special schools (kindergartens) do not take the recruitment examination, they can still work at private kindergartens or nursery schools.

A special movement about special education practices in Korea is called Eco-early childhood education (EECE), which urges that education for children with special needs should 'help children grow to be confident and happy' (Chung et al., 2008). Referring to EECE, there are 11 strategies in inclusive education:

i. All children belong
ii. Sense of community
iii. Diversity is valued
iv. Natural proportion
v. Realistic environment
vi. Services based on need rather than labels
vii. Support provided in general education classrooms
viii. Interdependence and support networks
ix. Support facilitators or collaborating teachers

x. Curriculum adapted when needed
xi. Resources combined.

These strategies reveal that there is a massive change of the approach used to support special needs children to develop their potentials. 'All children belong' is the representation of equality for all children in the classroom, in which the teacher needs to create the learning atmosphere where children do not feel any discrimination; 'sense of community' is the strategy that emphasises the activities that arises the feeling of togetherness; 'diversity valued' is the strategy in which differences should be positively valued by all class community; 'natural proportion' refers to strategies of learning that provide what needed in a proportional and natural way; 'realistic environment' is to do with efforts to establish learning environment and facilities that are realistic to students' conditions and needs; 'services based on needs rather than labels' is the strategies that put aside the unnecessary labels and provides the necessary supports when needed; 'support provided in general education classroom', which means that students with different abilities and needs are placed into the same learning group in the same classroom; 'interdependence and support network' refers to teaching and learning strategies that lead the students to support one another; 'support facilitators or collaborating teachers' means that the educator team work together to support special needs students gained their optimal development; 'curriculum adapted when needed' is to do with the flexibility of the curriculum that every special need students may need an individual curriculum to meet their needs; 'resource combined' is the strategies for teachers to make sure that learning is supported by appropriate learning resources.

The EECE movement requires that teachers of inclusive classes must undergo specialised training in special education. They need to complete an undergraduate programme in special education which is distinct from the general education licensure programme. Additionally, those who possess teaching licenses in other content areas should pursue a graduate-level special education teacher preparation programme (Kang & Shin, 2022).

Notwithstanding advancements in special education legislation, regulations and teacher training in South Korea, the implementation of inclusion techniques has faced numerous hurdles. A significant issue is the lack of awareness among teachers about the inclusion of children with impairments. In South Korea's competitive educational environment, teachers often prioritise high-achieving students, hence hindering the effective execution of inclusion. The referral procedure for at-risk pupils is usually protracted (Kwon, 2005) and educators often harbour negative attitudes and low expectations for students with impairments. Consequently, establishing inclusive educational environments is difficult without first enhancing educators' perceptions of children with special needs (Seo, 2020). Consequently, the requisite solution is to enhance teachers' knowledge and skills in addressing the needs of students with disabilities to facilitate the implementation of inclusive education.

3.5.3 The Implementation of Inclusive Education in Thailand

The legislative milestones indicate that the movement towards inclusive education in Thailand has been evolving since the mid-1990s, culminating in significant legal frameworks by 2008 that support the inclusion of students with disabilities in mainstream education (Kaur et al., 2016; Martinez & Fabella, 2023). The legal framework that supports inclusive education, including references to the Compulsory Education Act (2002) and the Persons with Disabilities Empowerment Act (B.E. 2550) (Vibulpatanavong, 2017).

There are three main policies as the cornerstones of the development of inclusive education in Thailand, which include (1) National Special Education Plan (1995). This plan focuses on developing educational services for children with disabilities, marking an early commitment to inclusive practices in the country. (2) It is the National Education Act (1999). This act emphasises the rights to education for individuals with disabilities and integrates them under the constitutional right to education, which further promotes the idea of inclusion. (3) Education Provision for People with Disabilities Act (2008): This law empowers children with special educational needs by granting them the right to education from early years to higher education. It declares it unlawful for schools to refuse admission to students with disabilities in mainstream classrooms, solidifying the commitment to inclusive education in Thailand (cited in Kaur et al., 2016).

Vibulpatanavong (2017) and Martinez and Fabella (2023) added several key policies related to education for persons with disabilities in Thailand, particularly focusing on inclusive education. Here are the main policies highlighted: (1) Compulsory Education Act (2002). This act mandates that all Thai children aged between 7–16 years must be enrolled in school, with provisions for special education for children with various disabilities. It emphasises the need for appropriate programmes and accommodations to ensure equal educational opportunities. (2) Persons with Disabilities Empowerment Act (B.E. 2550). This act entitles persons with disabilities to education as outlined in national education legislation. It emphasises the need for suitable facilities and services in both general and special education settings. (3) International Commitments. It considers global frameworks like the Salamanca Statement and the Convention on the Rights of Persons with Disabilities, which advocate for inclusive education and equal access to high-quality education for everyone, including individuals with disabilities.

Kaur et al., (2016) explain three fundamental approaches which underpin inclusive education in Thailand. Those are (1) Inclusive Education Philosophy. This approach is rooted in the belief that all students, regardless of their abilities or disabilities, should have access to quality education in mainstream settings. It promotes the idea of acceptance, respect and equal opportunities for all students. (2) Individualised Educational Plans (IEPs). The Thai Ministry of Education supports the development of IEPs for students with special needs, ensuring that their unique educational requirements are met through tailored strategies. (3) Collaborative Teaching, which emphasises collaboration among teachers, special educators and support staff to create a supportive learning environment for SEN students.

In addition, Vibulpatanavong (2017) and Martinez and Fabella (2023) expound several other approaches, namely (1) Education for All. This approach emphasises that all children, regardless of their abilities or disabilities, should have access to quality education. It advocates for the removal of barriers that prevent students with disabilities from participating in regular educational settings. (2) Universal Design for Learning (UDL). This approach involves designing educational environments and curricula that are accessible to all students from the outset. UDL aims to accommodate diverse learning needs by providing multiple means of engagement, representation and action/expression. (3) Professional Development for Teachers. Training regular teachers to understand and implement inclusive practices is crucial. This includes equipping them with the skills to modify curricula and teaching methods to accommodate students with diverse needs. (4) Community and Parental Involvement: Engaging families and communities in the educational process is essential for promoting inclusive education. This approach encourages collaboration between schools, parents and community organisations to support students with disabilities.

Kaur et al., (2016) also clarified some principles dealing with inclusive education, namely (1) Normalisation. This principle caters for individuals with disabilities to live and be educated alongside their peers, promoting social integration and acceptance. (2) Human Rights and Social Justice. Inclusive education is framed as a matter of human rights, ensuring that all students have equal access to educational opportunities and full participation in school life. (3) Diversity and Inclusion. Inclusive education aims to create an environment where all students feel valued and respected, regardless of their backgrounds or abilities. (4) Support and Training for Teachers. Effective inclusion requires that teachers receive adequate training and support to adapt their teaching practices and curricula to meet the needs of all students.

The strategies used in inclusive education, as explained in the case-study research in Thailand, are categorised into socio-emotional, cognitive and physical strategies (Kaur et al., 2016). Among those three categories, there are six strategies as follows:

(1) Socio-emotional Support: Teachers give support and encourage a positive environment for students with special educational needs (SEN), such as a teacher listening to a student with a delayed development by responding to her questions attentively.
(2) Peer Support and Collaboration: Using buddies to give help. By this way, the students with delayed development can improve peer interaction to promote learning. Buddies can also collaborate with teachers and senior students to monitor and support SEN students during school activities.
(3) Structured Interaction: It provides students with SEN an opportunity to take part in group activities, which enhance a sense of inclusion and value.
(4) Responsibility and Participation: The students with SEN are given tasks such as classroom duties to develop a sense of competence and self-esteem. They can feel appreciated and integrated in the classroom community.

(5) Non-instructional Support: Planning a schedule to give additional help during lunchtime and breaks to ensure that students with SEN get the necessary support.
(6) Use of Gestures and Expressions: Using physical gestures or facial expressions to show understanding and support when words may not work well.

Dhedchawanagon et al., (2024) elaborate the six strategies implemented for a successful programme of inclusive education:

(1) Curriculum Management Adjustments: Teachers are encouraged to modify the curriculum to include more visual learning tools, especially for students with hearing impairments, to facilitate better understanding of complex subjects.
(2) Individualised Education Programmes (IEPs): Teachers develop IEPs tailored to the unique needs of each student, which include specific goals and teaching methods designed to support their learning.
(3) Use of Supplemental Curriculum: Teachers create supplemental materials and lessons that focus on essential skills, such as life skills and social skills, particularly for students with autism.
(4) Peer-to-Peer Learning: Implementing collaborative activities where students can learn from each other, which helps in building social skills and acceptance among peers.
(5) Regular Assessment and Modification of Teaching Plans: Teachers are advised to evaluate and adjust IEPs at least twice a year based on student progress, ensuring that the teaching strategies remain effective and relevant.
(6) Community and Parental Involvement: Engaging parents and community members in the educational process to create a supportive network for students with special needs.

During its implementation, the teachers faced challenges. Several problems of inclusive education in Thailand raised by Kaur et al., (2016) are:

(1) Lack of Governmental Support. There is often insufficient governmental backing for inclusive education policies, leading to ineffective implementation and a gap between policy and practice.
(2) Inadequate Training for Teachers. Many teachers lack the necessary training and skills to effectively support students with special educational needs (SEN), which can lead to negative attitudes towards inclusion.
(3) Insufficient Resources and Infrastructure. Schools frequently face challenges related to inadequate funding, resources and infrastructure to support inclusive practices, making it difficult to accommodate SEN students.
(4) Unsystematic Curriculum Modifications. The absence of systematic modifications in school curricula and teaching approaches can hinder the effective inclusion of SEN students.
(5) Negative Attitudes and Perspectives. Teachers' attitudes towards inclusion can be influenced by a lack of support and training, leading to resistance against implementing inclusive practices.

Additionally, Martinez and Fabella (2023) explained six challenges of inclusive education, namely:

(1) Lack of Teacher Training: Many teachers are not adequately trained to handle children with disabilities. This lack of training affects their ability to adapt curricula and teaching methods to meet the diverse needs of students with special educational needs (SEN).
(2) Cultural Attitudes: Cultural perceptions of disability can hinder the implementation of inclusive education. The article mentions that many Thai students' express feelings of 'songsarn' (sympathy or pity) towards peers with disabilities, which can lead to stigmatisation rather than acceptance.
(3) Inconsistent Policy Implementation: Although there are laws and policies supporting inclusive education, their implementation is often inconsistent. The article points out that the government does not always execute these policies coherently, which undermines efforts to promote inclusion.
(4) Physical Barriers: There are physical challenges related to accessibility in schools and transportation. The transition from home to school can be particularly difficult for children with physical disabilities, especially in urban areas like Bangkok.
(5) Resource Limitations: Schools may lack the necessary resources, such as specialised materials and support staff, to effectively support students with disabilities. This can create an environment where inclusive education is difficult to achieve.
(6) Attitudinal Barriers: There are also attitudinal barriers among educators and the community that can impede the acceptance and integration of students with disabilities into mainstream classrooms.

3.6 Future Directions

The investigation of the approaches and strategies employed in teaching inclusive education indicates similarities among the four nations: Indonesia, Thailand, Korea and China. Differentiated instruction serves as the foundation for tailoring the teaching and learning process to meet the diverse requirements of pupils. An IEP is considered essential. Collaborative engagement with stakeholders is essential for the treatment of students. The implementation of the various strategies has demonstrated favourable outcomes and evident enhancements in behaviour modification and communication abilities. The study's findings in Bali align with those of Vasoya and Vansdadiya (2023) and Murtagh et al. (2022), indicating that the implementation of varied strategies and programmes via meaningful projects and play-based activities enhances students' communication skills and behaviour, while also fostering their empathy and awareness of inclusivity.

Inclusivity plays a crucial role in the context of heterogeneous classes, which involve students with diverse backgrounds who come from different countries of origin and different learning styles and abilities that involve students with ASD. By encouraging inclusivity, classes can create a learning environment that supports the growth and development of every student regardless of differences (Nelis et al.,

2023; Szumski et al., 2022). Children with disabilities will grow in self-esteem if given as many learning chances as possible in a loving and supportive atmosphere (Damianidou & Phtiaka, 2018).

The inclusive approach allows recognition of individual diversity, providing opportunities for every student to feel valued and recognised (Wibowo & Muin, 2018). It is not only about creating social justice but also providing benefits academically by enabling collaboration between students with different backgrounds (Thomas, 2016; Woodcock et al., 2022). By strengthening inclusivity, heterogeneous classrooms can become platforms that motivate, support and inspire students to achieve their full potential. This supports Anupriya and Salim (2014), who confirm that the primary goals of inclusive education are to help students with disabilities achieve independence and to ensure that everyone has access to educational materials, activities and practices.

This study also found that the three approaches of teaching and learning (Differentiated Instruction, The Whole-Person Approach and Constructivism) remained effective before and after the pandemic in classes in the school in Bali that included students with ASD. The differences are those applied in the classroom of students with special needs (ASD), which requires teachers' adaptive strategies that meet students' needs and characteristics (Khayati et al., 2020) and higher understanding (Sipayung & Hsu, 2023).

The challenges faced by the teacher were when students could not study at school with teachers, such as during holidays or at home. Without consistent teacher intervention, students' abilities and attitudes return to the baseline. Teachers must restart managing students' SEL so they are prepared to go on to the next stage of learning. The significance of SEL is not only in the framework of formal education but also as an investment in developing persons who are balanced, caring and ready to tackle the complexity of modern life (MacCann et al., 2019; Schonert-Reichl, 2017). The three approaches to teaching and learning (i.e. differentiated instruction, constructivism and the whole-person approach) play an essential role in generating teaching methods and strategies.

Tomlinson (2011) states that differentiated instruction involves teaching students with individual differences in skill levels, interests and learning styles. The students with ASD were taught using different individual education plans designed based on each student's needs. Constructivism stimulates the chance for the students to have direct experiences and social interaction from which new knowledge can be formulated, and cooperative learning is empowered (Slavin, 2015). Besides, constructivism could also improve student's motivation to learn (Do et al., 2023; Tsai et al., 2023). At the same time, the whole-person approach recognises students' complexity as human beings with physical, mental and psychomotor dimensions. By combining the three approaches holistically, teachers can create an inclusive, relevant and responsive environment for every student, fulfilling the complexity of their needs for the future.

During the transition period, students need to be provided with SEL as a support system to manage their feelings and social skills. SEL programmes help students build their social connections, improve their empathic feelings and manage their conflicts, all becoming necessary skills to adapt to the changing learning environment (Rymanowicz et al., 2020; Silke et al., 2024). SEL also has a vital role in managing the students' mental health and assisting students with the social acculturation process in teaching and learning (Ağırkan & Ergene, 2022; MacCann et al., 2019). By prioritising emotional and social aspects through SEL, we can ensure that this transition period focuses on academic and overall recovery, creating a supportive and inclusive learning environment. According to Bell (2016) addressing social and emotional competencies is crucial in twenty-first-century education to provide sustainable education and even more sustainable personality development for instructors and students.

After the transition process, the focus was on students' meaningful academic learning. Using relevant approaches as foundations, teachers design teaching strategies that accommodate all the differences and heterogeneity in the classroom, enhance the special needs students' achievements in their behaviour modifications and communication abilities, and empower inclusivity (Bryant et al., 2008). Although it took a while, the post-pandemic advancements of the students show a noticeable improvement. They all develop at different rates, even though their rates are distinct. Despite the transitional process that has been provided, many students experience learning loss. Therefore, the school decided to provide a holistic review and remedial treatment for at least the first three months before moving forward to regular teaching processes and also provide multiculturalism to maintain inclusivity.

Inclusivity in education creates an environment where every student, including those with special needs, is recognised, valued and supported in achieving their potential. Therefore, applying the principles of multiculturalism can help create an inclusive education system where diversity is seen as an asset that enriches the learning experience and allows every student, including children with special needs, to develop optimally. It supports Sipayung and Hsu (2023), who reconfirms that inclusivity in the curriculum reform in the context of Indonesia is very relevant to achieving SDG 4 because an educational approach that includes all students without discrimination makes an essential contribution to ensuring fair, equitable and quality access to education for all, in line with global efforts to achieve education inclusive and sustainable. Inclusive multiculturalism allows students to feel fully included and linked while acknowledging their individuality and differences (Ferdman, 2017).

Besides sharing similar approaches and strategies, Indonesia, China, Thailand and Korea also share commonalities in the challenges faced by teachers. The implementation of inclusive education in China, Thailand and Korea presents significant challenges for achieving a genuinely inclusive educational landscape. The primary issue is the scarcity of resources. A lack of teachers proficient in inclusive teaching methodologies exacerbates the inadequacy, leading to a disconnect between students' needs and instructional delivery (Bualar, 2015; Ji, 2024; Kim, 2013). This phenomenon also occurs in Indonesia (Padmadewi et al., 2024).

Examining the implementation of inclusive education in Indonesia, China, Thailand and Korea reveals inadequate training and resources, which are crucial for its success. Teachers frequently lack the specialised training necessary to deliver effective teaching and learning experiences for students with special needs, thereby hindering the fulfilment of inclusive education (Li et al., 2022). Consequently, educators require ongoing professional development in this area; they must understand the diverse developmental needs of students and demonstrate proficiency in classroom management. They should possess expertise in modifying and adapting curriculum and instruction and engage in meaningful collaborations for effective integrated education.

3.7 Conclusion

The implementation of inclusive education in Indonesia, China, Thailand and Korea shares much in common. In these countries, special needs educators employ various strategies for the purpose of supporting special needs children's optimal development. The strategies include differentiated instruction, individualised learning, collaboration and Universal Design for Learning, alongside the incorporation of social-emotional learning, to ensure equitable and high-quality education for diverse students. Most educators in these nations require specialised training to provide optimal learning experiences for effective inclusive education. Public awareness of the significance of inclusive education is essential for its successful implementation in schools. The enhancement of public knowledge about inclusive education and increased support will influence the acceptance of diversity, eliminate stigma and facilitate the supply of resources necessary to guarantee equitable access to quality education for all students.

In light of future considerations and the comparable challenges encountered by educators across Asia, it is imperative for Asian nations to collaborate in the pursuit of establishing high-quality inclusive education within the region to manifest the principles of Education for All. This chapter gives research-based evidence about how well-managed inclusive education affects children's personal and social development. The close reading of inclusive education implementation in China, Thailand and Korea enhances our comprehension of the necessary maintenance and improvement to ensure equal opportunities for quality education for all children, regardless of their circumstances.

Highlights

- Differentiated instruction, the whole-person approach and constructivism are foundational strategies for fostering inclusivity and addressing the diverse needs of students, including those with ASD, in classrooms across Indonesia, Thailand, Korea and China.
- Social and Emotional Learning (SEL) is critical in supporting students' mental health and social acculturation, particularly during transitions, and helps students build social connections and empathy.

- Post-pandemic, the focus shifted to remedial support and holistic reviews to address learning losses, ensuring that students received the necessary emotional and academic recovery before returning to regular teaching.
- A shortage of trained teachers and resources across these nations underscores the need for ongoing professional development to ensure effective delivery of inclusive education for all students.

References

Ağırkan, M., & Ergene, T. (2022). What do social and emotional learning interventions (SEL) tell us? A meta-analysis. *Revista de Psicodidáctica (English Ed.), 27*(2), 97–108. https://doi.org/10.1016/j.psicoe.2022.02.002

Amthor, R. F., & Roxas, K. (2016). Multicultural education and newcomer youth: Re-imagining a more inclusive vision for immigrant and refugee students. *Educational Studies, 52*(2), 155–176. https://doi.org/10.1080/00131946.2016.1142992

Anupriya, M. S., & Salim, D. (2014). *Including children with special needs at the primary stage: Department of education of groups with special needs*. National Council of Education Research and Training.

Bell, D. V. J. (2016). Twenty-first century education: Transformative education for sustainability and responsible citizenship. *Journal of Teacher Education for Sustainability, 18*(1), 48–56. https://doi.org/10.1515/jtes-2016-0004

Bryant, D. P., Smith, D. D., & Bryant, B. R. (2008). *Teaching students with special needs in inclusive classrooms*. Pearson Education, Inc.

Bualar, T. (2015). What has gone wrong with inclusive education in Thailand? *Journal of Public Affairs, 16*(2), 156–161. https://doi.org/10.1002/pa.1563

Chung, K.-S., Lee, S.-Y., & Goh, E.-K. (2008). Inclusive education: Teacher's experiences in South Korean eco-early childhood education. *AARE Annual Conference.*

Damianidou, E., & Phtiaka, H. (2018). Co-operating with parents for equal opportunities in education. *International Journal about Parents in Education, 10*, 91–102.

Dhedchawanagon, K., Pitsamai, S., & Nak-in, N. (2024). Development of management guidelines of inclusive education in Special Education Center, Phetchabun Province. *Community and Social Development Journal, 25*(2), 156–172. https://doi.org/10.57260/csdj.2024.268040

Do, H.-N., Do, B. N., & Nguyen, M. H. (2023). How do constructivism learning environments generate better motivation and learning strategies? The design science approach. *Heliyon, 9*(12), e22862. https://doi.org/10.1016/j.heliyon.2023.e22862

Ferdman, B. M. (2017). Paradoxes of inclusion: Understanding and managing the tensions of diversity and multiculturalism. *Journal of Applied Behavioral Science, 53*(2), 235–263. https://doi.org/10.1177/0021886317702608

Freeman, S., Eddy, S. L., McDonough, M., Smith, M. K., Okoroafor, N., Jordt, H., & Wenderoth, M. P. (2014). Active learning increases student performance in science, engineering, and mathematics. *Proceedings of the National Academy of Sciences, 111*(23), 8410–8415.

Jansen, W. S., Otten, S., & Zee, K. I. (2015). Being part of diversity: The effects of an all-inclusive multicultural diversity approach on majority members' perceived inclusion and support for organizational diversity efforts. *Group Processes and Intergroup Relations, 18*(6), 817–832. https://doi.org/10.1177/1368430214566892

Ji, A. (2024). The study on the dilemma of teachers in inclusive education in China. *Transactions on Social Science, Education and Humanities Research, 5*, 340–346. https://doi.org/10.62051/496d3g73

Kang, J., & Shin, M. (2022). Special education in South Korea: Policies and issues. *Journal of Special Education Preparation, 2*(2), 70–78. https://doi.org/10.33043/josep.2.2.70-78
Kaur, A., Noman, M., & Awang-Hashim, R. (2016). Exploring strategies of teaching and classroom practices in response to challenges of inclusion in a Thai school: A case study. *International Journal of Inclusive Education, 20*(5), 474–485. https://doi.org/10.1080/13603116.2015.1090489
Khayati, N. A., Muna, F., Oktaviani, E. D., & Hidayatullah, A. F. (2020). Teacher's role program in inclusive education to achieve Sustainable Development Goals (SDGs). *KOLOKIUM Jurnal Pendidikan Luar Sekolah, 8*(1), 1–6. https://doi.org/10.24036/kolokium-pls.v8i1.382
Kim, Y. W. (2013). Inclusive education in Korea: Policy, practice, and challenges. *Journal of Policy and Practice in Intellectual Disabilities, 10*(2), 79–81. https://doi.org/10.1111/jppi.12034
Kim, Y. W. (2014). Inclusive education in South Korea. *International Journal of Inclusive Education, 18*(10), 979–990. https://doi.org/10.1080/13603116.2012.693402
Kim, K. H., & Kim, J. (2020). Transition to higher education in South Korea: Current status and issues. *Intervention in School and Clinic, 55*(5), 319–324. https://doi.org/10.1177/1053451219881718
Kusimo, A. O., & Chidozie, F. C. (2019). Inclusive education and sustainable development goals: A study of the physically challenged in Nigeria. *Cogent Arts & Humanities, 6*(1), 1–13. https://doi.org/10.1080/23311983.2019.1684175
Kwon, H. (2005). Inclusion in South Korea: The current situation and future directions. *International Journal of Disability, Development and Education, 52*(1), 59–68. https://doi.org/10.1080/10349120500071910
Li, D., Gavaldà, J. M. S., & Badia Martín, M. (2022). Listening to students' voices on inclusive teaching strategies in Chinese primary schools. *International Journal of Chinese Education, 11*(2). https://doi.org/10.1177/2212585X221120971
MacCann, C., Jiang, Y., Brown, L. E. R., Double, K. S., Bucich, M., & Minbashian, A. (2019). Emotional intelligence predicts academic performance: A meta-analysis. *Psychological Bulletin, 0000219*, 1–98. https://doi.org/10.1037/bul0000219
Martinez, G. L., & Fabella, F. E. T. (2023). Educational attainment and years of teaching children with special needs as factors in teachers' attitudes toward inclusive education in Thailand. *International Journal For Multidisciplinary Research, 5*(6), 1–15. https://doi.org/10.36948/ijfmr.2023.v05i06.9408
McCabe, H., & Jiang, T. (2024). Inclusive education in China: From policy to implementation to on-the-ground experiences. *Review of Disability Studies: An International Journal, 19*(1), 1–42. http://www.rdsjournal.org/index.php/journal/article/view/119
Murtagh, E. M., Sawalma, J., & Martin, R. (2022). Playful maths! The influence of play-based learning on academic performance of Palestinian primary school children. *Educational Research for Policy and Practice, 21*(3), 407–426. https://doi.org/10.1007/s10671-022-09312-5
Nelis, P., Pedaste, M., & Šuman, C. (2023). Applicability of the model of inclusive education in early childhood education: A case study. *Frontiers in Psychology, 14*(July). https://doi.org/10.3389/fpsyg.2023.1120735
OECD. (2018). *Education at a glance 2018: OECD indicators.* OECD Publishing. https://doi.org/10.1787/eag-2018-en
Padmadewi, N. N., Artini, L. P., & Sukadana, I. M. S. A. (2024). Teaching an inclusive education in EFL setting: A phenomenological study. *The Art of Teaching English as a Foreign Language, 5*(2), 186–195. https://doi.org/10.36663/tatefl.v5l2.905
Rambla, X., & Langthaler, M. (2016). The SDGs and inclusive education for all: From special education to addressing social inequalities. *Briefing Paper, Austrian Foundation for Development Research* (ÖFSE, 14). https://doi.org/10.36663/tatefl.v5l2.905
Rapp, A. C., & Corral-Granados, A. (2021). Understanding inclusive education – a theoretical contribution from system theory and the constructionist perspective. *International Journal of Inclusive Education*, 1–17. https://doi.org/10.1080/13603116.2021.1946725

Rymanowicz, K. A., Moyses, K. J., & Zoromski, K. S. (2020). School readiness. In J. B. Benson (Ed.), *Encyclopedia of Infant and Early Childhood Development* (Second Edition) (pp. 55–64). Elsevier. http://hdl.handle.net/10419/142739

Schonert-Reichl, K. A. (2017). Social and emotional learning and teachers. *Future of Children, 27*(1), 137–155.

Seo, J. (2020). Initial teacher education for inclusion: A literature review of special educational needs and disabilities in the United Kingdom and in South Korea. *International Journal of Special Education, 35*(2), 83–92. https://doi.org/10.52291/ijse.2020.35.16

Shin, M., & Choi, N. (2021). The invisible population of students with learning disabilities in South Korea. *Intervention in School and Clinic, 57*(3), 50–53. https://doi.org/10.1177/10534512211014883

Silke, C., Davitt, E., Flynn, N., Shaw, A., Brady, B., Murray, C., & Dolan, P. (2024). Activating social empathy: An evaluation of a school-based social and emotional learning programme. *Social and Emotional Learning: Research, Practice, and Policy, 3*, Article 100021. https://doi.org/10.1016/j.sel.2023.100021

Sipayung, E. R., & Hsu, C.-H. (2023). Indonesian teachers' beliefs and competencies in implementing the Indonesian prototype curriculum to promote SDG Goal 4. *Educative: Jurnal Ilmiah Pendidikan, 1*(3), 118–126. https://doi.org/10.37985/educative.v1i3.220

Slavin, R. E. (2015). Cooperative learning in elementary schools. *International Journal of Primary, Elementary and Early Years Education, 43*(1). https://doi.org/10.1080/03004279.2015.963370

Slee, R. (2018). *Defining the scope of inclusive education: Think piece prepared for the 2020 Global Education Monitoring Report*. UNESCO.

Szumski, G., Smogorzewska, J., & Grygiel, P. (2022). Academic achievement of students without special educational needs and disabilities in inclusive education—Does the type of inclusion matter? *PLoS ONE, 17*(7), Article e0270124. https://doi.org/10.1371/journal.pone.0270124

Thomas, G., Walker, D., & Webb, J. (1998). *The making of the inclusive school*. Taylor & Francis.

Thomas, L. (2016). Developing inclusive learning to improve the engagement, belonging, retention, and success of students from diverse groups. In M. Shah, A. Bennett & E. Southgate (Eds.), *Widening Higher Education Participation* (pp. 135–159). Chandos Publishing. https://doi.org/10.1016/B978-0-08-100213-1.00009-3

Tomlinson, C.A. (2011). *How to differentiate instruction in mixed-ability classrooms*. Association for Supervision and Curriculum Development.

Tonegawa, Y. (2023). Education in SDGs: What is inclusive and equitable quality education? In S. Urata, K. Kuroda, & Y. Tonegawa (Eds.), *Sustainable development disciplines for humanity: Breaking down the 5Ps—people, planet, prosperity, peace, and partnerships* (pp. 117–138). Springer Nature Singapore. https://doi.org/10.1007/978-981-19-4859-6

Tsai, C.-A., Song, M.-Y.W., Lo, Y.-F., & Lo, C.-C. (2023). Design thinking with constructivist learning increases the learning motivation and wicked problem-solving capability—An empirical research in Taiwan. *Thinking Skills and Creativity, 50*, Article 101385. https://doi.org/10.1016/j.tsc.2023.101385

Vansdadiya, R. P., Vasoya, N. H., & Gondaliya, P. R. (2023). Beyond the classroom walls: Activity-based learning for a real-world math experience. *Asian Journal of Education and Social Studies, 43*(1), 1–9. https://doi.org/10.9734/ajess/2023/v43i1930

Vasoya, N., & Vansdadiya, R. (2023). Effective strategies for promoting foundational literacy and numeracy in early childhood education. *Journal of Social Sciences, 19*(1), 92–95. https://doi.org/10.3844/jssp.2023.92.95

Vibulpatanavong, K. (2017). Inclusive education in Thailand. *International Conference of Early Childhood Education*, 67–70. https://doi.org/10.2991/icece-17.2018.16

Wang, D., Huang, L., Huang, X., Deng, M., & Zhang, W. (2024). Enhancing inclusive teaching in China: Examining the effects of principal transformational leadership, teachers' inclusive role identity, and efficacy. *Behavioral Sciences, 14*(3), 175. https://doi.org/10.3390/bs14030175

Weissberg, R. P., Durlak, J., Domitrovich, C. E., & Gullotta, T. P. (2015). Social and emotional learning: Past, present, and future. In J. A. Durlak, C. E. Domitrovich, R. P. Weissberg, & T. P.

Gullotta (Eds.), *Handbook of Social and Emotional Learning: Research and Practice* (pp. 3–19). The Guilford Press.

Wibowo, S. B., & Muin, J. A. (2018). Inclusive education in Indonesia: Equality education access for disabilities. *KnE Social Sciences, 3*(5 SE-Articles). https://doi.org/10.18502/kss.v3i5.2351

Woodcock, S., Sharma, U., Subban, P., & Hitches, E. (2022). Teacher self-efficacy and inclusive education practices: Rethinking teachers' engagement with inclusive practices. *Teaching and Teacher Education, 117*, Article 103802. https://doi.org/10.1016/j.tate.2022.103802

Zhang, H. (2010). Cultivating an inclusive individuality: Critical reflections on the idea of quality education in contemporary China. *Frontiers of Education in China, 5*(2), 222–237. https://doi.org/10.1007/s11516-010-0015-0

Zhao, S., & Kokhan, S. T. (2022). Accessibility of education for persons with disabilities in China: Transforming special and inclusive education. *Professional Education in the Modern World, 12*(2), 370–376. https://doi.org/10.20913/2618-7515-2022-2-20

Ni Nyoman Padmadewi finished her undergraduate degree in English education at Udayana University, Denpasar, Bali, Indonesia. She earned a master's degree in Education from the School of Education at Macquarie University, NSW, Australia, in 1993, and a doctorate degree in Linguistics from Udayana University, Denpasar, Bali, Indonesia, in 2005. Padmadewi is also the founder of an inclusive bilingual kindergarten and primary school in North Bali, Indonesia, as well as a consultant for curriculum designs in teaching English for students with special needs and a reviewer for many international journals.

Luh Putu Artini is a professor in English language education at Universitas Pendidikan Ganesha, Bali, Indonesia. She earned her bachelor degree from Udayana University, Bali (1987); master degree from La Trobe University, Australia (1994) and Ph.D. in Education from Newcastle University, Australia (2006). She attended a non-degree training on Bilingual Education in Leuven Education College, Belgium (2009); and Early Literacy in Indiana University (2010). Artini is the member of three international professional organisations: Teaching English as a Foreign Language in Indonesia (TEFLIN), ASIA TEFL and International Society for Development & Sustainability.

Ni Made Ratminingsih is a senior lecturer at the English Education Study ProgrammeUniversitas Pendidikan GaneshaBali. She earned her master's degree from Deakin University, Australia in 1997 and her doctorate from Jakarta State University in 2011. Her expertise includes ELT, particularly TEYL, assessment and literacy skill development in EFL Context. She has published in internationally reputable journal and serves as a reviewer for Scopus-indexed journals, including Q1 journals such as the International Journal of Instruction (Turkey), Studies in English Language Education (Indonesia), and Studies in Educational Evaluation (Elsevier, UK) and the Q4 journal Forum for Linguistics Studies (Singapore).

Chapter 4
Exploring Early Education for Sustainable Development: Insights from a Japanese Kindergarten

Fung Lan Loo, Mahani Abdul Malik, Aliza Ali, Md Rosli Ismail, and Nurhusna Abdul Malek

Abstract In an era marked by growing concerns for global sustainability, the role of education in shaping the future for everyone is paramount. However, a fundamental question arises: Are we adequately preparing future generations with the essential skills and resources? This critical inquiry fuels the global conversation on sustainability. Responding to the need for educational solutions to address sustainability challenges, Education Sustainable Development (ESD) has gained prominence. This study explores kindergarten practices in Kagoshima Prefecture, Japan, revealing how ESD elements are seamlessly integrated into the curricula, teaching practices and daily routines. The research identifies critical components of high-quality preschool education that aim to instil knowledge, awareness and action while embedding ESD values in early childhood development. This exploration highlights its relevance and potential influence across Asia, urging educators, researchers and policymakers to reconsider the transformative role of early childhood education in fostering a sustainable future. The study offers novel perspectives on nurturing future leaders with a dedicated focus on sustainability, emphasising the importance of culturally grounded practices and their scalability across diverse Asian contexts.

Keywords Curriculum · Education for sustainable development · Preschool · Early childhood · Japan

---

F. L. Loo (✉) · M. A. Malik · A. Ali · M. R. Ismail · N. A. Malek
Faculty of Education, Open University Malaysia, Petaling Jaya, Malaysia
e-mail: loofunglan@oum.edu.my

M. A. Malik
e-mail: mahani_abdulmalik@oum.edu.my

A. Ali
e-mail: aliza_ali@oum.edu.my

M. R. Ismail
e-mail: mdrosli@oum.edu.my

N. A. Malek
e-mail: nurhusna@oum.edu.my

A. Pandian et al. (eds.), *Future of Education in Asia*,
https://doi.org/10.1007/978-3-032-07285-6_4

4.1 Introduction

Education is becoming increasingly important in an era of growing concern for the sustainable life on earth for everyone now and in the future. However, are we educating future generations with the skills and resources they need to thrive and ensure our precious planet's survival? Education for Sustainable Development (ESD) arose from the need for education to meet growing sustainability challenges. This chapter explores the complex structure of ESD within preschool education, a foundational stage that serves as the seed of future societal attitudes and behaviours. It begins by exploring the integration of ESD principles into early childhood education, a crucial period for instilling sustainability and environmental responsibility values. This contextual exploration is within Japan's National Curriculum Standard for Kindergarten, a pivotal policy document that shapes the nation's approach to nurturing young minds. These guidelines serve as a blueprint for embedding sustainability concepts into the fabric of early learning and development.

The chapter then examines specific pedagogical practices instrumental in fostering sustainable societies. These practices are not just theoretical constructs but are actionable strategies educators employ to cultivate an awareness of and commitment to sustainable living from a tender age. Dissecting these methodologies provides insight into how educators can effectively translate the abstract concept of sustainability into concrete learning experiences for preschool children. Through this analysis, the chapter aims to illuminate the decisive role that early childhood education plays in shaping a sustainable future. It underscores the importance of aligning educational practices with broader societal sustainability goals, preparing young learners for academic success and active participation in building more sustainable societies.

4.2 Education for Sustainable Development in Preschool

Education for Sustainable Development (ESD) in kindergarten represents an approach that seamlessly integrates principles, values and sustainable development practices into all facets of education and learning (UNESCO, 2017). This paradigm empowers young learners to make informed decisions and take responsible actions, fostering environmental integrity, economic viability and a just society for both present and future generations while embracing cultural diversity (UNESCO, 2014). ESD is realised in preschool through interactive activities and play-based learning, encouraging children to explore their natural and social environments. These activities, including storytelling, role-playing and nature exploration, foster environmental awareness and nurture empathy and essential social skills (Davis & Elliott, 2014).

The ESD curriculum covers themes such as biodiversity, water conservation, recycling and energy-saving, engaging children in impactful projects like classroom gardens that teach about plant life cycles and biodiversity (Pratt & Luby, 2016). Moreover, preschool educators play a pivotal role in modelling sustainable practices and guiding children to comprehend complex concepts tailored to their understanding (Siraj-Blatchford, 2010). Crucially, ESD in preschool emphasises cognitive development and the emotional and social aspects of learning. Children must develop a sense of connection and responsibility towards their environment and community from an early age. Integrating ESD principles into early childhood education lays a solid foundation for a more sustainable future, instilling stewardship and respect for the natural world in the next generation (McKeown & Hopkins, 2003).

This study highlights Japanese kindergartens as vital environments for cultivating sustainability, recognised for their holistic development focus and cultural heritage. Centring on Kagoshima Prefecture—a region renowned for its cultural diversity and abundant natural resources—the research explores the integration of for Sustainable Development ESD within early childhood education. The study explores how ESD elements are woven into preschool curricula, teaching methods and daily routines. This book chapter unveils the nuanced incorporation of ESD principles, showcasing how these practices nurture young learners capable of driving sustainable change in the future.

4.3 Holistic Approaches to Sustainability in Japanese Early Childhood Education

Japan's Early Childhood Care and Education (ECCE) system is a parallel structure composed of preschools and day nurseries. The National Curriculum Standard for Kindergarten governs Japan's preschools, while day nurseries adhere to the minimal requirements outlined in the National Guidelines for Care and Education. Previously, the two main early childhood education institutions, *Jochen (kindergarten)* (kindergarten) and *Haiku* (childcare) operated independently. However, in 1999, the Japanese Ministry of Education and Ministry of Welfare established a unified approach to integrate care and education. Both institutions were merged as *Boiku*, where care and education occur in the same environment.

In Japanese kindergartens, the curriculum comprises five areas: health, human relationships, environment, language and expression. The areas are presented to the children in a harmonious and integrated approach, considering the local culture, community, family and children's development. The curriculum also embeds sustainable development education, where the environment and sustainability education are linked. Good governance has a vital role in the success of the curriculum. Japanese teachers need to support children's social and emotional development, have an outlook on things based on Japanese culture, and have the ability to observe, care for and look at the world through a wide-angle lens. As we know, children learn from watching good examples, and values are learned from watching others.

Good pedagogical practices are required for the children to experience, become aware of, understand, reflect and make judgments through the curriculum. To embrace a zest for living, the kindergarten nurtures the following qualities and abilities:

- A foundation of knowledge and skills that enable individuals to feel, realise, understand and apply what they have learned through rich experiences.
- An essential ability to think, judge and express enables individuals to think, try, innovate and communicate using their skills and knowledge.
- Learning disposition and humanity enable individuals to embrace lifelong learning, striving for a better life while fostering the development of emotions, motivation and attitudes.

Play pedagogy in Japanese early childhood education is pivotal in fostering ESD through integrating *ganbari. Ganbari* is a concept deeply rooted in Japanese culture that embodies endurance, persistence, hard work and effort, essential for preparing children to face challenges with resilience (Hoffman, 2000; Yamamoto & Satoh, 2019). Children engage in creative problem-solving and teamwork through structured and unstructured play activities. For example, building structures with blocks or creating collaborative artwork requires sustained effort, patience and adaptability. These play-based experiences allow children to practice perseverance in an enjoyable and low-pressure setting. Moreover, the concept of *ganbari* is closely tied to *Shudan seikatsu* (group living), which emphasises the importance of collective responsibility and social harmony. Yamamoto and Satoh (2019) emphasise that group life offers a unique context for children to develop individual effort and a sense of interdependence. Through participation in group activities, children learn to balance their personal needs with the goals and expectations of the group, fostering empathy, cooperation, and a shared commitment to success (Aspinall, 2016; Hellman, 2016). By embedding *ganbari* into play pedagogy and group-based learning, Japanese kindergartens cultivate a generation equipped with the skills and values necessary for sustainable living.

Another crucial pedagogical practice in Japanese kindergartens involves using tasks and rules to deepen children's understanding of their roles and responsibilities within a group. According to Guo and Kuramochi (2019), engaging children in tasks and rules helps them internalise the roles of both themselves and their teachers, fostering an appreciation for collective responsibility. These practices enable children to align their learning behaviours with societal expectations, bridging the gap between individual actions and communal goals. By instilling cultural and social knowledge early on, these practices contribute to shaping responsible citizens who are attuned to the needs of a sustainable society. Through this alignment, children develop a sense of duty and accountability, which is essential for contributing to a harmonious and sustainable future.

Japan's commitment to sustainability is further reflected in its policy frameworks. As one of the few countries where the government has designated Education for Sustainable Development (ESD) as a core policy, Japan has taken a leading role in advancing global sustainability education (Nagata, 2017). Japan spearheaded the 2005–2014 United Nations Decade on Education for Sustainable Development (UNDESD), highlighting its proactive stance in embedding ESD principles across educational systems. While UNESCO provided overarching guidelines for ESD, each region, including Japan, was granted autonomy to adapt these principles to their local context and priorities (UNESCO, 2005). This autonomy allowed Japanese kindergartens, like the one examined in this study, to integrate sustainability into their pedagogy in culturally relevant ways. By integrating global ESD principles with local practices, Japanese kindergartens prepare children to understand and handle challenges in the world while encouraging them to support and build sustainable communities.

Figure 4.1 illustrates a framework for sustainable education in kindergarten to address the research question of how Japanese kindergarten integrates ESD into early childhood education. The foundation of sustainable education is built upon the physical environment, where the kindergarten is envisioned as an eco-school, prioritising health and safety. It features spacious and accessible open areas, large windows and effective ventilation, minimising the use of furniture in classrooms. Within this eco-school environment, the curriculum is a robust foundation that seamlessly integrates education, intertwining sustainable development, good governance and an interdisciplinary curriculum with a central focus on local society and culture. Through effective pedagogical practices, such as play, tasks and rules, the curriculum aims to cultivate individuals with strong and balanced characters nurturing both mind and body. These individuals are intended to embody the qualities of responsible Japanese citizens, contributing significantly to developing a sustainable society.

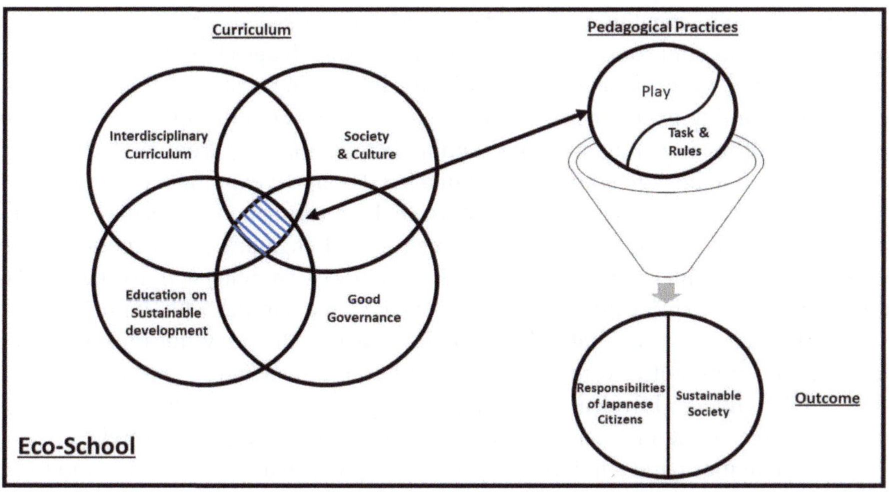

Fig. 4.1 Framework for sustainable development education in Japanese kindergarten linking eco-school infrastructure, interdisciplinary curriculum and culturally contextualised pedagogical strategies

4.4 Integration of ESD Practices in a Japanese Kindergarten

Pedagogical practices in Japanese kindergartens are designed to foster a sense of responsibility, community and sustainability from an early age. Japan places a strong emphasis on holistic development and instils values that contribute to the creation of environmentally conscious and socially responsible individuals. In this section, we will further discuss some ways that Japanese kindergarten pedagogical approaches support the development of sustainable societies.

4.4.1 Sustainable Teaching and Learning

The path towards sustainability is fundamentally linked to how societies educate the next generation. Addressing how teaching and learning can be sustainable for children and what they learn about the world is essential in navigating this path (Siraj-Blatchford et al., 2016). In the teaching and learning process, providing opportunities for each child to make meaning of their own experiences is crucial, as this is a pivotal aspect of the learning process. When children construct their own understanding instead of merely repeating what has been told, their knowledge becomes connected to their emotions, and consequently, children gain knowledge (Pramling et al., 2017). In this context, play is a seamless vehicle for integrating sustainable elements into children's learning experiences. See Fig. 4.2. Play or *Asobi* in Japanese becomes a powerful tool for instilling an understanding of sustainability from an early age by allowing them to engage with concepts in a hands-on and exploratory manner.

The kindergarten takes a proactive approach by prioritising various play activities and experiences. Throughout the year, three types of engaging play experiences are offered. Playing with English (英語遊び *Eigo Asobi*) is a weekly activity designed to allow children to explore different cultures and become familiar with English conversation. The activities facilitated by a foreign teacher, this play not only promotes language learning but also nurtures cross-cultural understanding and communication skills. Similarly, Physical Education Play (体育遊び *Taiiko Asobi*) is conducted weekly with external instructors leading exciting games and group exercises.

These activities build children's physical strength, enhance motor skills and foster teamwork and coordination. Another engaging play experience is the monthly swimming sessions in collaboration with the adjacent swimming school. See Fig. 4.3. This initiative aims to make children comfortable in aquatic environments, instil essential water safety and practical life skills and empower them with knowledge about climate change. Fostering water safety and survival skills can better equip children to deal with climate-related issues like floods and tsunamis.

Beyond play, children's values and participation are central to learning through their daily life experiences (Samuelsson & Park, 2017). The kindergarten incorporates four categories of activities to provide children with a holistic educational

Fig. 4.2 Role of play or *Asobi* in fostering participation, collaboration and engagement among children in a Japanese kindergarten

Fig. 4.3 Swimming class as a practical approach to developing water safety skills and resilience to climate-related challenges

experience. The first category, interaction with nature, includes horse riding, rice planting and harvesting and field trips to Mandarin and sweet potato gardens. These activities foster a connection with nature, hands-on agricultural experience, and an understanding of the importance of farming in the community. The second category, teamwork activities, comprises stage performances, a drum and flute team, Sports Day and sociodramatic play. These activities encourage creative expression, teamwork, self-confidence, public speaking skills and collaboration through music.

The third category includes seasonal events, such as rice cake making, Christmas parties and Bean-scattering ceremonies. These seasonal events are traditional activities that introduce engaging experiences embedded with educational and cultural values. For example, the Bean-scattering ceremony during *Setsubun* is an essential part of Japanese culture, and it teaches children about the changing seasons and the importance of good fortune. Lastly, the fourth category is facility experiences. Beyond traditional learning spaces, facility experiences add a dynamic dimension to education, allowing children to explore and connect with the natural world. For instance, visits to aquariums and zoos allow children to explore various animal species and their habitats, fostering an appreciation for wildlife and the conservation of resources.

Moreover, children are active learners and will learn best through direct engagement and experiences. A project-based learning environment offers children numerous opportunities to enhance their critical thinking and comprehension skills and develop a holistic understanding of sustainable principles (MacDonald, 2015). These projects can be likened to ongoing learning narratives that seamlessly integrate ESD elements through diverse strategies such as conversation, play, revisiting documentation and participation in related activities.

The kindergarten curriculum strongly emphasises active learning, considering it a fundamental approach for fostering sustainable education through project-based learning. The learning experience is designed to foster the development of various characteristics and abilities across different domains. Children are prompted to explore and showcase global changes and environmental phenomena within the classroom, facilitating their comprehension and awareness of the interconnectedness of all living things. For instance, children are specifically encouraged to depict the Sakurajima volcano and articulate their thoughts and perspectives about this volcano through their drawings. Notably, Sakurajima is Japan's most active volcano in Kagoshima prefecture.

4.4.2 Culture and Value in Practice

The Japanese have a long history that stresses the value of "cultivating young children's feelings, interests and motivations by providing a supportive, stimulating environment, which was initially championed by the founding father of Japanese early childhood, Sozo Kurahashi (Doyon, 2016). Bhuiyan (2019) states that Japanese children must be taught norms and etiquette early to ensure a good society.

Our research revealed that the kindergarten employed various approaches to establish culture and values. Teachers support children's learning through practices like *Mimamoru* (watchful waiting). It involves teachers observing children from a distance, allowing them to independently explore, solve problems and navigate social interactions without immediate adult intervention. The concept reflects a belief that children learn best when they can struggle, experiment and resolve challenges independently rather than being provided with direct solutions (Nakatsubo et al., 2022). A good example is a scene in the classroom where a girl was teased by her classmates for taking too much time to set her table for a meal and instead of intervening, the teacher just watched from afar. The Japanese believe children know how to resolve problems by themselves, which builds their character.

The second approach is developing emotional concepts like *sabishi* (loneliness or sadness) and *amae* (the desire to be taken care of). As the example above, when the little girl cried due to the teasing, it showed strength as crying builds strength in a child's character and her need for help. It also relates to empathy and the need to attend to the needs and feelings of others. The kindergarten settings that allow plenty of room for free play allow children to develop an emotional inter-subjectivity between and among children. Another approach we always missed out on as educators is what Hayashi and Tobin (2015) describe as the *gyarari* (gallery). We overlooked the children on the periphery of the scenes. When fighting and mediation occur in the class, there is a gallery of observers learning by observing. Teachers usually take this opportunity to ask children to reflect and analyse conflicts and learn from what is seen. Children must view the world through a wide-angle lens and see things in contexts.

Values such as respect and responsibilities are ingrained in children through tasks and rules. Children are taught that they have specific roles in the class, and their primary role is to respect and listen to the teacher. Respect is not taken for granted in kindergarten, and teachers must earn that respect. Here, children are treated as adults, never raising their voices or scolding them. If a child misbehaves, they are taken to another room where they will be advised and given room to reflect on their behaviour.

Children are always placed in groups to do tasks, where they learn to work as a team and the importance of living as a group. Japanese children are never compared, and soloists are never highlighted, as it does not produce positive results (Bhuiyan, 2019). One good example of this was observed during sports practice, where groups of three children were asked to use their bodies to build a complex formation of bridges. When the children did not perform as expected, the coach had a personal talk with the children and stressed the importance of teamwork and how one failure can cause the whole team to fail. Roles and responsibilities were constantly bombarded in the children until it became an understanding of an ingrained Japanese culture among them.

Children in kindergarten are also taught to value resources such as food by participating in planting and harvesting rice and picking fruits and vegetables. See Fig. 4.4. This indirectly teaches children the difficulty of obtaining food, hence understanding the importance of not wasting food. They are also responsible for deciding if the food given is too much for them to consume or return what they do not like to eat. Health and hygiene are other areas emphasised in kindergarten, where children are taught the importance of cleanliness and hygiene. For example, in kindergarten, children are required to change into special slippers when entering and leaving the school and a different slipper when entering the restroom. It shows the detailed emphasis given to hygiene.

Regarding health, the school encourages outdoor physical activities even during hot summer afternoons. See Fig. 4.5. It not only makes the children sweat but also teaches them about learning to endure hardship, improve their immune system and improve their focus and body coordination. Centralised air-conditioning is not utilised at all times but is only switched on after play and exercise to hasten the room's cooling down with open windows. It prevents the circulation of bacteria and viruses, hence preventing sickness. During one week in the kindergarten, no researchers observed any child sniffling with a cold or cough. The Japanese kindergarten is unique as culture and values are made a way of life from an early age.

Fig. 4.4 Children grow plants to learn environmental care, food appreciation and sustainability values through hands-on experiences

Fig. 4.5 Outdoor physical activities promote endurance, boost immunity and enhance children's focus and coordination

4.4.3 Daily Routines in Kindergarten

Routines play a central role in fostering meaningful connections, engagement and development in children and provide a safe and familiar environment for learning. According to Fitton (2013), a well-structured daily routine provides a sense of security and fosters children's curiosity by encouraging them to explore the world and develop their skills. Routines allow children to quickly complete everyday tasks required by the teacher and the children. Routines also help to smooth transitions between activities and, therefore, provide fewer opportunities for disruption (Burden, 2003). When children are expected to complete routine tasks, they also can take on more responsibility and learn more self-management skills.

The kindergarten has a structured daily routine and provides a comprehensive educational experience for kindergarten children through activities such as morning greeting, free play, group exercises, educational activities, lunch and outdoor playtime. The day at kindergarten begins with a warm welcome that creates a positive and inclusive atmosphere. Teachers then take attendance and set a detailed schedule to create expectations and routines that promote the children's sense of security and confidence throughout the day. The kindergarten emphasises physical activity and outdoor play, focusing on motor skill development and cooperative play. Free play is a specific time when the children can pursue their interests indoors and outdoors.

Teachers act as facilitators and intervene when necessary to support learning and social interactions, which is consistent with the findings of Flores et al. (2019) and Luo et al. (2022). Structured learning activities in language development, basic maths, science exploration and arts and crafts are carried out in groups. This approach promotes children's social interaction, communication skills and sense of community and prepares them for future academic challenges.

Mealtime at kindergarten is more than just a meal; it is a structured and organised event that provides a sense of community and shared responsibility. When the children are ready to take their meal, they place their cloth luncheon mat that they have brought from home and lay it out together with a small cup on their classroom table. See Fig. 4.6. Children bring these items from home daily in a little cloth pouch bag that they usually hang off this bag at the rear of the classroom.

Serving lunch in the classroom in kindergartens is a common practice influenced by cultural, educational and logistical factors. Eating together in the classroom fosters a sense of community and encourages social interaction among the children. It allows children to share meals, develop table manners and converse with their peers and teachers. See Fig. 4.7.

Japanese kindergartens may prioritise environmental sustainability by reducing the need for disposable items and minimising food waste. Serving meals directly in the classroom allows for better control over portions and waste management. A meal is normally served and eaten in the classroom, where children also clean up after themselves and serve one another. Children take turns to pick their food and serve it to their classmates. They take great pride in serving lunch in the classroom and frequently eagerly await their time. They take the lead and demonstrate their abilities to their friends and teachers. They learn what they consume and the health benefits of

Fig. 4.6 Child is responsible for placing a lunch mat and a small cup from a cloth pouch

Fig. 4.7 Serving meals directly in the classroom allows for better control over portions and waste management

particular foods and nutrients. Japan has the lowest obesity rates among developed countries' adult and child populations, partly due to the culture's emphasis on eating a healthy, balanced diet.

The children and the teachers actively eat together, promoting social skills and creating a supportive learning atmosphere. Personal hygiene practises such as hand washing, gargling and tooth brushing are emphasised to instil lifelong habits that promote overall well-being. Teachers strive to create a calm and positive atmosphere in the classrooms. Children collect their belongings, and teachers assist with a seamless transition for those children participating in after-school programmes. This extra time is supervised by teachers or other staff to ensure continuous support and help for the children. Children are encouraged to practice sustainability by not wasting the food they eat. Thus, they only take a small portion of food to encourage them to finish their meals without wasting it.

Safety is another aspect of sustainability in education. We observed that the the kindergarten was provided with environmentally safe school facilities such as bus service, particularly for families far from the school. The availability of a school bus in the kindergarten building offers children a dependable and secure way to get to and from kindergarten. There is indoor access for children to board the bus safely and conveniently. See Fig. 4.8. The school bus facility helps parents with logistical concerns and fosters inclusion by allowing children from different places to attend. Ensuring the safety of children during transportation requires a high priority, which includes implementing safety measures such as having certified bus drivers and making suitable seating arrangements.

Fig. 4.8 Indoor access in the school building for children to board the bus safely

Besides that, the kindergarten has a kitchen and cooking staff to prepare meals. Considering the importance of schools for providing sustainable food selections, school meals are prepared in fully equipped kitchens within the school building, leading to improved social, economic and environmental outcomes. See Fig. 4.9. The cooking staff prepared meals for the children using fresh ingredients rather than packaged frozen food. The kitchen staff then loaded up a cart with the food without any packaging that could contribute to waste for each class in the kindergarten. Within a few hours, the kitchen prepares all the meals to serve hot, fresh lunches for children.

4.5 Integration of ESD Practices in Asian Countries

The observation findings from the Japanese kindergarten revealed a model that embeds ESD into Early Childhood Care and Education. The model emphasises the holistic integration of ESD into interdisciplinary curricula, linking health, relationships, environment, language and expression. Practices such as *ganbari* (persistence), *shūdan seikatsu* (group living) and *mimamoru* (watchful waiting) foster autonomy, resilience and collective responsibility. Seasonal traditions and eco-school frameworks provide children with meaningful connections to nature and cultural heritage, aligning sustainability education with local and national values. Additionally, emotional concepts like *sabishii* (loneliness) and *amae* (the desire to be cared for), along with the *gyarari* (gallery) approach of observing and reflecting on interactions, foster empathy and emotional intelligence.

Fig. 4.9 Cooking staff working in the school kitchen prepares fresh meals from fresh ingredients without packaging that contributes to waste

Both Japan and Malaysia have integrated Education for Sustainable Development (ESD) into their national education frameworks, with Japan incorporating it into the National Curriculum Standard for Preschools to align with cultural traditions and holistic development. At the same time, Malaysia embeds ESD in its National Preschool Curriculum Standard with a focus on Sustainable Development Goals (SDGs), particularly SDG 4, on quality education (Yean, 2014). Both countries emphasise environmental sustainability, with Japanese kindergartens adopting eco-school frameworks and activities like rice planting and recycling and Malaysian preschools focusing on 3R (reduce, reuse, recycle) practices and outdoor education to foster ecological awareness (Heng et al., 2024; Mokshein, 2019).

Teachers play a pivotal role in both contexts; Japanese educators utilise culturally sensitive, child-centred approaches like *mimamoru* (watchful waiting) to instil sustainability values, whereas Malaysian teachers act as knowledge and co-constructors of sustainable behaviours. As knowledge transmitters, they educate young learners about sustainability concepts, environmental issues and sustainable practices through structured lessons and activities. As co-constructors, they engage children in interactive, hands-on activities, encouraging collaboration, critical thinking and the practical application of sustainability principles. However, significant challenges often hinder these efforts, including insufficient training in ESD pedagogy and a lack of resources for effective implementation (Heng et al., 2024; Mokshein, 2019).

Japan and China are committed to integrating ESD into early childhood education, but their approaches exhibit similarities and differences shaped by cultural and institutional contexts. Both countries prioritise environmental education, emphasising activities like recycling and nature-based projects to instil ecological awareness in young children. Community and family involvement are a shared strength. Japanese kindergartens engage families in seasonal activities such as rice planting and Chinese schools incorporate families into recycling initiatives and environmental storytelling. Recycling projects in China often involve children and their families sorting recyclable materials, crafting items from waste and participating in community recycling drives, fostering ecological responsibility through hands-on engagement (Zhou, 2021).

Thematic science activities in Chinese preschools introduce environmental topics such as air pollution, water cycles, or conservation through experiments and stories, helping children grasp basic scientific concepts and environmental challenges (Wang et al., 2019). Both nations also draw from global frameworks, such as UNESCO's ESD guidelines, to inform their practices. However, while evolving, China's ESD practices focus predominantly on environmental education, often through teacher-led approaches, with limited emphasis on sustainability's social and economic dimensions (Zhou, 2021). Initiatives such as recycling projects and thematic science activities are promising but lack the interdisciplinary integration seen in Japan (Cheng & Yu, 2022; Wang et al., 2019).

Education for Sustainable Development (ESD) practices in Korea gradually evolved, strongly emphasising environmental education and growing initiatives to integrate sustainability into early childhood education. At the national level, ESD is supported through government policies and programmes that align with global frameworks, such as the Sustainable Development Goals (SDGs). However, its integration into the curriculum needs to be more cohesive. While Korea's National Curriculum includes ESD components, implementation often relies on regional collaborations like the Tongyeong Regional Centre for Expertise (RCE), which fosters partnerships among local stakeholders, schools and community organisations (Sung & Choi, 2022). The collaborations aim to bridge the gap between policy and practice, promoting community-based environmental stewardship. Programmes like the 'Keeper of Green Earth' project exemplify Korea's focus on environmental education in early childhood. These projects involve hands-on activities such as planting, recycling and discussions about environmental conservation, aiming to instil ecological awareness and responsibility from an early age (Park & Yu, 2016).

Additionally, Early Childhood Education and Care (ECEC) programmes in Korea often emphasise the importance of experiential learning, with activities designed to immerse children in nature and encourage active participation in sustainability efforts. For instance, outdoor projects involve exploring local ecosystems, learning about biodiversity and understanding the impact of human actions on the environment. These activities complement formal lessons, creating a well-rounded approach to instilling environmental values (Sung & Choi, 2022). Despite these efforts, integrating social and economic dimensions of sustainability still needs to be improved, reflecting a narrower focus compared to Japan's holistic approach, which incorporates cultural, social and environmental elements into ESD practices. Expanding Korea's ESD initiatives to include these dimensions could enhance their impact, equipping young learners with a more comprehensive understanding of sustainability challenges and solutions (Park & Yu, 2016; Sung & Choi, 2022).

The challenges faced by Japan, Malaysia, China and Korea in implementing Education for for Sustainable Development (ESD) highlight shared obstacles, including policy gaps, limited teacher training and difficulties adapting to local and modern contexts. While ESD benefits from strong policy support and integration into early childhood education in Japan, traditional teaching methods often limit its transformative potential, resulting in 'shallow ESD' that struggles to meet dynamic sustainability goals, particularly in urban settings (Nagata, 2017). Malaysia faces similar barriers, with teachers positioned as knowledge transmitters and co-constructors of sustainable behaviours, but needs to be improved by gaps in training and insufficient practical application of ESD concepts, restricting its overall impact (Heng et al., 2024; Mokshein, 2019).

In China, the absence of targeted national ESD guidelines further compounds these issues, with rural–urban disparities and inadequate teacher preparation undermining consistent implementation. Meanwhile, in Korea, early childhood educators recognise the importance of ESD and express a willingness to implement it. However, a lack of comprehensive training and pedagogical strategies impedes their ability to effectively integrate sustainability into classrooms, underscoring the need for professional development and curriculum reform (Kyung & Choi, 2024; Sung & Choi, 2022). Collectively, these challenges emphasise the critical need for targeted policies, enhanced teacher training and context-specific approaches to strengthen ESD practices across these countries.

4.6 Future Directions

The study on ESD practices in Japanese kindergartens holds significant implications for early childhood education and sustainability across Asia. By emphasising holistic learning experiences that integrate cultural values, play-based learning and community involvement, the Japanese model serves as a compelling framework for fostering well-rounded development in young learners, particularly within the diverse sociocultural contexts of the region. This chapter provides critical insights into the effectiveness of Japan's curriculum guidelines related to ESD, offering a foundation for policy adjustments that align with the sustainability goals of neighbouring Asian countries.

Given the shared challenges across the region, such as limited teacher training, urban–rural disparities and balancing traditional and modern educational demands, the practices outlined in this study present scalable and adaptable strategies for Asian educators and policymakers. By contributing to a regional knowledge-sharing platform, this chapter facilitates the dissemination of practical

ESD approaches, equipping stakeholders across Asia with the tools to advance sustainability education and nurture a generation of sustainability-conscious citizens.

As education systems in Asia continue to adapt to the region's dynamic cultural, social and technological changes, the future of ESD will increasingly rely on integrating local cultural values, traditions and regional context into its Framework. Future directions in ESD research should focus on deepening the understanding of how different socio-political and cultural environments across Asian nations influence sustainable education practices. Emphasising the value of non-Western perspectives, future studies will shed light on the diverse ways sustainability is conceptualised and addressed within educational systems, allowing for region-specific models that respect cultural nuances and local priorities. Exploring these localised approaches will enable the creation of ESD frameworks that are globally relevant and contextually grounded in the needs and values of specific communities.

Future research should focus on the professional development of educators, emphasising building the competencies required to implement effective ESD practices. As the demand for ESD skills grows, teachers must be equipped with the knowledge, tools and adaptive teaching methods needed to address the diverse needs of students in varying educational contexts. Future initiatives should prioritise the development of cross-country collaborative training platforms, allowing educators from different Asian nations to share successful strategies and challenges. This approach will ensure that sustainable education practices are adaptable and scalable, meeting the evolving demands of both local and global educational goals.

4.7 Conclusion

In conclusion, this chapter significantly enhances our understanding of Education for for Sustainable Development (ESD) in the context of a Japanese kindergarten, with implications that extend across Asia. The findings emphasise the effectiveness of an integrated approach to ESD, blending instructional strategies with environmental and cultural supports to foster sustainability from an early age.

This chapter offers a detailed and contextually rich perspective on implementing ESD in diverse cultural settings by addressing educators' unique opportunities and challenges in Asia. The recommendations provide actionable insights for improving ESD practices in Japan and offer a framework adaptable to other Asian nations. By fostering regionally grounded yet globally aligned approaches, this work contributes to developing sustainable education systems that nurture ecological awareness, cultural responsibility and global citizenship among young learners.

Highlights

- Japan's ESD model integrates cultural values, community involvement and interdisciplinary curricula, promoting environmental sustainability and emotional intelligence in early childhood education.

- While Japan and Malaysia emphasise environmental sustainability in preschool curricula, with Japan focusing on eco-school frameworks and Malaysia integrating SDGs, both face challenges in teacher training and resource availability.
- China and Korea are integrating ESD into early childhood education with a strong focus on environmental education, but further development is needed to incorporate social and economic sustainability dimensions.
- Future ESD practices in Asia should focus on local cultural values, improving teacher training and developing cross-country collaborative platforms to enhance regional sustainability education.

References

Aspinall, R. W. (2016). Children's rights in a risk society: The case of schooling in Japan. *Japan Forum, 28*(2), 135–154. https://doi.org/10.1080/09555803.2015.1076871

Bhuiyan, M. Z. A. (2019). The early education in Japanese society: Morality and etiquette issues. Retrieved from https://doi.org/10.13140/RG.2.2.22328.80641

Burden, P. R. (2003). *Classroom management: Creating a successful learning community*. John Wiley & Sons, Inc.

Cheng, C., & Yu, Y. (2022). Early childhood educators' practices in education for sustainable development in China: Evidence from Shandong Province. *Sustainability, 14*(4), 2019. https://doi.org/10.3390/su14042019

Davis, J., & Elliott, S. (2014). *Research in early childhood education for sustainability: International perspectives and provocations*. Routledge.

Doyon, P. (2016). Teaching embodied: Cultural practices in Japanese preschools. *Teachers College Records*. Retrieved from http://www.tcrecord.org ID Number 19561.

Fitton, V. (2013). Attachment theory: History, research, and practice. *Child and Adolescent Psychotherapy and Psychoanalysis*, 121–143.

Flores, F. S., Rodrigues, L. P., Copetti, F., Lopes, F., & Cordovil, R. (2019). Affordances for motor skill development in home, school, and sport environments: A narrative review. *Perceptual and Motor Skills, 126*(3), 366–388. https://doi.org/10.1177/0031512519829271

Guo, K., & Kuramochi, K. (2019). Exploring kyoiku: Children's educational experiences in Japanese kindergartens. *Issues in Educational Research, 29*(1), 89–105.

Hayashi, A., & Tobin, J. (2015). *Teaching embodied: Cultural practice in Japanese preschools*. University of Chicago Press.

Hellman, A. (2016). Teaching reflective care in Japanese early childhood settings. *Early Child Development and Care, 186*(10), 1693–1702. https://doi.org/10.1080/03004430.2015.1124869

Heng, S. Y. B., Anna, P. W. L., & Siew, S. K. (2024). Preschool teachers' perceptions on classroom practices to sustain environmental education development. *Horizon Journal of Human and Social Science Research, 6*(1), 64–75. https://doi.org/10.37534/bp.jhssr.2024.v6.nS.id1246.p64

Hoffman, D. M. (2000). Pedagogies of self in American and Japanese early childhood education: A critical conceptual analysis. *The Elementary School Journal, 101*(2), 193–208. Retrieved from https://www.jstor.org/stable/1002342

Kyung Jang, S., & Choi, E. Y. (2024). A study on the perceptions of pre-service early childhood teachers on education for sustainable development (ESD) in South Korea. *Asia-Pacific Journal of Research in Early Childhood Education, 18*(2). https://doi.org/10.17206/apjrece.2024.18.2.199

Luo, W., Berson, I. R., & Berson, M. J. (2022). Enhancing preschool–home collaboration: Lessons from virtual home visit experiences during the COVID-19 outbreak in China. *Early Years, 42*(1), 119–135. https://doi.org/10.1080/09575146.2021.2008880

MacDonald, M. (2015). Early childhood education and sustainability: A living curriculum. *Childhood Education, 91*(5), 332–341.
McKeown, R., & Hopkins, C. (2003). Educating for a sustainable future. *Applied Environmental Education & Communication, 2*(3), 155–163.
Mokshein, S. E. (2019, June). Education for sustainable development (ESD) in Malaysia: Policy, program, and evaluation. In *3rd International Conference on Current Issues in Education (ICCIE 2018)*, 6–12. Atlantis Press.
Nagata, Y. (2017). A critical review of education for sustainable development (ESD) in Japan: Beyond the practice of pouring new wine into old bottles. *Educational Studies in Japan, 11*, 29–41.
Nakatsubo, F., Ueda, H., & Kayama, M. (2022). Why don't Japanese early childhood educators intervene in children's physical fights? Some characteristics of the Mimamoru approach. *Early Childhood Education Journal, 50*(4), 627–637. https://doi.org/10.1007/s10643-021-01184-3
Park, E., Kim, H., & Yu, S. (2016). Perceptions and attitudes of early childhood teachers in Korea about education for sustainable development. *International Journal of Early Childhood, 48*, 369–385. https://doi.org/10.1007/s13158-016-0176-y
Pramling, N., Doverborg, E., & Samuelsson, I. P. (2017). Re-metaphorising teaching and learning in early childhood education beyond the instruction–social fostering divide. In C. Ringsmose & G. Kragh-Müller (Eds.), *Nordic social pedagogical approach to early years* (pp. 205–218). Springer. https://doi.org/10.1007/978-3-319-42557-3_12
Pratt, R., & Luby, A. (2016). Early childhood environmental education: Making it mainstream. *Early Childhood News*.
Samuelsson, I. P., & Park, E. (2017). How to educate children for sustainable learning and for a sustainable world. *International Journal of Early Childhood, 49*, 273–285.
Siraj-Blatchford, J., Mogharreban, C., & Park, E. (2016). *International research on education for sustainable development in early childhood*. Springer.
Siraj-Blatchford, J., Smith, K. C., & Samuelsson, I. P. (2010). *Education for sustainable development in the early years*. OMEP, World Organization for Early Childhood Education.
Sung, J. H., & Choi, J. E. (2022). The challenging and transformative implications of education for sustainable development: A case study in South Korea. *Journal of Curriculum Studies Research, 4*(2), 1–14. https://doi.org/10.46303/jcsr.2022.8
UNESCO. (2005). *UNDESD international implementation scheme*. UNESCO Educational Sector.
UNESCO. (2014). *Shaping the future we want: UN Decade of Education for Sustainable Development (2005–2014) final report*. UNESCO.
UNESCO. (2017). *Education for sustainable development goals: Learning objectives*. UNESCO.
Wang, G., Zhou, X., & Cui, H. (2019). Exploring education for sustainable development in a Chinese kindergarten: An action research. *ECNU Review of Education, 2*(4), 497–514. https://doi.org/10.1177/2096531119897638
Yamamoto, Y., & Satoh, E. (2019). Cultivating perseverance and motivation in early childhood education in Japan. In *Contemporary perspectives on research in motivation in early childhood education* (pp. 229–245).
Yean, A. S. (2014). Education for sustainable development in Malaysia's national curriculum reformation: A theoretical exploration. *Journal of International and Comparative Education (JICE), 199–212*. https://doi.org/10.14425/00.73.61
Zhou, D. (2021). Explore the notion of education for sustainable development in early childhood education in China. *Asia-Pacific Journal of Research in Early Childhood Education, 15*(1). https://doi.org/10.17206/apjrece.2021.15.1.77

Fung Lan Loo is an experienced professional with over 20 years of experience in early childhood education. She is the programme director for Open University Malaysia's Bachelor of Early Childhood Education programme. She was the founder and principal of a private kindergarten and educational centre. She is actively involved in lecturing, research, training and consultation. Her areas of expertise encompass early literacy, child psychology, teaching pedagogy and curriculum development. She obtained a master's degree in Early Childhood Education from the University of National Malaysia and a Ph.D. from the University of Malaya.

Mahani Abdul Malik has been in early childhood education since 1992. She earned a Science degree from the University of Malaya in 1983 and later completed a Nursery Foundation Diploma at the London Montessori Centre to open her first kindergarten. She founded two Montessori preschools and became active in training, writing and designing children's educational materials. After earning a Master's in preschool Education in 2006, she lectured part-time and trained educators. She is now a senior lecturer and programme director at Open University Malaysia and an MQA assessor and external examiner for several universities.

Aliza Ali is the dean of the Faculty of Education at Open University Malaysia. She holds a Ph.D. and master's in education, specialising in Curriculum and Pedagogy. Her areas of expertise are curriculum, pedagogical approach, curriculum development, early childhood education, assessment of learning and teaching and learning. As a senior academic, she has completed supervising more than 20 master students and one doctoral student. She has published two academic books, a chapter in books, more than 10 children's books, and scholarly journals in SCOPUS and ISI. Her research area interest is early childhood pedagogy and curriculum.

Md Rosli Ismail enrolled on the Japanese Preparatory Special Course programme at the University of Malaya before pursuing his first degree at Keio University, Tokyo, for four years. Upon graduation, he was hired by the Bank of Tokyo-Mitsubishi headquarters in Tokyo as an international executive. He was attached to various corporations upon his return to Malaysia for 19 years. He has been a practitioner of early childhood education since 2005. He earned a master of Philosophy and a doctor of Philosophy from the National University of Malaysia.

Nurhusna Abdul Malek is a dedicated professional in early childhood education. She embarked on her educational journey at the University Pendidikan Sultan Idris, where she honed her expertise with a diploma, degree and master's degree in early childhood education. Since 2013, she has contributed actively to the field as a lecturer. Currently, she holds the position of Programme Director for Diploma in Early Childhood Education lecturer at Open University Malaysia, where she continues to inspire and educate future educators in early childhood development.

Chapter 5
Sustainable Cultural Transformations in Asia: Vietnamese Students' Intercultural Awareness Following a Malaysian Exchange Programme

Wun-Chiew Pung, Connie Keh-Nie Lim, and Su-Hie Ting

Abstract This empirical study examined the impact of a short-term student exchange programme on Vietnamese students' intercultural awareness, positioning it within the context of growing global interconnectedness. The goal of this chapter is to show how experiential learning through study abroad initiatives can foster intercultural sensitivity, particularly within the Asian region. To achieve this, a descriptive methodology was adopted, using an online questionnaire administered to 104 Vietnamese students who participated in an exchange programme at Universiti Malaysia Sarawak. The questionnaire examined the students' exposure to other cultures in their family and among friends, and communicative behaviours and attitudes towards other cultures. Key findings indicated that despite limited prior intercultural contact, participants experienced positive shifts in cultural perception and increased ethnorelativism, though they encountered challenges such as adjusting to local cuisine. These results show that even brief study abroad experiences can enhance students' appreciation for diverse worldviews, values and beliefs. The study suggests that formal intercultural education is essential to prepare students to engage constructively with cultural differences, an increasingly vital skill in a globally connected world.

Keywords Sustainable cultural transformations · Intercultural awareness · Ethnorelativism · Exchange programme · Study abroad · Vietnamese

W.-C. Pung · S.-H. Ting (✉)
Faculty of Education, Language and Communication, Universiti Malaysia Sarawak, Kota Samarahan, Malaysia
e-mail: shting@unimas.my

W.-C. Pung
e-mail: wcpung@unimas.my

C. K.-N. Lim
Faculty of Applied and Creative Arts, Universiti Malaysia Sarawak, Kota Samarahan, Malaysia
e-mail: lknconnie@unimas.my

A. Pandian et al. (eds.), *Future of Education in Asia*,
https://doi.org/10.1007/978-3-032-07285-6_5

5.1 Introduction

Today's world is marked by increasing global interconnectedness. The rapid development of technology in communication and transportation facilitates international collaborations. It facilitates student mobility or study abroad programmes in higher education. During their time abroad, students are immersed in the culture(s) of the host country which may be different from the home country. The brief duration of these programmes makes them a viable option for a broader spectrum of students seeking international educational experiences, catering especially to those unable to engage in extended stays overseas due to work or familial commitments (Carley et al., 2011).

For the benefits of student mobility programmes to be reaped, the students need some intercultural awareness to effectively navigate the diverse cultural environments they encounter. Mere exposure to cultural differences may not be sufficient, and there must be opportunities for cultural reflection (Tarchi & Surian, 2022). The engagement with both local and fellow international students allows students to improve their intercultural competence (Sobkowiak, 2019). Sierra-Huedo and Foucart (2022) point out the need for universities to shift focus from quantity (i.e. the number of international students and student mobility programmes) to quality of intercultural experiences which involves changing students' attitudes and developing their intercultural competence.

Past findings are inconsistent on the association between intercultural exposure and intercultural competence. Some studies show a link (Skefsrud, 2022; Sobkowiak, 2019) and researchers even found that short-term mobility programme is effective (San & Htwe, 2023). However, there are contradictory findings on whether ethnocentrism increased after the time abroad (Czura, 2017).

At this juncture, past studies in intercultural awareness often focus on students travelling between western and eastern countries, or within the western context such as the Erasmus+ mobility programme (Sierra-Huedo & Foucart, 2022). However, very few studies investigate the intercultural awareness of students involved in mobility programmes within Asian cultures. The interest in academic exchange between universities in Asian countries has been driven by the desire to have international linkages. Having a large foreign student population is one indicator of university prestige (Slipchuk et al., 2021). Furthermore, universities are pressured to boost their number of international students not only to pursue branding and reputation (rankings) but also to generate revenue (de Wit, 2019). The MOUs signed for such purposes often include staff and student exchange activities. However, the students' experiences may be neglected in the pursuit of points for the university ranking.

As far as Malaysia is concerned, the spotlight should be turned on Vietnam because although diplomatic ties have been established for only half a century, since 30 March 1973, movement between the two countries for education is just beginning. Tourism-related travels between Malaysia and Vietnam may look promising. As of September 2023, Malaysia recorded 272,852 Vietnamese arrivals (Panaroma Destination, 2023). Malaysian tourist arrivals in Vietnam totalled 470,000 in 2023 (Phong,

2023). However, Malaysia and Vietnam are not education destination hubs for either country. Joint academic activities such as training and conferences are merely beginning after the COVID-19 pandemic eased in 2022. Examples are MOUs between Ho Chi Minh City University of Technology and the University Kebangsaan Malaysia (HUTECH, 2024) and Universiti Malaysia Sarawak and Hoa Sen University (HSU, 2023).

Thus far, there is a shortage of research and scant empirical evidence on intercultural awareness of students involved in mobility programmes in the Asian region, and our literature review did not find any studies on student exchange programmes and intercultural awareness of Vietnamese students when they spend time abroad. The past studies were on Vietnamese learners of English in Vietnam (Kiet, 2016; Nguyen, 2023). Kiet (2016) found that students' active engagement in intercultural learning resulted in better understanding of the English culture and cultural self-awareness, while Nguyen (2023) concluded that cultural knowledge should be given the same emphasis as linguistic knowledge in English language teaching at university level in Vietnam. Learning English in mobility programmes abroad provides the advantage of being in a different learning environment. Extending language education beyond the home country can broaden students' cultural awareness through interacting with peers from different linguistic and cultural backgrounds (Kong & Rajandram, 2023), allowing them to develop better interpersonal skills and cultural appreciation. Due to the lack of studies on gains from short-duration international experiences in the Asian region, further exploration is much needed to understand the impact of the exposure on intercultural awareness.

The study investigated Vietnamese students' intercultural awareness after an exchange programme in Malaysia. The research questions are:

(1) What are the Vietnamese students' exposure to other cultures prior to their arrival in Malaysia?
(2) In what ways has participation in the mobility programme impacted the cultural awareness of Vietnamese students?

5.2 Literature Review

5.2.1 Intercultural Awareness

In general, intercultural awareness involves understanding the similarities and differences between one's culture and other cultures (Council of Europe, 2001). This encompasses the ability to recognise, appreciate and respect cultural differences between people in different contexts (Kim, 2001). Adler (1987) posits that intercultural awareness involves a process of internalising insights about group-specific common understandings that shape an individual's predominant values, attitudes, beliefs and outlooks. Hanvey (1987) categorises intercultural awareness as encompassing:

(1) superficial cultural traits which involve recognising the most visible and surface-level aspects of a culture;
(2) significant and subtle cultural traits which involve understanding deeper and different traits of other cultures; and
(3) insider's perspective involves understanding how insiders experienced the culture.

Closely linked to intercultural awareness is Bennett's (1993) Developmental Model of Intercultural Sensitivity. According to Bennett (1993), intercultural competence develops along a continuum from ethnocentrism to ethnorelativism. Ethnocentric individuals who have little exposure to other ethnic groups carry the attitude that one's own group, ethnicity, or nationality is superior. Ethnorelativistic individuals believe that no one culture is superior to another. This progression begins with a denial of cultural differences, then moves through defence, minimisation, acceptance, adaptation and finally culminates in integration of cultural diversity. However, Ting-Toomey (1999) emphasises that intercultural competence does not happen in isolation, but through interactions with and in response to individuals from diverse backgrounds.

5.2.2 *Intercultural Awareness in Student Mobility Programmes*

Student mobility programmes can reduce ethnocentrism as they enable students to gain overseas experiences, improve cultural and linguistic skills, foster tolerance and broaden perspectives (Skefsrud, 2022). These programmes also enhance intercultural communication skills and competence (Sobkowiak, 2019). More importantly, mobility programmes enable hands-on intercultural experience gained from living and collaborating with people from diverse backgrounds (Behrnd & Porzelt, 2012).

However, results from past studies on intercultural awareness in student mobility programmes are not consistent. Krajewski's (2011) study showed a positive link between intercultural experience and competence, with participation in these programmes alone being a form of learning. Czerwionka et al. (2015) discovered that students who went abroad showed increased knowledge related to daily life, food and drink, values and politics of the host country. Engle and Engle (2004) found participants' intercultural sensitivity improved, even when time spent abroad was as short as one semester. Similarly, Fang and Baker (2017) also reported that even a short-term mobility programme resulted in changes in students' attitudes and behaviours whereby they became more understanding of other cultures, developed a more inclusive mind, and were more open to diversity. San and Htwe (2023) found that Myanmar students who were in a one-semester study abroad programmes in Thailand, Vietnam, Sweden and the Netherlands reported enhanced intercultural awareness. The Myanmar students reported open-mindedness, ethno-relative views, attitude changes, skill development and the ability to exhibit appropriate and effective intercultural behaviours following their study abroad experience.

However, there were also studies which showed that positive outcomes from mobility programmes were not automatic, supporting Ting-Toomey's (1999) contention. For instance, Czura (2017) reported that students on mobility programmes return home being more ethnocentric and less willing to communicate with people of other cultures. As such, the findings of these studies imply that merely mixing domestic and international students does not necessarily contribute to intercultural competence. This underscores the need for universities to create opportunities for domestic and international students to engage in meaningful interaction. Otherwise, international students may spend most of their time with peers from the same culture. As a result, their relationships with domestic students, local community members and other international students can become superficial, brief, or even negative and combative (Geibel, 2020). Hence, factors such as frequency and levels of intercultural engagement, as well as facilitating contexts which require individuals to reflect on their intercultural interactions are important requisites for intercultural competence (Sobkowiak, 2019).

Intercultural competence is also influenced by a person's previous intercultural experiences and personality traits. These factors affect one's ability to interact with others, behave appropriately, and handle new situations when entering different cultures. Rings and Allehyani (2020) found positive associations between personality traits like openness and conscientiousness and intercultural competence. Therefore, educational institutions need to provide opportunities and support for students to experience and participate in diverse cultural experiences so that they step out of their comfort zones (Penbek et al., 2012).

5.2.3 Higher Education Student Mobility in Asia

Traditionally, Western universities were the preferred destinations for international students. However, recent trends show a growing interest in Asian universities, especially among Southeast Asian students. In 2019, China hosted 28,600 Thai students, 15,000 Indonesian students, 11,300 Vietnamese students and 9,500 students from other Southeast Asian countries, making it the top destination for many Southeast Asian students (ICEF Monitor, 2024). Japan aims to increase its foreign enrolment to 400,000 students, South Korea targets 300,000 by 2027, and Taiwan aims for 320,000 by 2030 (ICEF Monitor, 2024).

There are several reasons why intraregional student mobility is popular. Among them are familiarity with cultural customs, a sense of physical and cultural safety, and cost savings due to proximity (ICEF Monitor, 2024). Government policies also play a significant role. Initiatives like the University Mobility in Asia and the Pacific Association, the Association of Southeast Asian Nations University Network and the Association of Southeast Asian Institutions of Higher Learning promote better educational opportunities for member countries (Pfaff-Czarnecka, 2020). Additionally, the number of top-ranked universities in Asia is rising. For example, 33 Asian universities ranked within the 2024 Times Higher Education top 200, up from 28 in 2023 (ICEF Monitor, 2024). Furthermore, Asia is recognised as the world's fastest-growing regional economy, and students understand that attending Asian universities can provide opportunities for industry internships and jobs in the region (ICEF Monitor, 2019).

Despite the advantages of intraregional student mobility in that geographic and cultural proximity can ease challenges like homesickness and adjustment difficulties when studying abroad, students still face various issues. Students often form in-groups based on nationality, ethnicity and educational background (Pfaff-Czarnecka, 2020), which defeats the purpose of fostering intercultural competence and integration. These divisions can exacerbate existing prejudices and stereotypes. There are reports of local prejudices, such as Taiwanese locals' views of Indonesian students as migrant workers (Paramitaningrum, 2013), Chinese students in Singapore being stereotyped as prostitutes (Yang, 2014), Chinese students in Japan being discriminated against due to their nationality (Taniguchi et al., 2022), and Malaysian students in South Korea being looked down upon due to their country of origin and the misconception that Malaysians are poor (Dos Santos, 2022).

Although intraregional mobility in Asia is a growing trend, at this juncture, little is known about the experiences and intercultural awareness of Asian students within the region (Pfaff-Czarnecka, 2020). Most existing literature focuses on intercontinental mobility between Western and non-Western countries, but the impact of intraregional mobility on students' intercultural awareness and competence may differ from these perspectives.

5.3 Background on Vietnam and Malaysia

5.3.1 Vietnam

Vietnam primarily comprises the Kinh (Viet) ethnic group (85.3%), has Vietnamese as its official language and is increasingly favouring English as a second language (Central Intelligence Agency, 2024b). Like the rest of the world, Vietnam is also actively internationalising its education system, following the Higher Education Reform Agenda introduced in 2005 and the announcement by the Prime Minister to enhance the quality of higher education from 2019 to 2025 (Luong et al., 2023). Vietnam's future direction and move towards internationalisation are largely spurred by its low education quality which lags behind global standards. This could be attributed by the prevalence of traditional teaching methods, which focus heavily on theory with little practical application (Nguyen, 2021), as well as insufficient qualified staff (Uoc et al., 2023). This affects the quality of graduates, who often lack the skills, adaptability and competitiveness needed in the global market.

Key steps in improving the quality of education through internationalisation efforts include adopting English as a medium of instruction, incorporating foreign curricula and forming international partnerships for transnational education initiatives (Tran et al., 2018). It is expected that by 2025, at least 50% of tertiary institutions will integrate foreign curricula, either partially or wholly, into their programmes (Luong et al., 2023). The English language mobility programme in Malaysia, which was structured as credit transfer, is an example of Vietnam's integration of foreign curricula into their existing educational structure. Initiatives like this have paved the way for stronger educational ties between the two nations. On October 17, 2023, Vietnam and Malaysia signed a Memorandum of Understanding on cooperation in higher education, formalising their commitment to enhancing educational cooperation (VietnamPlus, 2023).

Student exchange programmes are crucial in preparing Vietnamese students for global competitiveness and enhancing their intercultural competence. Findings from past studies on student mobility programmes for Vietnamese students have been positive. Cuong's (2020) research revealed that Vietnamese students show a high level of readiness for international mobility programmes. Their primary motivations include learning about different cultures, practising English with locals, broadening their knowledge, experiencing a new learning environment, developing personal skills and making friends. However, students also expressed a need for more orientation activities and faculty support to better prepare them for the different environments they would encounter, indicating concerns about adjusting to new settings.

Similarly, Diem et al. (2023) found that student teachers had positive perceptions of student exchange programmes. They noted advantages such as access to new knowledge, enhanced teaching skills and valuable support from foreign partners and mentors. These benefits are crucial for Vietnam, particularly in improving teacher training strategies and curricula to elevate education standards. However, the study also revealed challenges related to language differences and cultural norms, suggesting that student teachers would benefit from better orientation and preparation before going abroad.

Overall, findings on student mobility programmes for Vietnamese students highlight their importance not only for internationalisation efforts to improve the quality of education in Vietnam, but also for fostering intercultural awareness and competence. However, there is a need for adequate preparation and support to maximise the benefits and address potential challenges. This need may stem from Vietnam's relatively homogeneous population, with the majority being of the Kinh (Viet) ethnic group (Central Intelligence Agency, 2024b). Additionally, although English is increasingly being used as a second language in Vietnam, English proficiency among Vietnamese students remains a barrier, impacting their academic performance and social integration in the host country (Le et al., 2017).

These findings have significant implications for the future of education in Vietnam. Efforts must be directed towards acquiring essential skills that equip students with the mindset and capabilities to thrive as global citizens, and these include having greater intercultural awareness and adaptability to navigate new environments. It is also crucial to address the barriers posed by limited English proficiency in order to ensure that students are well-prepared for cultural and academic adjustments.

5.3.2 Malaysia

Unlike Vietnam, Malaysia has a more multicultural population with Bumiputera comprising Malays and indigenous groups such as Dayaks, Kadazans and Orang Asli (62.5%), Chinese (20.6%), Indians (6.2%) and others (Central Intelligence Agency, 2024a, 2024b). Malaysia also has a greater diversity of languages, such as English, Chinese, Indian, indigenous languages and the national language, *Bahasa Melayu*.

In recent years, Malaysia has intensified efforts to internationalise higher education. The aim is to host 250,000 international students by 2025 (Ministry of Education Malaysia, 2015). Key initiatives for internationalisation include integrating international elements into course syllabi, hiring more international staff, establishing student mobility programmes, and offering twinning programmes with institutions in the UK, US, Australia and New Zealand (Muthu & Chelliah, 2020). Malaysia is also a host for international branch campuses such as Newcastle University Medicine Malaysia, University of Southampton Malaysian Campus, Netherlands Maritime Institute of Technology, University of Reading Malaysia and University of Nottingham Malaysia Campus (ICEF Monitor, 2016). These efforts are part of a broader strategy to position Malaysia as a world-class education provider and a hub for education in Southeast Asia (Chang & Abdullah, 2021).

International students are drawn to Malaysia for various reasons, including its culture, the use of English as the medium of instruction (Singh, 2019), and the combination of quality and affordability in education (Khalid et al., 2020). This appeal is particularly strong among Asian students. The top three sending markets for Malaysia are China, Indonesia and Bangladesh (ICEF Monitor, 2023), but Vietnam is also a significant contributor. Vietnamese students find the lower costs and cultural similarities between Malaysia and their home country advantageous (Thoo et al., 2022). In recent years, extensive collaborations between Vietnam and Malaysia have resulted in approximately 1,000 Vietnamese students studying in Malaysia (VietnamPlus, 2023).

Despite Malaysia being a popular destination for Asian students, they still face significant challenges. Tuerxun et al. (2020) found that Muslim Chinese international students in Malaysia faced problems with the hot humid and unpredictable climate, spicy food and language. Although both China and Malaysia are Asian countries, the students also reported social and cultural challenges. They found that Malaysian students kept to themselves. The students' lack of proficiency in English and the local language, Malay, led them to socialise more with other international students rather than with locals. This indicates that better language support and planned opportunities for intercultural interactions are important.

A study by Rathakrishnan et al. (2021) involving 541 international undergraduate students at a public Malaysian university highlighted similar issues. The international students experienced homesickness and cultural stress, primarily due to language adaptation challenges. They faced language barriers that led to feelings of depression, homesickness and an inability to adjust to the new environment. Even those proficient in English struggled to understand the English spoken by locals due to the heavy Malay accent. This prevented them from building and maintaining good relationships with local people, resulting in social anxiety and a sense of not fitting in.

The findings have significant implications for student mobility programmes and the development of intercultural competence. Firstly, the challenges faced by international students in Malaysia highlight the necessity for universities to provide comprehensive support systems that address both academic and non-academic needs. This includes orientation programmes that prepare students for the local climate, cuisine and social norms, as well as ongoing language support to help them overcome communication barriers. Encouraging local students to engage with their international peers can help reduce homesickness, cultural stress and social anxiety and bridge social gaps. By doing so, universities can foster a more inclusive environment.

5.4 Methodology

The study involved 104 first-year inbound mobility students from Vietnam (80 male, 24 female). Most participants came from families with modest incomes, typically earning less than 14,999,999 VND per month (equivalent to about USD610). The students were in a one-month intensive English language mobility programme in one of the universities in Sarawak, Malaysia. It was designed to enhance the students' English language skills: listening, speaking, reading and writing. The programme was a credit transfer programme.

The programme comprised a blend of both English lessons and cultural immersion. During the weekdays, the students spent five hours each day in intensive English classes. During weekends, their educational experience extended beyond the classroom to prominent local attractions such as museums and Sarawak Cultural Village to enrich the students' cultural understanding. The programme was also augmented with on-campus activities that allowed students to engage with local people and learn local cultures. This included opportunities to learn traditional Malaysian dances, an experience facilitated by expert traditional Malaysian dancers. In addition, throughout their stay, the mobility students were accommodated at a residential college within the university campus, giving them opportunities to interact with other students at the college.

In investigating students' intercultural awareness, an online questionnaire consisting of four sections (background information, initial intercultural exposure, intercultural awareness and elaboration of experiences) was constructed in Google Forms. The mobility students were told about the study on intercultural awareness, voluntary participation and confidentiality of responses. Informed consent was secured from all participants. The link was sent to students by the second researcher on the last day of the mobility programme through their WhatsApp group. The students were encouraged to complete the questionnaire.

For the questionnaire, the participants' responses were downloaded in the form of an Excel sheet. The responses were coded numerically (e.g. 1 for female, 2 for male, 1 for strongly disagree, 2 for agree, 3 for neutral, 4 for agree, 5 for strongly agree). Then the Excel statistical functions were used to compute means, standard deviations and percentages. For the responses to the open-ended questions on students' accounts of their experiences, their responses were read a few times to understand their context and content. Thematic analysis was conducted based on Cresswell and Cresswell (2018) to determine key themes and sub-themes. Categories were developed based on the aspects of the experiences such as culture, language and practical issues. The responses were then assigned to the developed categories. Finally, the themes and sub-themes were counted because the number indicate whether a particular theme or sub-theme was of common concern of interest to the students, or the unique experiences of a small subset of the mobility student experiences.

5.5 Findings

5.5.1 Participants' Initial Exposure to Other Cultures

All participants had minimal direct engagement with other cultures before coming to Malaysia. Nearly 70% indicated they did not have an ethnically diverse family background, 59.23% had fewer than three international friends and 76.93% had visited fewer than three countries (Table 5.1).

5.5.2 Participants' Intercultural Awareness After Student Mobility Programme

The other culture referred to here is the Malaysian culture. The participants reported respect for people from different cultures in terms of their ways ($M = 4.41$) and values ($M = 4.36$). When interacting with people from other cultures, the participants tried to give positive responses ($M = 4.13$) and obtain as much information as they could ($M = 4.09$). They were keen to find opportunities for such interactions by attending events ($M = 4.07$). The participants had an interest in cultural exchange and a high regard for cultural diversity. Table 5.2 shows that the programme fostered positive intercultural awareness, based on the Vietnamese students' self-reports.

Next, in terms of adaptability in intercultural situations, the participants' mean scores were around the neutral of three, and the larger standard deviations showed variability in responses. As a group, the Vietnamese participants were marginally positive about the importance of having agreement from friends of different cultures

Table 5.1 Participants' intercultural exposure (N = 104)

Intercultural exposure	Level of agreement	Frequency	Percentage
Ethnic diversity within family members	Strongly disagree	71	68.27
	Disagree	1	0.96
	Neutral	8	7.69
	Agree	16	15.38
	Strongly agree	8	7.69
Number of friends from other culture groups	0	35	33.98
	1	7	6.80
	2	19	18.45
	3	14	13.59
	4	2	1.94
	5	1	0.97
	6	25	24.27
Number of countries	0	31	29.81
	1	39	37.50
	2	10	9.62
	3	12	11.54
	4	3	2.88
	5	0	0
	6	9	8.65

Table 5.2 Participants' experiences of intercultural awareness (N = 104)

Item	Mean	SD
1. I respect the ways people from different cultures behave	4.41	0.89
2. I respect the values of people from different cultures	4.36	0.97
3. I often give positive responses when interacting with people from other cultures	4.13	1.03
4. I try to obtain as much information as I can when interacting with people from different cultures	4.09	1.07
5. I like to attend events where I might get to know people from different cultures	4.07	1.05
6. It is important that friends from different cultures agree with me on issues	3.31	1.20
7. Getting to know someone of another culture is generally uncomfortable for me	3.15	1.41
8. I am sensitive to the subtle meanings conveyed by people from other cultures	2.87	1.20
9. I get upset easily when interacting with people from different cultures	2.33	1.46

(M = 3.31) and the discomfort of the acquaintance process (M = 3.15). However, the participants disagreed that they were sensitive to the subtle meanings conveyed by Malaysians but they do not get upset easily in such intercultural interactions (M = 2.33). These findings hint at the complexity of reconciling personal beliefs with the respect for diverse cultural perspectives, and that the participants acknowledged the challenges in fully understanding subtle communication across cultures.

5.5.3 Changes in Cultural Views Towards Malaysia and Vietnam

The Vietnamese students were asked to directly reflect on the changes in cultural views towards both Malaysia (host country) and Vietnam (native country). Table 5.3 shows the results.

Over 70% of the participants reported a considerable increase in their cultural knowledge of Malaysia (71.15%) and ease in interacting with Malaysians (70.19%). The open-ended item which allowed them to give additional responses revealed the cultural exchange events were meaningful for the participants. One of the participants explained that he or she 'participated in many cultural exchange events and learned a lot about food and life'. Examples of cultural exchange events are excursions to cultural sites ("Here I went to some places like museums, mosques, cultural villages") and intercultural activities like sports ("I have some Malaysian friends play basketball with me") and cultural performances ("Get on stage and dance"). The cultural exchange events enabled the participants to gain an understanding and appreciation of different cultures. The acquired cultural knowledge was translated into actions, with several participants reporting better adaption to the local culture ("trying to adapt to the culture is important", "I learned how to behave with people around me").

Most of the participants also experienced greater ease when interacting with Malaysians as only 6.73% reported that their ease level remained the same. Many of them viewed Malaysians as friendly ("People here friendly"). A few of them had

Table 5.3 Changes in cultural views towards Malaysia and Vietnam (N = 104)

Item	Increased a lot	Increased a little	Remained the same
1. Cultural knowledge of Malaysia	74 (71.15%)	25 (24.04%)	5 (4.81%)
2. Ease in interacting with Malaysians	73 (70.19%)	24 (23.08%)	7 (6.73%)
	Became more positive	Became more negative	Remained the same
3. View of Malaysia	84 (80.77%)	1 (0.96%)	19 (16.35%)
4. View of Vietnam	58 (55.77%)	8 (7.69%)	38 (36.54%)

even formed personal relationships with the locals ("Have a friend from Kuching", "I have a girlfriend in Kuching").

The next part of the results in Table 5.3 show that after the one-month student mobility programme, 80.77% of the participants' cultural views also underwent positive changes. Only one of 104 participants (0.96%) reported their views becoming more negative while 16.35% retained their cultural views of Malaysia.

The open-ended responses revealed that their negative experiences were predominantly cuisine-related. They were not used to spicy food ("The food in Kuching very spicy") and having chicken for most meals ("I only eat chicken, I am very sad"). Although Malaysian food is generally spicier than Vietnamese food, Malaysian cuisine include lamb, beef and pork (consumed by non-Muslims). The Vietnamese students could have reached the conclusion about chicken because of the food catering selection during their stay. The implication of the findings is that a more diverse range of cuisines that accurately represent Malaysian culture should be offered. This would address misconceptions such as the belief that 'Malaysians do not eat pork' due to Muslim dietary laws. Offering a broader variety of local cuisines that accurately reflects the host culture will not only break down stereotypes, but also leads to better understanding of the culture.

The Vietnamese students also were not used to the prayer calls for Muslims to pray. They commented to the organisers about the singing, and wondered why there was singing at different times of the day. The organisers explained to the students who queried, but this experience indicated that the exchange students should be informed about various religious practices in Malaysia.

As for cultural views of Vietnam, the results showed that the time in Malaysia enabled 55.77% to view their own country more positively while 7.69% reported a negative shift in their views of Vietnam and 36.54% retained their views. Some participants stated they developed respect for a variety of cultures while still taking pride in their own heritage ("Respect different cultures and be proud of your nation"). This finding suggests an advanced understanding of intercultural sensitivity where individuals appreciate the richness of other cultures without diminishing the value of their own cultural identity.

5.6 Discussion

This study highlights three noteworthy points regarding student mobility programmes and intercultural education in Asia in terms of their intercultural awareness, social and personal development and challenges in cross-cultural adaptation.

Firstly, we showed that even a brief student mobility programme can significantly impact students' intercultural awareness and sensitivity. Before the programme, the Vietnamese students had limited exposure to cultural diversity, be it in terms of their family background, social circle, or travel experience. By the end of the programme, they reported positive shifts in intercultural awareness. This finding aligns with previous studies by Fang and Baker (2017) and San and Htwe (2023) which also concluded that short-term mobility programmes are effective in promoting students' intercultural awareness.

Our findings indicated that the inclusive, hands-on cultural experiences that can bring about increased intercultural awareness include excursions, participation in cultural events, conversations with locals and learning local dances. The Vietnamese students participated actively in various informal activities as well. The programme structure which included both formal talks on culture and informal interactions with locals allow for deeper cultural engagement. The hands-on cultural experiences led to considerable positive shifts in their cultural perceptions towards both Malaysia and Vietnam, advancing towards greater ethnorelativism. This progression is a critical stage in Bennett's (1993) Developmental Model of Intercultural Sensitivity. The Vietnamese students exhibited a proactive and open approach to understanding other cultures. This finding thus underscores the need for an educational structure that promotes inclusion, engagement and cultural synergy to achieve intercultural competence. This finding echoes Sobkowiak's (2019) assertion that factors like the frequency and depth of intercultural engagement are essential for developing intercultural competence.

Secondly, the benefits of studying abroad extended beyond academic achievement (improving English proficiency) to include social and personal development. Immersed in Malaysia, where English is widely used as a second language, the students not only improved their language skills through increased exposure but also used English to communicate with local students and build new friendships, broadening their perspectives on their own and other cultures. As noted by Skefsrud (2022), such programmes are instrumental in helping students develop a deeper appreciation of diverse cultures and improve their global competencies.

Finally, despite participating in immersive activities, the Vietnamese participants had difficulties adapting to local cuisine. They expressed disappointment with the limited variety of food, specifically the frequent serving of chicken, and struggled with the spiciness of the meals. This issue shows the need for better dietary options in mobility programmes to accommodate diverse student needs while allowing them to experience the local food. This finding concurs with Tuerxun et al.'s (2020) findings, where Muslim Chinese international students in Malaysia also reported difficulties with spicy food. As such, it is crucial for student mobility programmes to have greater empathy towards foreign students' needs and design mobility programmes that reflect the full range of the host culture, including cuisine and religion. Even when the programme is based in Asia and aimed at Asian students, there are still cultural differences which can pose challenges and adversely affect their memory of their stay. Furthermore, universities should facilitate open dialogues where students feel supported and encouraged to express their concerns, ensuring their issues are addressed.

Another problem faced by the Vietnamese participants was their difficulties in establishing deep cross-cultural connections. They reported uncertainty in forming cross-cultural friendships. They were unsure about intercultural communication nuances, suggesting that a one-month programme may be too brief. This finding echoes Jon's (2013) findings in Korea, where short programme length limited students' opportunities for forging meaningful social bonds. These findings indicate that while short-term programmes offer initial intercultural exposure, students may benefit from additional support, such as virtual exchanges or follow-up collaborations, to help them maintain relationships and extend their intercultural engagement beyond the formal programme period. Mobility programmes should be seen as the starting point for fostering meaningful humanistic benefits, rather than the end goal.

The difficulties Vietnamese students faced in forming cross-cultural friendships also highlight the need to reconsider the pedagogical framework of the programme. Focusing solely on international students is insufficient. Local students must also be engaged and take on active roles. In short, effective student mobility programmes should foster positive relationships that lead to mutual improvement and growth for all participants, not just international students.

5.7 Situating the Findings Within the Broader Asian Context

The study was on the impact of a Malaysian short-term mobility programmes on intercultural awareness of Vietnamese students. The study generated new findings as similar studies have not been conducted in other Southeast Asian countries. In short, Vietnamese students with limited prior exposure to diverse cultures showed increased intercultural sensitivity after a brief exchange in Malaysia. This outcome reflects findings from Jon's (2013) study, where Korean students in short-term programmes reported gains in intercultural competence and confidence. Similarly, Hofmeyr (2021) found that even brief, multicultural interactions raised cultural awareness among Japanese students. As such, the findings of these studies suggest that a well-structured short-term mobility programme can provide meaningful cultural learning experiences.

Student mobility programmes had conventionally involved bringing students across the eastern and western hemispheres. Our study showed that there are intercultural gains from intraregional exchanges within Asia. Shared cultural elements can facilitate adaptation for students. For instance, Jon and Yoo (2021) found that regional mobility programmes in Korea enhanced mutual understanding and encouraged students' interest in neighbouring Asian cultures. This implies that intraregional student exchanges provide an accessible entry point for intercultural engagement, particularly for students without prior experience in culturally diverse settings.

Our study provided an indication of informal, experiential activities that are important for fostering intercultural contact with locals, such as cultural events and social gatherings. Similarly, a study on a one-month teacher exchange programme in the Philippines (Agustin & Montebon, 2018) found that Thai teachers' intercultural awareness increased through activities like homestays, where host families cooked local dishes and guided visits to cultural heritage sites. These findings emphasise the importance of experiential learning tailored to cultural context, which is especially effective when the locals are involved, as they offer first-hand cultural perspectives. This kind of interactive engagement helps foster intercultural sensitivity and enables students to connect more authentically.

Other studies on student mobility programmes are on those with a longer duration. For example, the Mobility Programme of University Student in Asia (CAMPUS Asia) initiative among Japan, South Korea and China demonstrates that extended exchanges with integrated academic and extracurricular activities are more effective at fostering deep, lasting intercultural connections. For example, Yung et al. (2017) observed that CAMPUS Asia's extended duration encouraged sustained engagement, enabling students to form meaningful cross-cultural relationships.

By contrast, shorter programmes, while valuable for initial intercultural exposure, often face challenges in building such bonds, as seen in this study where Vietnamese students reported difficulties forming close friendships within the programme's brief duration. This also aligns with Jon's (2013) findings in Korea, where limited programme length constrained students' ability to establish strong social connections. As such, to enhance the impact of short-term exchanges, incorporating virtual exchanges or follow-up collaborations could allow students to maintain relationships and deepen their intercultural understanding beyond the programme's duration.

Finally, the challenges Vietnamese students faced in adapting to Malaysian cuisine reflect the complex role of food in cultural immersion. Tuerxun et al. (2020) reported that Chinese Muslim students in Malaysia also struggled with local spicy foods, illustrating how unfamiliar cuisine can act as a barrier to engagement. However, some programmes have successfully used food as a bridge to cultural understanding. For instance, Umemiya et al. (2024) described a Bruneian student's experience in Korea, where a traditional sweet cooking class became a meaningful way to connect with the host culture. These contrasting experiences suggest that while regional proximity can ease some cultural adjustments, differences in food preferences often persist within Asia. To enhance students' comfort and engagement, mobility programmes should balance cultural immersion with sensitivity to dietary preferences.

5.8 Future Directions

Looking ahead, the long-term evolution of short-term intercultural exchange programmes will increasingly be shaped by the global interconnectedness of Asia. As students' intercultural sensitivity is enhanced during these exchanges, it is essential to examine how these shifts in cultural awareness continue to evolve over time.

This exploration should focus not only on immediate changes but also on whether these experiences lead to sustained shifts in attitudes, behaviours and cross-cultural engagement once students return to their home countries. Integrating post-exchange activities, such as virtual exchanges or alumni networks, could enhance the sustainability of intercultural learning, addressing the challenges of brief mobility programmes while supporting the continuous development of intercultural competence. This expansion of follow-up opportunities highlights how short-term exchanges can contribute meaningfully to the lifelong cultivation of global citizenship.

As the role of local cultural context becomes even more critical, future work will need to explore how experiential learning activities—such as homestays, cooking classes, or collaborative community projects—affect intercultural understanding across different cultural settings in Asia. A comparative analysis across Asian nations could reveal how various cultural contexts shape the experience of intercultural exchange and help identify the most effective methods for fostering lasting intercultural competence. Expanding the types of cultural immersion offered will provide a more holistic view of intercultural exchange and support the development of genuine, enduring relationships.

In parallel, integrating these experiential elements into the broader academic curriculum could further elevate the educational value of exchange programmes. Formalising intercultural learning within academic structures will provide greater depth and continuity, enabling students to connect their exchange experiences with global issues such as sustainable development, cultural diversity and the complexities of an interconnected world. By embedding these themes into structured academic programmes, future studies can better equip students with the critical thinking tools needed to engage with the evolving challenges of global interconnectedness.

5.9 Conclusion

This study validates the value of short-term student mobility exchanges as an effective entry point for intercultural awareness and sensitivity, especially within an intraregional Asian context. The future of education in the Asian region lies in formal intercultural education in well-structured student mobility programmes that focus on both short-term and long-term benefits. This means that the programmes must prioritise not only academic achievements but also personal growth through culturally immersive activities. This approach will enable students to form meaningful friendships, gain broader cultural perspectives and foster greater tolerance and competencies.

Nevertheless, effective programmes will also require the active participation of both international and local students to ensure meaningful cultural integration. These programmes must also accurately represent host countries' cultural practices to avoid misunderstandings and be sensitive to the diverse needs of international students. Preparing students with orientation sessions on cultural norms and language before they go abroad can also help them adapt and thrive more easily in a new environment. Formal intercultural education in student mobility programmes will become vital components of future education, serving as essential tools for preparing students, both international and domestic, to thrive in a globally interconnected world.

Highlights

- Short-term student mobility programmes significantly enhance intercultural awareness and sensitivity, with Vietnamese students reporting positive shifts in cultural perceptions after their exchange in Malaysia.
- Challenges include difficulties in adapting to local cuisine and forming deep cross-cultural connections, suggesting the need for better dietary options and longer programmes for meaningful relationship-building.
- Intraregional exchanges within Asia, particularly among neighbouring countries, provide valuable intercultural learning opportunities, highlighting the potential of regional mobility programmes for promoting mutual understanding.
- Future directions for student mobility programmes should include follow-up activities, experiential learning and integration of intercultural education into academic curricula to ensure sustained intercultural development.

References

Adler, P. S. (1987). Culture shock and the cross-cultural learning experience. In E. C. Smith & L. F. Luce (Eds.), *Toward internationalism: Readings in cross-cultural communication* (pp. 24–35). Newbury.

Agustin, R., & Montebon, A. (2018). An assessment of the project teacher exchange for ASEAN teachers (TEACH) program. *International Journal of Evaluation and Research in Education, 1*(1), Article 7635. https://files.eric.ed.gov/fulltext/EJ1174874.pdf

Behrnd, V., & Porzelt, S. (2012). Intercultural competence and training outcomes of students with experiences abroad. *International Journal of Intercultural Relations, 36*(2), 213–223. https://doi.org/10.1016/j.ijintrel.2011.04.005

Bennett, M. J. (1993). Towards ethnorelativism: A developmental model of intercultural sensitivity. In M. Paige (Ed.), *Education for the intercultural experience* (pp. 21–71). Intercultural Press.

Carley, S., Stuart, R., & Dailey, M. P. (2011). Short-term study abroad: An exploratory view of business student outcomes. *Journal of Management Policy and Practice, 12*(2), 44–53.

Central Intelligence Agency. (2024a). *The world factbook: Malaysia.* https://www.cia.gov/the-world-factbook/countries/malaysia/

Central Intelligence Agency. (2024b). *The world factbook: Vietnam.* https://www.cia.gov/the-world-factbook/countries/vietnam/#people-and-society

Chang, D. W., & Abdullah, D. (2021). Internationalisation of Malaysian universities: Policies, practices, and the SDGs. *International Journal of Comparative Education and Development, 23*(3), 212–226.

Council of Europe. (2001). *Common European framework of reference for languages: Learning, teaching, assessment*. Cambridge University Press.

Cresswell, J. W., & Cresswell, J. D. (2018). *Research design: Qualitative, quantitative, and mixed methods approaches* (5th ed.). SAGE.

Cuong, L. T. (2020). Students' readiness for international academic mobility programs: Evidence from Faculty of English Linguistics and Literature, University of Social Sciences and Humanities, VNUHCM. *Ho Chi Minh City Open University Journal of Science, 10*(2), 52–67. https://doi.org/10.46223/HCMCOUJS

Czerwionka, L., Artamonova, T., & Barbosa, M. (2015). Intercultural knowledge development: Evidence from student interviews during short-term study abroad. *International Journal of Intercultural Relations, 49*, 80–99. https://doi.org/10.1016/j.ijintrel.2015.06.012

Czura, A. (2017). Student teachers' recollections of a short-term study abroad experience: Critical incident analysis. *Konin Language Studies, 5*(1), 135–156.

de Wit, H. (2019). Internationalization in higher education, a critical review. *SFU Educational Review, 12*(3), 9–17.

Diem, H. T. T., Thinh, M. P., & Mung, T. T. (2023). An investigation into the benefits and challenges of international student exchange programs: Perspectives from student teachers. *International Journal of Learning, Teaching and Educational Research, 22*(7), 258–280. https://doi.org/10.26803/ijlter.22.7.14

Dos Santos, L. M. (2022). Study abroad and student exchange experiences of international students: The sociocultural and psychological racism perspective. *Academic Journal of Interdisciplinary Studies, 11*(1), 15–27. https://doi.org/10.36941/ajis-2022-0002

Engle, L., & Engle, J. (2004). Assessing language acquisition and intercultural sensitivity development in relation to study abroad program design. *The Interdisciplinary Journal of Study Abroad, 10*, 151–163.

Fang, F., & Baker, W. (2017). A more inclusive mind towards the world: English language teaching and study abroad in China from intercultural citizenship and English as a lingua franca perspective. *Language Teaching Research, 22*, 1–17.

Geibel, W. R. (2020). A pedagogy of student mobility: Facilitating humanistic outcomes in internationalization and student mobility. *Journal of Comparative and International Higher Education, 12*, 67–77.

Hanvey, R. G. (1987). Cross-culture awareness. In E. C. Smith & L. F. Luce (Eds.), *Toward internationalism: Readings in cross-cultural communication* (pp. 13–33). Newbury.

Hofmeyr, A. S. (2021). Intercultural competence development through co-curricular and extracurricular at-home programs in Japan. *Journal of Studies in International Education, 27*, 363–386.

HSU. (2023, May). *Hoa Sen University cooperates with the University of Malaysia Sarawak (UNIMAS) to organize the ICGD international conference*. Hoa Sen University. Retrieved November 13, 2024, from https://www.hoasen.edu.vn/en/hoa-sen-university-cooperates-with-the-university-of-malaysia-sarawak-unimas-to-organize-the-icgd-international-conference/

HUTECH. (2024, March). *HUTECH and the University Kebangsaan Malaysia signing MoU and MoA*. HUTECH University of Technology. Retrieved November 13, 2024, from https://www.hutech.edu.vn/english/training-news/14617850-hutech-and-the-university-kebangsaan-malaysia-signing-mou-and-moa

ICEF Monitor. (2016, Aug 22). *Malaysia competing for a greater share of international students*. ICEF Monitor. Retrieved May 31, 2024 from https://monitor.icef.com/2016/08/malaysia-competing-greater-share-international-students/

ICEF Monitor. (2019, Dec 4). *Closer to home: Intra-regional mobility in Asia*. ICEF Monitor. Retrieved May 31, 2024 from https://monitor.icef.com/2019/12/closer-to-home-intra-regional-mobility-in-asia/

ICEF Monitor. (2023, Feb 1). *Malaysia exceeds target for new international student applications in 2022*. ICEF Monitor. Retrieved May 31, 2024 from https://monitor.icef.com/2023/02/malaysia-exceeds-target-for-new-international-student-applications-in-2022/

ICEF Monitor. (2024, Feb 7). *Southeast Asian students increasingly considering "studying abroad" close to home*. ICEF Monitor. Retrieved May 31, 2024 from https://monitor.icef.com/2024/02/southeast-asian-students-increasingly-considering-studying-abroad-closer-to-home/

Jon, J. E. (2013). Realizing internationalization at home in Korean higher education: Promoting domestic students' interaction with international students and intercultural competence. *Journal of Studies in International Education, 17*(4), 455–470. https://doi.org/10.1177/1028315312468329

Jon, J., & Yoo, S. S. (2021). Internationalization of higher education in Korea: Policy trends toward the pursuit of the SDGs. *International Journal of Comparative Education and Development, 23*(2), 120–135.

Khalid, J., Nordin, N. M., Ali, N. J., Iftikhar, U., & Jamil, A. (2020). Exploring factors influencing student study abroad choice of destination in the contemporary globalized era. *Akademika,90*(Special Issue 2), 17–28. https://doi.org/10.17576/akad-2020-901

Kiet, H. S. T. (2016). Developing intercultural awareness and attitudes of Vietnamese learners of English: Case studies. *Tạp chí Khoa học và Công nghệ-Đại học Đà Nẵng, 109*(12), 80–84.

Kim, Y. Y. (2001). *Becoming intercultural: An integrative theory of communication and cross-cultural adaptation.* Sage.

Kong, K., & Rajandram, K. V. (2023). *Global learning: Virtual exchange between Malaysia and the US*. Retrieved November 14, 2024 from https://diversityabroad.org/public/DIVaPublic/GIE-Archives/GIE-2023/GIE-Sp2023/GIE-Sp23-Article-14.aspx

Krajewski, S. (2011). Developing intercultural competence in multilingual and multicultural student groups.*Journal of Research in International Education,10*(2), 137–153. https://doi.org/10.1177/1475240911408563

Le, H., Koo, F. K., Arambewela, R., & Zutshi, A. (2017). Voices of dissent: Unpacking Vietnamese international student experience. *International Journal of Educational Management, 31*(3), 280–292. https://doi.org/10.1108/IJEM-07-2015-0099

Luong, P. M., Tran, L. T., Nguyen, H. T. T., Ngo, N. T. H., Nguyen, T. T. M., Dang, G. H., & Nguyen, H. T. (2023). Student agency for intercultural adaptability in international programs: Insights into internationalization at home in Vietnam. *International Journal of Intercultural Relations, 96*, Article 101855. https://doi.org/10.1016/j.ijintrel.2023.101855

Ministry of Education Malaysia. (2015). *Malaysia education blueprint 2015–2025 (higher education).* Ministry of Education Malaysia. Retrieved January 20, 2024, from https://www.um.edu.my/docs/um-magazine/4-executive-summary-pppm-2015-2025.pdf

Muthu, K., & Chelliah, S. (2020). Perceived factors influencing internationalisation of Malaysian public higher learning institutions. In *European proceedings of social and behavioural sciences* (pp. 517–526). European Publisher. https://doi.org/10.15405/epsbs.2020.10.46

Nguyen, B. T. T. (2023). *Cultural awareness and intercultural awareness in ELT practices from global Englishes perspective in higher education in a Vietnamese university* (Doctoral dissertation, University of Southampton).

Nguyen, M. T. (2021). Developing education in Vietnam in the context of international integration. In *Proceedings of the 17th international conference of the Asia association of computer-assisted language learning (AsiaCALL 2021)* (pp. 234–240). Atlantis Press. https://doi.org/10.2991/ASSEHR.K.210226.029

Panaroma Destination. (2023, December 7). *Tourism Malaysia participates in the Batik Air show.* Panorama Destination. Retrieved November 14, 2024, from https://www.panorama-destination.com/malaysia-news/tourism-malaysia-participates-in-the-batik-air-roadshow/

Paramitaningrum, P. (2013). Enhancing people-to-people cooperation between ASEAN and East Asia countries through counterparts: The case of Indonesian students in Taiwan. *Journal of ASEAN Studies, 1*(2), 164–176. https://doi.org/10.21512/jas.v1i2.69

Penbek, S., Yurdakul, D., & Cerit, G. (2012). Intercultural communication competence: A study about the intercultural sensitivity of university students based on their education and international experiences. *International Journal of Logistics Systems and Management, 1*, 232–252.

Pfaff-Czarnecka, J. (2020). Shaping Asia through student mobilities. *American Behavioral Scientist, 64*(4), 1–15. https://doi.org/10.1177/0002764220947753

Phong, H. (2023, December 28). *Vietnam wraps up 2023 with 12.6 million foreign visitors.* VnExpress International. https://e.vnexpress.net/news/travel/vietnam-wraps-up-2023-with-12-6-million-foreign-visitors-4694869.html#:~:text=Thailand%20(489%2C000)%2C%20Malaysia%20(,rounded%20off%20the%20top%2010

Rathakrishnan, B., Singh, S. B., Kamaluddin, M. R., Ghazali, M. F., Yahaya, A., Mohamed, N. H., & Krishnan, A. R. (2021). Homesickness and socio-cultural adaptation towards perceived stress among international students of a public university in Sabah: An exploration study for social sustainability. *Sustainability, 13*(9), Article 4924. https://doi.org/10.3390/su13094924

Rings, G., & Allehyani, F. (2020). Personality traits as indicators of the development of intercultural communication competence. *International Journal of Curriculum and Instruction, 12*(1), 17–32. https://ijci.wcci-international.org

San, C. K., & Htwe, K. M. (2023). Exploring the impact of study abroad on students' intercultural competence: The case of Myanmar. *Journal of Comparative & International Higher Education, 15*(3), 81–97.

Sierra-Huedo, M. L., & Foucart, J. (2022). Intercultural and professional skills in student mobility to boost employability. *Journal of Intercultural Communication, 22*(3), 1–12. https://doi.org/10.36923/jicc.v22i3.68

Singh, M. K. M. (2019). Academic reading and writing challenges among international EFL master's students in a Malaysian university: The voice of lecturers. *Journal of International Students, 9*(4), 972–992.

Skefsrud, T. (2022). Why student mobility does not automatically lead to better understanding: Reflections on the concept of intercultural learning. In D. Cairns (Ed.), *The Palgrave handbook of youth mobility and educational migration* (pp. 63–73). Palgrave Macmillan. https://doi.org/10.1007/978-3-030-99447-1_7

Slipchuk, V. L., Yuzkiv, H., Batechko, N., Pisotska, M., & Klymenko, L. (2021). The development of academic mobility among the students of universities. *Linguistics and Culture Review, 5*(S3), 224–236. https://doi.org/10.21744/lingcure.v5nS3.1514

Sobkowiak, P. (2019). The impact of studying abroad on students' intercultural competence: An interview study. *Studies in Second Language Learning and Teaching, 9*(4), 681–710. https://doi.org/10.14746/ssllt.2019.9.4.6

Taniguchi, N., Takai, J., & Skowronski, D. (2022). Intra-cultural and inter-cultural: Contact orientation of international students in Japan. *Journal of International Students, 12*(4), 843–866. https://doi.org/10.32674/jis.v12i4.3747

Tarchi, C., & Surian, A. (2022). Promoting intercultural competence in study abroad students. *European Journal of Psychology Education, 37*, 123–140. https://doi.org/10.1007/s10212-021-00554-0

Thoo, A. C., Lim, M. B. P., Huam, H. T., & Sulaiman, Z. (2022). Increasing destination loyalty of international students towards Malaysian higher educational institutions. *International Journal of Evaluation and Research in Education, 11*(1), 31–41. https://doi.org/10.11591/ijere.v11i1.21619

Ting-Toomey, S. (1999). *Communication across cultures.* Guilford Press.

Tran, L. T., Phan, H. L. T., & Marginson, S. (2018). The 'advanced programmes' in Vietnam: Internationalising the curriculum or importing the 'best curriculum' of the west? In L. Y. Tran & S. Marginson (Eds.), *Internationalisation in Vietnamese higher education* (pp. 55–76). Springer International Publishing AG.

Tuerxun, S., Hisham, N. A., Abdulla, R., & Khalid, N. S. (2020). Psychological adjustment of Muslim Chinese international students in Malaysia. *International Journal of Education, Psychology and Counselling, 5*(34), 60–79. https://doi.org/10.35631/IJEPC.534005

Umemiya, N., Sugimura, M., Kosaikanont, R., Nordin, N. M., & Ahmad, A. L. (2024). Impact of a consortium-based student mobility programme: The case of AIMS (Asian International Mobility for students). *Journal of International Cooperation in Education, 26*(1). https://doi.org/10.1108/jice-08-2023-0020/full/html

Uoc, T. M., Ai, L. H., & Binh, N. T. T. (2023). Educational philosophy for Vietnam in the present era. *Multidisciplinary Science Journal, 5*. https://doi.org/10.31893/multiscience.2023062

VietnamPlus. (2023, July 18). *Malaysian PM's Vietnam visit hoped to develop substantive bilateral ties.* VietnamPlus. https://en.vietnamplus.vn/malaysian-pms-vietnam-visit-hoped-to-develop-substantive-bilateral-ties-post259562.vnp

Yang, P. (2014). Privilege, prejudice, predicament: "PRC scholars" in Singapore—An overview. *Frontiers of Education in China, 9*(3), 350–376. https://doi.org/10.1007/bf03397027

Yung, A. C. H., Hill, C., Chen, K. H., Tsai, H., & Chen, V. (2017). A comparative study of student mobility programs in SEAMEO-RIHED, UMAP, and Campus Asia: Regulations, challenges, and impacts on higher education regionalization. *Higher Education Evaluation and Development, 11*(1), 12–24. https://doi.org/10.1108/HEED-08-2017-003

Wun-Chiew Pung is a lecturer at the Faculty of Education, Language and Communication, Universiti Malaysia Sarawak. She graduated from the University of Malaya with a Ph.D. in Pragmatics. Her research interests include communication strategies and gendered discourse.

Connie Keh-Nie Lim is a senior lecturer in the Music Programme at the Faculty of Applied and Creative Arts, Universiti Malaysia Sarawak (UNIMAS). In addition to her academic role, she serves as an assistant programme coordinator for the mobility programme at UNIMAS Business School. She earned her Ph.D. in Music from Universiti Putra Malaysia. Her research interests encompass ethnomusicology and cultural musicology.

Su-Hie Ting is currently a professor and she teaches research methodology at the Faculty of Education, Language and Communication, Universiti Malaysia Sarawak. She graduated from the University of Queensland with a Ph.D. in Applied Linguistics in 2001. She has published extensively on language use, attitudes and identity, communication strategies and health communication.

Chapter 6
AI's Impact on Higher Education and Academic Publishing in Asia: Insights from China, South Korea and Singapore

Márton Gosztonyi

Abstract Navigating the intricate intersection of artificial intelligence (AI), higher education and academic publishing in Asia requires a keen understanding of both technology and pedagogy. This chapter highlights the transformative potential of AI within Asian higher educational landscapes, set against the rich tapestry of the continent's diverse academic traditions. It also explores the impact of artificial intelligence on scholarly publications, a vital component of the academic ecosystem. The growing influence of AI in Asia has prompted a shift in educational paradigms, fostering new opportunities for student engagement, instructional methods and the democratisation of knowledge. In this chapter, the readers are provided with a concise summary of these profound transformations. By drawing on insights from China, South Korea and Singapore as illustrative examples, this chapter examines how AI is integrated into the research process and academic publishing workflows, with a particular focus on how these countries adapt to technological change while maintaining cultural and academic traditions Grounded in rigorous research, this chapter presents a multi-dimensional perspective, synthesising technology, tradition and practice, encouraging readers to reimagine the contours of learning and scholarship within the Asian context. As AI continues to evolve, the intersection between technology, ethics and pedagogy becomes increasingly complex, requiring a balance between innovation and the preservation of academic integrity.

Keywords Artificial intelligence (AI) · Higher education innovations · Asian academic traditions · Scholarly publishing · Digital pedagogy · Ethical implications

M. Gosztonyi (✉)
Faculty of Social Sciences, Eötvös Loránd University, Budapest, Hungary
e-mail: gosztonyi.marton@gmail.com

A. Pandian et al. (eds.), *Future of Education in Asia*,
https://doi.org/10.1007/978-3-032-07285-6_6

6.1 Introduction

In the dynamic realm of higher education and academic publishing in Asia, a remarkable transformation is underway. This region, steeped in rich educational traditions, is now embracing cutting-edge artificial intelligence (AI) technologies. These advances are not merely technological feats; they represent a paradigm shift in the way education is delivered and knowledge is produced, disseminated and evaluated. AI's role in this context is pivotal, offering innovative tools for automated manuscript screening, AI-assisted peer review, content curation and even the generation of scholarly content. As AI becomes increasingly embedded in academic publishing ecosystems across Asia, balancing technological innovation with culturally rooted values presents both opportunities and ethical dilemmas. The landscape is thus a tapestry of tradition and modernity, evolving in response to scholarly priorities, national research agendas and global standards of academic excellence.

This chapter delves into the evolving relationship between AI and academic publishing in higher education across Asia, with a particular emphasis on national approaches in China, South Korea and Singapore. It seeks to examine how AI is reshaping publishing workflows, research evaluation and editorial practices, highlighting both the potential benefits, such as increased efficiency and transparency and the risks, including algorithmic bias and ethical ambiguity. Rather than presenting empirical research, the chapter offers a conceptual synthesis of policy, practice and scholarly literature to support critical reflection and informed innovation. In doing so, the chapter contributes to ongoing regional dialogues on the future of education in Asia, offering insight into how AI-driven publishing practices can evolve in ways that honour both academic integrity and cultural diversity.

6.1.1 Asian Academic Traditions and Contemporary Challenges

Asian higher education is deeply rooted in a rich historical and cultural tapestry, characterised by distinct educational philosophies that have evolved over centuries. Central to these traditions is the influence of Confucian philosophy, particularly in Chinese, Korean and Japanese and South-East Asian education systems, where respect for teachers and the pursuit of moral and intellectual development is emphasised (Kim & Park, 2013; Lee, 2000). The Gurukul system in India exemplifies another unique approach, prioritising holistic education derived from life experiences under the guidance of a guru, a stark contrast to the compartmentalisation seen in Western education (Sharma, 2017).

The impact of colonialism, as noted by Nguyen (2015), introduced Western educational methods to many Asian countries, creating a blend of traditional and Western educational paradigms. This historical intermingling has shaped the distinctive educational landscape in Asia. Thus, western education, characterised by a focus on critical thinking, creative problem-solving and learner autonomy, has significantly influenced Asian higher education (Johnson, 2019; Smith, 2003). Asian educational systems have been evolving to incorporate these aspects, particularly in higher education, leading to a complex picture of contrast and convergence between Asian and Western academic models. The significant influence of Western academic models on Asian higher education is further elaborated by Altbach (1989), highlighting how these models have shaped educational structures and philosophies in Asia. This influence, coupled with the impact of Japanese colonialism in regions like Korea and Taiwan, and South-East Asia presents a dynamic interaction of local and external educational traditions (Altbach, 1989; Jung, 2018).

Recent studies by Welch (2019), Yang (2019), Jarvis and Mok (2019) and Shahjahan (2020) have focused on the challenges Asian higher education faces in balancing local traditions with external influences. These challenges are amplified by the rapid economic transformations in Asia, affecting higher education governance and the construction of Asian higher education in global discourse. Dee (2008) emphasises the rapid expansion of higher education in Asia since the 1980s, noting the emergence of Asian universities as prominent global players. This growth signifies a shift in the global educational landscape, necessitating a complex understanding of the diverse paths these institutions have taken. Chan (2016) discusses the historical decline and recent resurgence of academic innovation in Asia. Despite the Westernisation of educational models, there is a growing recognition of Asia's unique contributions to the global academic landscape, particularly in technological advancements.

Yun (2015) examines the impact of globalisation on higher education in various East Asian contexts. The study highlights how regions like Singapore, Hong Kong, Taiwan, China, Korea and some South-East Asian nations are adapting to global educational standards, with a focus on the challenges and transformational efforts undertaken. Marginson (2011) introduces the concept of the Confucian Model in Asia–Pacific higher education, emphasising the role of the nation-state and the pursuit of world-class educational standards. This model, while fostering rapid development, also raises concerns about social equity and academic creativity. Moreover, Ziguras (2016) addresses the massification and internationalisation of STEM fields in Asian higher education. The study also highlights the achievements of Asian students in international assessments like PISA, underscoring the effectiveness of these education systems. Furthermore, Yonezawa et al. (2014) and Jung et al. (2018) emphasise the ongoing evolution of higher education in Asia, focusing on the international dimensions and the challenges in developing a robust higher education research community. These studies call for a more international outlook and the integration of diverse perspectives in educational research. Yang (2018) underscores the importance of incorporating indigenous perspectives to fully grasp the unique aspects of the region's educational development.

Clearly, Asian academic traditions, influenced by historical, cultural and external factors, present a rich and complex landscape. The interplay of traditional values, Western influences and contemporary challenges creates a dynamic educational environment in Asia. Understanding these multifaceted influences is crucial for grasping the potential and challenges of higher education in the region, especially in the context of integrating emerging technologies like AI.

6.2 Literature Review

6.2.1 *A Brief History of Higher Educational Technologies and Teaching Methodologies Leading up to AI*

The evolution of higher educational technologies, intertwined with shifts in teaching methodologies, represents a fundamental transformation within higher education. This progression from traditional tools to advanced artificial intelligence systems illustrates not only a technological revolution but also a pedagogical paradigm shift, profoundly impacting teaching and learning processes at the higher educational level.

The innovation of the printing press by Johannes Gutenberg in the fifteenth century marked a new era in higher education, facilitating a transition from oral to visual and written learning. This democratisation of knowledge significantly impacted academic scholarship (Eisenstein, 1979). The 19th-century introduction of the chalkboard further diversified instructional methodologies, moving towards more interactive and student-centric approaches (Tyack & Cuban, 1995). The twentieth century saw a surge in electronic and digital educational tools in higher education. Overhead projectors, for instance, became essential for displaying complex academic content, fostering a more engaging and participative classroom environment (Saettler, 2004). The introduction of personal computers and the internet later in the century revolutionised access to academic resources, enabling a global exchange of knowledge and collaboration (Bates, 2000). This era also marked a pedagogical shift from lecture-based to more active and collaborative learning methods, aligning with constructivist theories that emphasise student-centred learning (Jonassen, 1991).

The emergence of learning management systems (LMSs) like Blackboard and Moodle signified a major leap in digital education, offering sophisticated platforms for course management and online learning (Watson & Watson, 2007). The proliferation of MOOCs further extended the reach of higher education, making quality learning resources accessible globally (Pappano, 2012). The integration of interactive technologies like whiteboards and tablets has enhanced pedagogical strategies, promoting collaborative and experiential learning in higher education (Beetham et al., 2013).

The latest advancement in higher education was the integration of artificial intelligence, a development that reshaped the educational landscape in many ways. AI's adaptive learning systems and virtual assistants had provided personalised educational experiences, aligning with the academic emphasis on individualised learning, as highlighted by Pane et al. (2017) and Graesser et al. (2018). AI also played a crucial role in data analytics, enhancing teaching strategies and course design through insights into student performance (Siemens & Gasevic, 2012). Popenici and

Kerr (2017) explored AI's diverse applications in teaching, learning, student support and administration, demonstrating the transformative potential of technologies like IBM's Watson. Moreover, Xia (2020) viewed AI as essential for the training of innovative talents, personalised learning and the integration of VR technology, crucial for the development of assistant teaching and intelligent cloud schools. During the quarantine period, Syzdykbayeva et al. (2021) highlighted AI's growing importance in distance learning. Furthermore, Huang et al. (2021) discussed AI's influence on teaching methodologies and smart campus development, acknowledging both its benefits and challenges. Xia and Li (2022) focused on AI's strategic role in reshaping teaching models and fostering lifelong learning, linked to economic development. Dilmurod and Fazliddin (2021) underscored AI's role in enhancing university operations and selecting learning strategies aligned with market demands. In line of this, Ge and Hu (2020) advocated for the construction of an AI ecosystem in higher education to innovate teaching management. These developments indicated AI's significant impact on educational approaches and the changes of higher education.

The journey from Gutenberg's printing press to the current era of AI and digital technologies in higher education is a testament to the continuous changes of educational tools and teaching methodologies. This technological advancement has been paralleled by a pedagogical shift towards more interactive, student-centred and personalised learning experiences. As higher education continues to embrace these innovations, it faces the challenge of effectively integrating these technologies while maintaining academic rigour and quality.

6.3 Methodology

This chapter adopts a conceptual and interpretive approach to examine the evolving role of artificial intelligence in higher education and academic publishing across Asia. Rather than presenting formal case studies or primary data, it draws upon a synthesis of existing literature, policy frameworks and national experiences, particularly those of China, South Korea and Singapore, as illustrative examples. These examples serve to highlight how AI technologies are being locally interpreted and integrated within diverse educational traditions, governance structures and academic cultures.

The chapter is situated within an interpretivist-constructivist paradigm, assuming that knowledge and practices surrounding AI are socially constructed and contextually embedded rather than universally generalisable (Creswell, 2013; Denzin & Lincoln, 2011). This paradigm is appropriate for a conceptual study that aims to explore how different national systems define, adopt and frame AI in relation to scholarly communication, authorship practices and publishing ethics. Rather than seeking generalisable patterns or outcomes, the chapter interprets culturally embedded meanings, policies and responses across settings. This orientation allows for reflection on the cultural, ethical and institutional dimensions of AI adoption in academic publishing. China exemplifies a large-scale, state-driven model with ambitious AI development strategies. South Korea reflects a technologically advanced system that prioritises ethical concerns and academic integrity. Singapore represents a pragmatic, policy-driven approach with a strong emphasis on innovation, global benchmarking and multicultural inclusivity (Kim & Park, 2021; Marginson, 2011; Tan & Ng, 2023).

This chapter does not aim to generate empirical findings but rather to develop a conceptual understanding of how AI intersects with scholarly communication, institutional reform and publishing ethics in the Asian higher education context. By drawing connections across diverse national settings, the chapter contributes to a broader understanding of the opportunities and tensions shaping AI's impact on academic futures in the region.

Data for this study were gathered exclusively through desk-based research. Sources included peer reviewed academic journal articles, scholarly monographs, national policy documents and institutional reports from higher education agencies and international organisations. The rationale for using a desk-based approach lies in the availability of a rich body of secondary data, which allows for synthesis and comparison across contexts without the logistical limitations of fieldwork (Bowen, 2009). Particular emphasis was placed on literature examining AI applications in academic publishing, including authorship practices, peer review systems, research evaluation frameworks and digital ethics. In addition, national policy initiatives, such as China's 'Next Generation AI Development Plan' and Singapore's 'Smart Nation' strategy, were analysed alongside scholarly debates on AI ethics and the evolving landscape of knowledge production in Asia.

The analytical process followed a qualitative thematic content analysis. Documents and academic literature were read closely and interpreted through inductive coding to identify recurring themes and categories relevant to the chapter's guiding questions. These themes included AI's role in automated peer review, authorship attribution, research integrity and the cultural translation of digital publishing platforms. Within each country example, texts were analysed to understand how AI is being conceptualised and implemented within national and institutional contexts. Following this, a cross-case comparative analysis was conducted to highlight both converging trends and divergent approaches across the three countries. This form of analysis allowed for the identification of shared regional patterns, such as the increased use of AI for efficiency and transparency, as well as national distinctions in editorial governance, academic values and publishing norms.

Throughout the analysis, cultural specificity and reflexivity were prioritised. The interpretivist framework necessitated careful attention to the ways that Confucian traditions, collectivist values and policy-driven innovation shaped each country's educational and technological trajectory (Kim & Park, 2013; Lee, 2000). At the same time, the constructivist orientation required ongoing reflexivity regarding the role of the researcher in constructing meaning from the texts and documents consulted. The chapter does not claim objectivity but instead seeks to produce credible, contextualised interpretations based on rigorous reading, triangulation of sources and analytical consistency (Creswell & Poth, 2017).

It is also important to note that this conceptual and interpretive methodology, grounded in an Asia-centric lens and supported by qualitative desk-based analysis, provides a coherent foundation for understanding the evolving role of AI in academic publishing and higher education across the region.

6.4 Key Concepts and Issues

6.4.1 Basics of AI Technology in Higher Education

Artificial intelligence in higher education represents a significant evolution, transitioning from early computational methods to advanced, multifaceted learning and data analysis tools. This progression is a testament to AI's transformative impact on the educational landscape, particularly at the university level. The incorporation of AI into higher education has experienced profound changes, beginning with basic computational tools and evolving into sophisticated, personalised learning systems.

The late twentieth century marked the initial phase of AI in higher education, focusing primarily on data management and automated administrative tasks. This era laid the groundwork for future advancements in AI-driven educational tools. In the early twenty-first century, machine learning (ML) began revolutionising higher education. This was characterised by the development of algorithms capable of providing personalised learning experiences. A seminal study in this area is Xie and Zhu's (2020) work on utilising ML to predict student performance, a significant departure from traditional educational models.

The evolution of natural language processing (NLP) from basic text analysis to aiding complex research activities also marks a significant milestone. Smith and Hamilton's (2021) study on the application of NLP in automated content analysis demonstrates the utility of this technology in enhancing research efficiency.

The introduction of robotics into university laboratories has been a more recent development. Lee et al.'s (2019) research highlight how robotic assistants in labs can augment the learning experience, allowing students to engage in more sophisticated experiments.

Parallel of these changes, the role of data mining in higher education has evolved significantly. Johnson and Sam's (2018) research emphasise how data mining techniques are now utilised to identify gaps in curricula and suggest data-driven improvements.

These groundbreaking changes bring the consideration of ethical issues in AI applications in education in the forefront. Patel and Smith's (2022) work underscore the necessity of establishing ethical frameworks to guide AI deployment, ensuring fairness and privacy. These modifications are poised to significantly transform not only the educational landscape but also the paradigms of academic assessment. The increasingly accessible entry points to artificial intelligence and machine learning technologies present distinctive challenges to students within higher education. This shift necessitates a re-evaluation of traditional assessment methodologies employed by educators in these advanced learning environments.

The historical evolution of AI in higher education illustrates a journey from basic computational tools to advanced, dynamic learning systems. This progression has been marked by significant technological advancements and a growing awareness of ethical implications. As AI technology continues to develop, its potential to further revolutionise higher education remains vast, requiring ongoing exploration, adaptation and responsible implementation. The introduction of AI in higher education has the potential to revolutionise the learning process, with applications in adaptive learning, assessment and administrative services (Ali, 2021; Dilmurod & Fazliddin, 2021; Syzdykbayeva et al., 2021). However, its implementation is not without challenges, including the need for further exploration of pedagogical, ethical and cultural dimensions (Ali, 2021). Although, the use of AI in higher education is still in its early stages, with a need for more comprehensive research and a focus on the opportunities and challenges it presents (Hinojo-Lucena et al., 2019; Tahiru, 2021). Despite these challenges, AI has the potential to enhance learning and research environments in higher education (Kuleto et al., 2021).

6.4.1.1 Early Implementations and Case Studies

The integration of artificial intelligence in higher education across Asia has been a journey of transformative change, reshaping the academic landscape. For example, in China, platforms such as Squirrel AI have been used to personalise learning across large-scale educational systems. South Korean universities have introduced AI tools for academic advising and emotional well-being, while Singaporean institutions have implemented predictive learning analytics aligned with labour market needs (Kim & Park, 2021; Li & Xu, 2021; Tan & Ng, 2023). These diverse approaches reflect broader regional trends and challenges in adapting AI to different educational contexts. This examination is crucial to understanding how emerging technologies like AI are revolutionising teaching methodologies, student evaluation processes and the overall learning experience.

Asian universities have been at the forefront of innovatively applying AI technologies. In China, universities have leveraged AI to create personalised learning environments, significantly improving student engagement and performance, a development well-articulated by Zhang and Zhou (2019). Japanese institutions have utilised AI in language education, enhancing English language proficiency among students (Saito et al., 2020). South-East Asian universities has seen AI's implementation in academic advising, where systems like ChatGPT have been instrumental in analysing student performance data and assisting students in making informed academic and career decisions (Kim & Park, 2021). The emergence of AI in Asian higher educational paradigms has seen a significant rise in the use of AI applications, particularly in learning management systems (Manhiça et al., 2022; Tahiru, 2021). These applications, such as personalised learning experiences, adaptive testing and predictive analytics, have been found to improve learning efficiency and provide customised educational support (Wang et al., 2023). However, the adoption of AI in education has been accompanied by challenges and ethical issues, including the lack of association between theoretical pedagogical perspectives and the need for exploration of pedagogical, ethical, social, cultural and economic dimensions (Ali, 2021). The potential impact of AI on international students in higher education has also been explored, with the identification of significant risks and limitations (Wang et al., 2023). Despite these challenges, the use of AI in education is expected to continue to grow, with a focus on the development of applications for higher education (Pedró, 2020).

Thus, the successes of AI implementation in higher education across Asia are manifold. Chinese universities have witnessed enhanced personalisation in learning through AI, leading to better student outcomes. Japanese universities have seen AI bring efficiency to language education, addressing long-standing challenges in language proficiency. In South Korea, AI's role in academic advising has facilitated more informed decision-making processes for students. However, these advancements have not been without their challenges. A significant issue has been the disparity in AI resource availability, leading to uneven educational experiences across different institutions. Adapting AI tools like ChatGPT to fit local educational and linguistic contexts has been a complex task. Ethical concerns, especially around data privacy and the responsible use of AI, have emerged as critical issues that need stringent oversight and policies. Additionally, the integration of AI into existing curricula and the financial aspects, including cost and resource allocation, have posed substantial challenges.

The early implementation of AI in higher education within Asia, particularly with tools like ChatGPT, offers valuable insights into both the potential and the pitfalls of integrating advanced technology into educational settings. These case studies from various Asian countries highlight the importance of addressing technological inequalities, ensuring cultural adaptability and managing ethical considerations. They also emphasise the need for faculty training, curriculum integration and careful financial planning. As AI continues to evolve, the experiences and lessons learned from these Asian higher educational institutions will be crucial for the global academic community in adapting to the changing educational landscape.

6.4.1.2 Enhancing Learning Experiences—How AI Can Personalise Learning and Cater to Diverse Learning Styles?

Artificial intelligence in higher education marks a paradigm shift, redefining the pedagogical landscape by personalising learning experiences to accommodate diverse learning styles (Pedró, 2020). AI's adaptability to individual student needs offers a distinct perspective from traditional, uniform educational models, fostering an environment where students can engage with material in a manner that is most beneficial to their learning success (Wang et al., 2023). AI-driven personalisation in higher education is predicated on the use of adaptive learning technologies, which employ machine learning algorithms to tailor the educational path of each learner (Kim & Park, 2021). These technologies can discern a student's proficiency and learning style by analysing their interactions with the content, thereby creating a customised experience that evolves in real time (Ali, 2021). Insights gleaned from AI systems regarding student learning behaviours can lead to predictive models that anticipate academic outcomes, facilitating timely and effective interventions (Tiwari, 2023). This predictive capability enables educators to offer targeted support, ensuring that learning is both effective and efficient (Kuleto et al., 2021).

Furthermore, the administrative potential of AI is equally transformative, offering higher education institutions the ability to automate processes such as enrolment and scheduling, thereby streamlining operations and allowing faculty to concentrate on core educational responsibilities (Tahiru, 2021). Thus, the transformative potential of AI in higher education is evident in its ability to personalise learning experiences and cater to diverse learning styles (Manhiça et al., 2022; Pinto et al., 2023). AI can enhance learning experiences by adapting to individual student needs, predicting academic performance and providing personalised feedback (Lydia, 2023; Zaman, 2023). However, the ethical and societal implications of AI in education, such as privacy, data security and bias, need to be carefully considered (Ali, 2021; Pinto et al., 2023). Despite these challenges, the potential benefits of AI in higher education, including more efficient educational processes and enhanced student outcomes, are significant (Manhiça et al., 2022).

Thus, the ethical deployment of AI in education is paramount, with a focus on fostering inclusivity and preventing the exacerbation of existing inequalities (Ali, 2021). Ensuring data privacy, mitigating algorithmic bias and promoting digital literacy are critical to the responsible integration of AI in educational contexts (Manhiça et al., 2022). Realising the full potential of AI in enhancing higher education experiences necessitates rigorous interdisciplinary research that merges AI technological advancements with educational theory (Kim & Park, 2021).

6.4.1.3 Revolutionising How AI Changes Teaching Methodologies in Asia

At its core, AI in higher education is anchored in the principles of adaptive learning and personalised instruction in Asian universities as well. This paradigm shift is rooted in the constructivist theory, which posits that learning is an active, contextualised process of constructing knowledge rather than acquiring it (Piaget, 1954; Vygotsky, 1978). AI systems, through their capacity for data processing and pattern recognition, epitomise this approach by offering tailored learning experiences based on individual student profiles (Zhou et al., 2020).

In Asia, a region characterised by its rapid adoption of technology and high value on education, AI has been instrumental in reshaping educational practices. For instance, in China, AI-driven platforms like Squirrel AI have revolutionised the concept of after-school tutoring, providing personalised learning paths based on student performance (Li, 2019). Similarly, Japan's utilisation of AI in universities, for language learning and student support systems, showcases a shift towards more interactive and responsive educational environments (Tanaka, 2021). South Korea's EBSi system, utilises AI to provide customised learning materials and assessments, demonstrates a significant shift towards technology-driven education (Jung & Lee, 2022). In Singapore, AI tools are being used in higher education to predict student performance and provide early interventions, thus enhancing student outcomes (Nguyen et al., 2021).

Thus, AI's impact on teaching methodologies manifests in various forms nowadays. Leveraging AI's analytical capabilities, educators can now offer more personalised instruction, adapting to the needs, learning styles and pace of each student (Wang, 2018). AI tools analyse student engagement through metrics like participation and online behaviour, providing educators with insights to enhance teaching effectiveness (Zhang et al., 2021). AI tools have streamlined the assessment process, enabled more timely and accurate evaluations of student work, and provided immediate feedback (Chen, 2020). Furthermore, AI facilitates the creation of dynamic, interactive learning environments that engage students through simulations and virtual reality, enhancing experiential learning (Liu et al., 2022a, 2022b). Augmented reality powered by AI is creating immersive learning experiences, particularly in fields like medical and engineering education (Patel & Patel, 2019). Thus, AI systems are now aiding in the development and optimisation of curricula, ensuring they are more relevant and responsive to student needs (Huang & Soman, 2018).

Despite these advancements, integrating AI into Asian higher education is not without challenges. Issues such as data privacy, algorithmic bias and the digital divide pose significant concerns (Kim, 2020). Moreover, there is a risk that over-reliance on AI may diminish the role of human interaction in education, which is crucial for developing critical thinking and social skills (Park, 2021). While AI's benefits in education are substantial, there are critics who argue that AI's emphasis on technology might erode traditional educational values like critical thinking and humanistic learning (Choi, 2020). Furthermore, there is a concern that over-dependence on AI could lead to a lack of problem-solving and analytical skills in students, as they might rely too much on AI for answers (Kim & Lee, 2021).

AI's role in transforming higher education teaching methodologies in Asia is undeniable. It offers a more personalised, efficient and engaging learning experience. However, it is imperative to address the accompanying challenges to ensure that this technological integration benefits all stakeholders in the educational ecosystem.

6.4.2 AI in Academic Publishing

Artificial intelligence has revolutionised university education in Asia, profoundly impacting not just the pedagogical approaches but also significantly influencing the scholarly publications. These publications, integral to both the academic framework and the professional trajectories of university members, have undergone substantial changes in response to AI advancements. In the ensuing discussion, we will provide a concise overview of these transformative effects.

The landscape of academic publishing in Asia, steeped in history and tradition, presents a unique paradigm within the global scholarly communication system. Rooted in a diverse cultural and academic environment, Asian academic publishing has evolved over time, adapting to the changing needs of the scholarly community while retaining certain distinctive characteristics. Historically, the traditional academic publishing model in Asia has been characterised by a strong emphasis on print-based media. Journals and books, predominantly published in local languages, have been the primary channels for disseminating research findings. This model has often been associated with a regional focus, where the majority of the published work catered to local academic communities. As Li and Agha (2019) observed, the reliance on print media has been a defining feature of the Asian academic publishing landscape, often driven by the cultural and linguistic diversity within the region (Li & Agha, 2019). Furthermore, the academic publishing scene in Asia has traditionally operated within a highly hierarchical and rigorous peer review system, which, has its roots in the long-standing academic traditions of the region (Wang & Zhang, 2021). This system, while ensuring the quality and integrity of academic work, has also been criticised for its slow pace and potential biases, as discussed in the works of Chen and Hsieh (2018).

Despite these characteristics, the emergence of digital technologies and the increasing push towards open access publishing are transforming the Asian academic publishing model. Kim and Park (2020) highlighted the growing trend towards digital publication platforms, which are challenging the traditional print-centric model (Kim & Park, 2020). Moreover, initiatives like the Plan S movement, as analysed by Liu et al. (2022a, 2022b), are increasingly influencing Asian publishers to adopt open access policies, thereby altering the dynamics of scholarly communication in the region (Liu et al., 2022a, 2022b).

Academic journal publications serve as a cornerstone in the realm of higher education, primarily because they are a key indicator of an educator's research prowess and intellectual contribution. As argued by Thompson and Lee (2021), publications in reputed journals are often viewed as a testament to a scholar's expertise and originality in their field (Thompson & Lee, 2021). This perception is deeply ingrained in the academic culture, where publishing frequency and the prestige of journals are often directly linked to career advancement, funding opportunities and professional recognition. The significance of publications in academic journals is further emphasised by Morrison and Smith (2019), who point out that these publications play a critical role in shaping an educator's identity and authority in their respective disciplines (Morrison & Smith, 2019). However, this focus on journal publications is not without its critics. One major counterargument, as outlined by Patel and Jackson (2020), is the 'publish or perish' culture in academia. This phenomenon, where educators feel compelled to publish regularly irrespective of the quality or impact of their research, can lead to a quantity-over-quality dilemma (Patel & Jackson, 2020). Furthermore, as argued by O'Neil and Martin (2018), an overemphasis on journal publications can marginalise other forms of scholarly output, such as books, conference presentations, educational materials and teaching itself which are also vital to the academic ecosystem. Moreover, there is a growing concern about the relevance and accessibility of academic research. Critics like Singh and Gupta (2019) argue that the focus on publishing in high-impact journals often sidelines research that is regionally significant but may not attract a global readership (Singh & Gupta, 2019). This can lead to disconnect between academic research and societal needs, especially in the context of regional issues in Asia.

At this juncture, it is important to note that while the traditional academic publishing model in Asia has been deeply influenced by regional cultural and linguistic phenomena, it is currently undergoing a significant transformation. The integration of digital technologies and the shift towards open access are pivotal in shaping the future trajectory of academic publishing in this region. Academic journal publications remain a key element in shaping a higher educator's career and reputation in the field of higher education, although there are valid concerns and critiques regarding the current publishing paradigm. These concerns include the pressure to publish, the marginalisation of other scholarly contributions, and the potential disconnect between research and societal needs. Addressing these challenges requires a balanced approach that values diverse forms of academic contributions while maintaining the integrity and relevance of scholarly research.

6.4.2.1 AI-Driven Changes in Publishing and Future Perspective

In academic publishing, artificial intelligence is a catalyst for profound transformation. At its core, AI in publishing around enhancing efficiency, accuracy and accessibility. It streamlines editorial processes, making them more robust and less time-consuming. AI algorithms, adept at handling large volumes of data, are now increasingly employed to assist in initial manuscript screening (Smith & Johnson, 2021). This is a significant shift from traditional methods, where human editors solely shouldered this responsibility. However, critics like Anderson (2022) argue that AI may overlook the creativity of human writing, potentially dismissing innovative but unconventional manuscripts. The utilisation of NLP techniques enables AI systems to perform preliminary evaluations of submissions. They assess the relevance of a manuscript based on its adherence to the journal's scope and its originality (Brown, 2020). Although, sceptics like Evans (2023) caution against over-reliance on AI, pointing out that such systems can generate false positives and lack the subtlety to understand context. AI tools such as iThenticate are commonly used for plagiarism detection, ensuring the integrity of academic publications (Miller & Davis, 2022). However, Barnes and O'Neill (2022) highlight that AI cannot fully grasp the complexities of certain research methodologies, possibly leading to inaccurate assessments.

AI also revolutionised the peer review process, which is a cornerstone of scholarly communication. Traditional peer review processes often grapple with challenges like reviewer bias and time delays. AI introduces an element of objectivity and efficiency in this field (Williams, 2019). For instance, tools like StatReviewer conduct automated statistical checks, ensuring the statistical integrity of manuscripts (Peterson &

Green, 2021). Additionally, AI can assist in the identification and selection of appropriate peer reviewers. By analysing the content of a manuscript and comparing it with the scientific literature database, AI can suggest potential reviewers who have expertise in relevant fields (Taylor, 2022). This not only accelerates the peer review process but also enhances its quality by matching manuscripts with the most suitable experts. Although, critics such as Hughes (2021) argue that AI algorithms might perpetuate biases, favouring certain research topics or methodologies over others.

AI's impact extends to the dissemination of research as well. The traditional model of academic publishing has often been criticised for its limited reach and accessibility. AI-driven platforms are challenging this by facilitating broader dissemination of research findings (Clark & Hamilton, 2023). For instance, platforms like ResearchGate and Academia.edu use AI to recommend articles to users based on their interests and previous interactions (Moore & Patel, 2021). This personalised approach to content delivery ensures that research reaches a wider and more relevant audience. However, Robinson and Lee (2022) contend that such algorithms could create echo chambers, limiting exposure to diverse perspectives.

From the side of the authors the application of AI in scientific research extends beyond publishing processes to the generation of data itself. Scientists are increasingly employing AI to simulate experiments and predict outcomes, thereby augmenting traditional methods of data collection and analysis. AI's ability to process and analyse vast amounts of information rapidly makes it a valuable tool in fields where data collection is either too time-consuming or complex for human researchers (Johnson & Lee, 2021). However, the reliance on AI-generated data is not without its critics. Scholars like Adams (2022) caution against over-dependence on AI, noting that machine-generated data might miss aspects that human researchers would catch.

Another groundbreaking application of AI in scientific research from an author point of view is in manuscript writing. AI tools are now capable of drafting research papers, including literature reviews, methodology and even discussions based on data inputs. These tools, such as OpenAI's GPT models, use advanced NLP techniques to generate coherent and contextually relevant text. This development is particularly beneficial in drafting repetitive or standardised sections of scientific papers, allowing researchers to focus on more creative and analytical aspects of their work (Singh & Patel, 2022). However, there are concerns about the authenticity and originality of AI-generated content. Critics like Thompson (2023) argue that AI-written papers may lack the critical thinking and unique insights that come from human researchers. There's also a concern regarding the potential for AI to inadvertently perpetuate biases found in its training data.

The journey of AI-assisted manuscript writing in academic journals has been marked by significant milestones. In its early stages, AI's role in academic writing was largely confined to basic data analysis and simple text generation. The early 2000s witnessed the advent of AI tools capable of performing basic language processing tasks. However, these applications were rudimentary, primarily focusing on syntax rather than semantics or context (Davis & Taylor, 2004). The real breakthrough came with the development of advanced NLP techniques, particularly with the introduction of machine learning algorithms that could understand and generate human-like text.

This evolution was marked by a notable event in 2016 when an AI-authored paper was submitted to a scientific conference, opening the door to the possibilities of AI in academic writing (Wang & Zhang, 2016). Recent years have seen the emergence of more sophisticated AI models, which have revolutionised manuscript writing. These tools employ deep learning and large language models to produce text that is not only grammatically correct but also contextually relevant and coherent. They can draft various sections of research papers, from literature reviews to discussions, based on data inputs and pre-existing literature (Singh & Patel, 2022).

The use of AI in data generation and manuscript writing brings to light several ethical and practical considerations. One major concern is the transparency of AI-generated data and text. Ensuring that AI-assisted publications are clearly marked and differentiated from purely human-generated research is crucial for maintaining scientific integrity (Brown & Harris, 2023; Martinez & Rodriguez, 2021). Moreover, there is an ongoing debate about authorship when AI significantly contributes to a research paper. The academic community is grappling with questions about how to credit AI tools and the extent of their contribution (White & Zhao, 2022). Thus, the benefits of AI in manuscript writing are multifaceted.

AI can significantly reduce the time and effort involved in drafting repetitive or standardised sections of scientific papers, such as methodology descriptions and literature reviews (Johnson & Lee, 2021). Additionally, AI tools can assist in overcoming language barriers, enabling non-native English speakers to produce well-written manuscripts, thus democratising access to scientific publication (Lopez & Hernandez, 2022). Despite these a concern is the authenticity and originality of AI-generated content. Critics argue that AI-written papers may lack the depth of critical thinking and unique insights that come from human researchers (Thompson, 2023). There is also a risk that over-reliance on AI could lead to a homogenisation of scientific discourse, with many papers mirroring the style and limitations of the AI's training data.

Looking forward, the integration of AI in publishing is poised to evolve further. With advancements in AI, we can anticipate more sophisticated algorithms capable of providing even deeper insights into manuscript quality (Kapoor & Singh, 2023). There is potential for AI to assist in predicting the impact of research, thereby guiding editors and authors in their decision-making processes. However, it is crucial to balance AI's capabilities with ethical considerations. Ensuring the fairness and transparency of AI algorithms is imperative to maintain trust in the academic publishing process (Lee & Kim, 2021). As AI continues to reshape the landscape of academic publishing, it presents both exciting opportunities and significant responsibilities.

The integration of AI in manuscript writing necessitates a re-evaluation of ethical guidelines in scientific publishing. Issues such as authorship credit for AI contributions and the transparency of AI use in manuscript preparation need clear guidelines (White & Zhao, 2022). Looking ahead, it is likely that AI will continue to play an increasingly integral role in manuscript writing. However, the academic community must address the ethical and practical challenges to ensure that the use of AI enhances rather than undermines the integrity and diversity of scientific research.

AI-driven changes in publishing are revolutionising the way academic content is created, evaluated and disseminated. While these advancements promise efficiency and broader accessibility, they also call for careful consideration of ethical implications. Furthermore, AI's role in data generation and manuscript writing represents a paradigm shift in scientific research. While it offers significant efficiencies and capabilities, it also presents challenges that need careful consideration, particularly in terms of ethics, transparency and the preservation of human insight in scientific discovery. Thus, the integration of AI in academic publishing is not just a technological upgrade, it is a step towards a more inclusive and efficient scholarly communication system. The integration of AI in academic publishing is a complex interplay between technological advancements and the preservation of scholarly integrity.

6.4.2.2 Ethical and Cultural Considerations, Balancing Technology and Tradition—Asian Educational Values in the AI Era

In the evolving landscape of higher education, especially within Asian educational contexts, the integration of artificial intelligence brings to the fore distinctive ethical and cultural dilemmas. These systems, steeped in traditional values such as respect, discipline and a communal approach to learning, are now at a crossroads, where the imperative to embrace AI's potential must be harmoniously balanced with the preservation of cultural ethos. This can bring numerus challenges for the academia as in 'The Technological Singularity', Murray Shanahan (2015) presents an in-depth exploration of the concept of a future era where technological advancement spirals beyond human control and fundamentally transforms civilisation. The exponential progression in AI fields might catalyse a singularity event. AI system fixated on maximising a reward function might inadvertently trigger undesired outcomes. This point accentuates the unpredictability of AI behaviour and its far-reaching consequences for society (Shanahan, 2015). That is why necessity of careful consideration and ethical reflection in the advancement of AI technologies.

Historically, the Asian educational paradigm has placed a significant emphasis on teacher-centric methods and rote memorisation. This methodological approach, while often lauded for its rigour, faces challenges in the AI era, which promotes personalised and adaptive learning strategies. This shift necessitates a re-examination of the traditional roles of teachers and learners in Asian education systems.

The integration of AI in educational settings also raises concerns about data privacy and algorithmic bias. Asian societies, typically characterised by their collectivist nature, are confronted with the individualistic leanings of data-centric AI systems. It is critical that these systems are developed and implemented in a manner that respects cultural norms and safeguards individual privacy. Additionally, the problem of algorithmic bias poses a significant challenge in ensuring equitable and culturally sensitive AI applications in education (Wong & Chu, 2018). AI systems, often developed with data that may not fully represent Asia's diverse cultural and socio-economic landscapes, risk reinforcing educational disparities (Chen et al., 2020).

To address these challenges, it is vital to establish a synergy between technological innovation and traditional values. This involves reimagining the educator's role to one that integrates AI tools effectively while upholding the cultural integrity of the learning environment. Furthermore, the development of policy frameworks to guide the ethical application of AI in education is imperative. Such frameworks should be grounded in principles of fairness, transparency and accountability, tailored to align AI applications with cultural values and educational objectives. The creation of these frameworks should be a collaborative process, involving a broad spectrum of stakeholders including educators, technologists, policymakers and the society. These ideas are straitened in Boddington's (2017) book that delves into the multifaceted ethical challenges presented by AI. Boddington argues for the necessity of ethical codes in AI, acknowledging its extensive applications and inherent risks. She proposes strategies that supplement ethics codes, such as epistemic and technological strategies, alongside moral strategies in AI development.

AI, ethical considerations have increasingly become a pivotal area of discourse, particularly in the context of higher education and research. These epistemic and technological strategies (Boddington, 2017) include but are not limited to issues of data privacy, inherent biases in AI systems, and the accessibility of these technologies. Data privacy, issue arises when AI training data includes sensitive personal information, which can be vulnerable to misuse or unauthorised access. This concern is amplified in educational settings where student data is often involved. Scholars such as Shmueli et al. (2021) highlight the need for stringent data governance policies to ensure that personal data is handled responsibly and ethically in AI applications (Shmueli et al., 2021). Another significant ethical challenge is the presence of bias in AI systems. These biases often stem from the data used to train AI models, reflecting existing prejudices in society.

For instance, Buolamwini and Gebru (2018) provided a seminal analysis of gender and skin-type biases in AI facial-analysis systems, revealing substantial disparities in accuracy (Buolamwini & Gebru, 2018). Such biases can perpetuate and even amplify societal inequalities, making it imperative for AI developers and researchers to implement strategies that identify and mitigate biases in AI systems. Accessibility is also a critical ethical concern. AI technologies hold the potential to greatly enhance learning experiences and research capabilities in higher education. However, there exists a risk of creating or exacerbating digital divides. Students and institutions with limited resources may find themselves at a disadvantage, unable to access the latest AI tools and technologies. This aspect has been explored in depth by scholars like Watters (2020), who argues for inclusive AI design that considers diverse user needs and contexts to ensure equitable access to AI-driven educational technologies (Watters, 2020).

Addressing these ethical challenges, it is crucial to foster a multidisciplinary approach that includes not only technologists and AI developers but also ethicists, sociologists, legal experts and educators. Such collaborations can lead to the development of AI systems that are not only technologically advanced but also socially responsible and ethically fine. Additionally, the integration of ethical considerations into AI curricula in higher education is vital. By educating the next generation of AI professionals about these ethical challenges, we can nurture a culture of responsibility and ethical awareness in the field of AI.

The incorporation of AI into Asian higher education systems presents a complex mix of technological progress and cultural fidelity. The ethical implications of AI in higher education and beyond are multifaceted and complex. Addressing these issues requires a concerted effort from various stakeholders to ensure that AI technologies are developed and used in ways that are respectful of privacy, free from bias and accessible to all. As we progress into this new era, striking a balance that respects traditional values while leveraging the transformative capabilities of AI is essential. This equilibrium is not only crucial for the ethical and effective use of AI in higher education and scholar publications but is also pivotal in nurturing a generation that is technologically proficient and culturally attuned.

6.5 Discussion of Implications

Comparative Insights on AI's Transformative Role in Asian Higher Education and Academic Publishing: Illustrative National Perspectives from China**, South Korea and Singapore**

This section provides detailed case studies from China, South Korea and Singapore to illustrate how AI has been integrated into higher education and academic publishing in these countries. Through these examples, the chapter explores the unique approaches each nation has taken, shaped by their cultural, educational and policy contexts.

6.5.1 China—Scaling AI for Performance and Innovation

China has rapidly embraced AI in education, aligning it with its national ambitions to become a global leader in technology and innovation. The Chinese government's commitment to AI is exemplified in initiatives such as the 'Next Generation AI Development Plan', which prioritises AI integration across sectors, including higher education.

In the realm of education, China has implemented large-scale adaptive learning platforms, like Squirrel AI, that leverage AI to provide tailored educational experiences. According to Li and Xu (2021), Squirrel AI's system has led to a 13% improvement in academic performance and an 18% increase in retention rates across participating schools and universities. The platform adapts learning content to individual students' needs by analysing performance and adjusting the difficulty and type of material presented. This approach not only enhances academic outcomes but also aligns with China's strategic goal of cultivating technologically skilled graduates who can compete internationally (Li & Zhang, 2021).

AI's role in academic publishing within China is similarly ambitious. Chinese academic publishers, supported by the Chinese Academy of Sciences, are employing AI tools to streamline editorial processes. Zhao and Chen (2022) report that AI-driven screening technologies have reduced manuscript review times by 40%, allowing editors to process a larger volume of submissions efficiently. However, the rapid adoption of AI in publishing has raised ethical concerns, particularly regarding data privacy and intellectual property. Chinese academic institutions and publishers are beginning to establish guidelines to ensure that AI use in research and publishing adheres to ethical standards, balancing the efficiency gains AI provides with the need to protect authors' rights and maintain academic freedom.

China's approach to AI in education and publishing demonstrates the country's dual focus on enhancing educational outcomes and reinforcing its global competitive edge. The scale and scope of AI applications reflect China's ambition, yet they also raise ethical questions, highlighting the importance of developing robust governance frameworks to manage these issues.

6.5.2 South Korea—Focusing on Ethics and Student Well-Being

South Korea's adoption of AI in higher education and publishing is characterised by an emphasis on ethical use and student well-being. South Korea's 'AI National Strategy' integrates AI across various sectors, with a distinct focus on supporting students academically and emotionally within the competitive Korean educational environment.

In academic support, South Korean universities utilise AI-based advising platforms to provide personalised academic guidance and mental health support. Kim and Lee (2022) observed that these platforms have reduced academic probation rates by 22% by identifying students at risk and offering real-time, tailored support. The AI systems evaluate student data to detect patterns associated with academic difficulties or emotional stress, enabling early intervention. This focus on mental health reflects South Korea's recognition of the high-pressure educational culture and the government's commitment to holistic student development, an approach that combines academic assistance with well-being resources.

In academic publishing, South Korean institutions have widely adopted AI tools to uphold research integrity. The Korean Council for University Education has established ethical guidelines for AI applications in publishing, which prioritise transparency and fairness. For instance, plagiarism detection tools like iThenticate are standard across major academic publishers, contributing to a 97% decrease in detected plagiarism cases (Choi & Park, 2023).

However, South Korea's academic community debates whether AI tools may favour more conventional research, potentially limiting innovative work that does not fit typical patterns recognised by AI algorithms. This debate highlights South Korea's ethical concerns and dedication to maintaining academic integrity and diversity in research output.

South Korea's approach underscores the country's emphasis on student well-being and ethical standards. By integrating mental health support and ethical guidelines into AI applications, South Korea exemplifies a balanced approach that values academic integrity and the overall welfare of students, setting it apart in the region.

6.5.3 Singapore—Prioritising Workforce Readiness and Transparency

Singapore's strategy for AI integration in higher education and academic publishing focuses on preparing students for the future workforce and maintaining transparency in AI use. As part of its National AI Strategy, Singapore has prioritised education as a key area for AI deployment, with goals aligned towards economic relevance and employability.

In higher education, Singapore's universities use AI-driven predictive analytics for academic advising and career planning. The National University of Singapore, for instance, has implemented systems that analyse student performance data to identify optimal course pathways and align students' academic experiences with labour market needs. Tan and Ng (2023) report a 28% improvement in timely graduation rates through these AI-driven advising systems, which are also designed to provide students with information about potential career outcomes based on their academic choices. The Singaporean government's support for these initiatives reflects its aim to ensure that graduates are equipped with the skills and knowledge required for a rapidly evolving job market.

In academic publishing, Singaporean publishers have adopted AI technologies that streamline editorial processes, from plagiarism detection to reviewer selection. Ong and Lim (2023) found that AI-assisted workflows in Singapore have reduced the time from submission to publication by an average of 25%. However, in line with the government's emphasis on responsible AI, Singaporean academic publishers are encouraged to disclose AI's role in the editorial process, ensuring that authors and readers are aware of AI's contributions. This focus on transparency is a response to growing concerns about the ethical implications of AI in research, aiming to build trust within the academic community.

Singapore's approach to AI in higher education and publishing highlights its commitment to economic relevance and responsible innovation. By aligning AI applications with workforce readiness and emphasising transparency, Singapore creates an environment where AI not only enhances academic processes but also ensures ethical accountability.

6.5.4 Reflections and Broader Implications

These case studies from China, South Korea and Singapore showcase both commonalities and distinctions in AI's integration across Asian higher education and publishing. China's large-scale deployment of AI for educational performance and research output mirrors its ambition for technological and economic leadership, while South Korea's ethical approach underscores student well-being and academic integrity. Singapore's focus on workforce readiness and transparency reflects a pragmatic approach to AI that aligns educational outcomes with economic goals.

These case studies also highlight the diverse strategies employed by each country in addressing the challenges and opportunities posed by AI. China's emphasis on large-scale AI adoption in education reflects a broader push for technological dominance, yet it faces concerns about data privacy and over-reliance on AI in decision-making. South Korea, with its strong focus on ethical AI use, prioritises safeguards against biases and ensures that AI applications do not undermine the holistic development of students.

Meanwhile, Singapore's balanced approach of aligning AI-driven educational practices with workforce needs emphasises the importance of adaptability and lifelong learning, preparing students for an ever-evolving job market. These varied strategies provide valuable insights for other Asian nations looking to integrate AI effectively into higher education and publishing, each with its own priorities and context.

The comparative insights reveal that while these countries share a commitment to enhancing education and academic publishing through AI, their strategies are influenced by distinct cultural values and policy priorities. China's focus on national competitiveness drives its investment in high-performance AI applications, South Korea's ethical and supportive approach reflects its commitment to comprehensive student development, and Singapore's transparent and employment-oriented model aligns AI with future labour market demands.

6.6 Future Directions

This chapter highlights the transformative potential of AI in higher education and academic publishing in Asia, with a particular focus on the experiences of China, South Korea and Singapore. These case studies demonstrate how AI applications in educational systems are evolving to become more culturally responsive and ethically sound. The experiences of these countries provide valuable lessons for other Asian nations considering AI integration, underscoring the need for policies that strike a balance between innovation and ethical responsibility.

As AI becomes further embedded in academic environments, the development of AI-driven tools for personalised learning, research automation and academic publishing will continue to expand. While China, South Korea and Singapore have set distinct paths for AI, it is expected that a convergence of strategies will emerge across the region as common challenges, such as data security, digital literacy and ethical considerations, become more pronounced. Moving forward, deeper exploration into how these various AI approaches interact will be essential, with the potential for shared solutions or frameworks that can be applied regionally.

AI also plays a pivotal role in fostering cross-cultural academic exchanges and collaborations. By enabling real-time translation, collaborative learning platforms and virtual classrooms, AI is facilitating greater international cooperation in research and education across Asia. As AI technologies improve, they are expected to increase accessibility and inclusivity in higher education, particularly for students in under-represented or rural areas. The integration of AI in higher education in these countries also brings up important discussions about the balance between human-driven education and AI-driven processes, especially regarding creativity, ethics and humanistic values in learning. The insights from China, South Korea and Singapore will likely guide broader conversations on how AI can be used to enhance academic outcomes while respecting cultural and ethical considerations.

In addition to academic collaboration, AI tools are playing an important role in promoting sustainable practices within higher education systems in Asia. By enabling educational institutions to develop curricula that integrate environmental responsibility and sustainable development, AI helps ensure that students are exposed to crucial sustainability issues. As countries like China, South Korea and Singapore continue to expand AI integration, the focus will likely shift towards creating AI systems that prioritise not only academic efficiency but also sustainable educational models, equipping students with the skills and awareness needed to address global environmental challenges.

6.7 Conclusion

This chapter has delved deeply into the transformative role of artificial intelligence in reshaping the educational and publishing landscapes of Asia, weaving a narrative that interlaces technological advancement with pedagogical development and knowledge creation in academia. The transition to AI in Asian higher education was marked by a detailed examination of its foundational elements and initial applications in the academic sphere. Special focus was given to AI-powered tools like ChatGPT, which have revolutionised academic work and created personalised learning experiences. Further, the chapter underscored the profound potential of AI in enriching learning experiences and transforming pedagogical methodologies. This transformation is evidenced by AI's role in facilitating personalised education through its analytical capabilities and in enhancing the efficiency of academic assessments.

The comparative case studies of China, South Korea and Singapore provided a nuanced perspective on how AI is shaping higher education and academic publishing in the Asian context. These case studies illustrated the unique approaches each country has taken, influenced by their distinct cultural, economic and policy priorities. China's large-scale, performance-oriented AI initiatives reflect its strategic goal of technological leadership, while South Korea's ethically focused, student-centred approach emphasises well-being and academic integrity. Singapore's balanced strategy combines AI applications for workforce readiness with a strong emphasis on transparency and accountability in academic publishing. Together, these examples underscore the diverse pathways through which Asian countries are navigating the AI transformation in academia, offering insights into how technology and tradition can be harmonised to advance educational and scholarly excellence.

Implementing interdisciplinary research about this area could serve as a crucial catalyst for future exploration, fostering sustained intellectual curiosity and could provide real-world practical insights. Research, which synergises AI's technological advancements with educational theories, could paves the way for pioneering innovations and discoveries. It is a field that not only drives forward academic inquiry but also shapes the future of educational practices and publishing paradigms. We also think that the explorations of the integration of AI into the Asian higher educational and publishing sectors invites readers to engage in a continuous dialogue about the opportunities and challenges presented by AI, urging a thoughtful consideration of how technology can be harmoniously blended with higher educational values and practices.

Highlights

- The chapter emphasises the need for a balance between AI-driven innovation and traditional Asian educational values, with a focus on ethical concerns like data privacy, algorithmic bias and student well-being.
- Case studies from China, South Korea and Singapore illustrate how AI is reshaping educational practices, enhancing personalised learning, and improving academic publishing efficiency.
- AI technologies, including real-time translation and collaborative platforms, are increasing accessibility and fostering cross-cultural academic exchanges, improving research and education cooperation across Asia.
- AI tools are being leveraged to integrate sustainability into curricula, optimise resources and create sustainable educational models, ensuring students are equipped to address global environmental challenges.

References

Adams, R. (2022). Challenges of AI-generated data in scientific research. *Journal of Computational Science, 14*(3), 154–162.

Ali, M., & Abdel-Haq, M. K. (2021). Bibliographical analysis of artificial intelligence learning in higher education: Is the role of the human educator and educated a thing of the past? In *Fostering communication and learning with underutilized technologies in higher education* (pp. 36–52). IGI Global.

Altbach, P. G. (1989). Twisted roots: The Western impact on Asian higher education. *Higher Education, 18*(1), 9–29.

Anderson, K. (2022). The limits of AI in recognizing academic creativity. *Journal of AI and Humanities, 3*(1), 45–59.

Barnes, L., & O'Neill, D. (2022). Challenges in AI-driven statistical review. *Journal of Quantitative Methods, 22*(4), 199–211.

Bates, A. W. (2000). *Managing technological change: Strategies for college and university leaders.* Jossey-Bass.

Beetham, H., & Sharpe, R. (Eds.). (2013). *Rethinking pedagogy for a digital age: Designing for 21st century learning*. Routledge.

Boddington, P. (2017). *Towards a code of ethics for artificial intelligence* (pp. 27–37). Springer.

Buolamwini, J., & Gebru, T. (2018). Gender shades: Intersectional accuracy disparities in commercial gender classification. *Proceedings of Machine Learning Research, 81*, 1–15.

Bowen, G. A. (2009). Document analysis as a qualitative research method. *Qualitative Research Journal, 9*(2), 27–40.

Brown, L. (2020). The role of AI in manuscript screening and selection. *Journal of Publishing Innovations, 12*(3), 15–29.

Brown, L., & Harris, J. (2023). AI and the risk of bias in scientific writing. *Ethics in AI Research, 8*(1), 45–57.

Chan, S. J. (2016). A journey back to the center: Asian higher education in global context. In *The Palgrave handbook of Asia Pacific higher education* (pp. 25–37). Palgrave Macmillan.

Chen, G. (2020). AI in education: A systematic literature review and future research directions. *Educational Technology & Society, 23*(1), 25–37.

Chen, G., & Hsieh, Y. (2018). The impact of regional focus on academic publishing: An analysis of publishing trends in Asia. *Asian Journal of Communication, 28*(1), 85–100.

Chen, X., Liu, Y., & Zhang, H. (2020). Cultural considerations in the application of AI in Asian classrooms. *Asian Journal of Educational Research, 18*(4), 22–37.

Choi, J., & Park, S. (2023). Integrating AI in academic publishing: Ethical implications in South Korean universities. *Journal of Scholarly Publishing, 54*(2), 130–145.

Choi, M. (2020). Critical perspectives on AI in education. *Educational Philosophy and Theory, 52*(14), 1489–1504.

Clark, J., & Hamilton, P. (2023). The impact of AI on academic publishing accessibility. *International Journal of Digital Literacy, 7*(1), 48–65.

Creswell, J. W. (2013). *Qualitative inquiry and research design: Choosing among five approaches* (3rd ed.). SAGE.

Creswell, J. W., & Poth, C. N. (2017). *Qualitative inquiry and research design: Choosing among five approaches* (4th ed.). SAGE.

Davis, R., & Taylor, J. (2004). Early applications of NLP in academic writing. *Journal of Computational Linguistics, 20*(2), 35–48.

Dee, J. R. (2008). Asian universities: Historical perspectives and contemporary challenges. *Quality Assurance in Education, 16*(2), 201–204.

Denzin, N. K., & Lincoln, Y. S. (2011). *The SAGE handbook of qualitative research* (4th ed.). SAGE.

Dilmurod, R., & Fazliddin, A. (2021). Prospects for the introduction of artificial intelligence technologies in higher education. *ACADEMICIA: An International Multidisciplinary Research Journal, 11*(2), 929–934.

Eisenstein, E. L. (1979). *The printing press as an agent of change.* Cambridge University Press.

Evans, J. (2023). Contextual limitations of AI in plagiarism detection. *Journal of Ethics in Publishing, 17*(1), 32–47.

Ge, Z., & Hu, Y. (2020, April). Innovative application of artificial intelligence (AI) in the management of higher education and teaching. *Journal of Physics: Conference Series, 1533*(3), 032089.

Graesser, A., et al. (2018). Advancing the science of collaborative problem solving. *Educational Psychology Review, 30*, 1531–1551.

Hinojo-Lucena, F. J., Aznar-Díaz, I., Cáceres-Reche, M. P., & Romero-Rodríguez, J. M. (2019). Artificial intelligence in higher education: A bibliometric study on its impact in the scientific literature. *Education Sciences, 9*(1), 51.

Huang, J., Saleh, S., & Liu, Y. (2021). A review on artificial intelligence in education. *Academic Journal of Interdisciplinary Studies, 10*(206).

Huang, R., & Soman, D. (2018). AI and curriculum development: An exploration in Singapore. *Journal of Curriculum Studies, 50*(4), 456–474.

Hughes, M. (2021). AI and bias in peer reviewer selection. *Ethics in Scientific Review, 15*(3), 100–115.

Jarvis, D. S., & Mok, K. H. (2019). The political economy of higher education governance in Asia: Challenges, trends, and trajectories. In *Transformations in higher education governance in Asia: Policy, politics, and progress* (pp. 1–46).

Jonassen, D. (1991). Objectivism versus constructivism: Do we need a new philosophical paradigm? *Educational Technology Research and Development, 39*(3), 5–14.

Johnson, E. (2019). The Socratic method and its influence on Western pedagogy. *Educational Review, 71*(2), 215–229.

Johnson, H., & Lee, A. (2021). AI in climate modeling: Data generation and analysis. *Journal of Environmental Science and Technology, 15*(4), 212–227.

Johnson, R., & Sam, C. (2018). Data mining in educational curriculum design: A case study. *Journal of Educational Data Mining, 10*(1), 1–18.

Jung, J. (2018). Higher education in Korea: Western influences, Asian values, and indigenous processes. *Journal of Asian Public Policy, 11*(1), 1–13.

Jung, H., & Lee, J. (2022). AI in education: South Korea's strategy and its implications. *Journal of Educational Technology & Society, 24*(2), 102–116.

Jung, J., Horta, H., & Yonezawa, A. (2018). Higher education research in Asia: History, development, and challenges. In *Researching higher education in Asia: History, development and future* (pp. 355–366).

Kapoor, A., & Singh, S. (2023). Predictive analytics in academic publishing: The future of AI. *Advances in Scholarly Communication, 11*(2), 102–116.

Kim, E. (2020). Ethical considerations of using AI in higher education. *International Review of Information Ethics, 28*(1), 85 92.

Kim, J., & Park, H. (2020). Digital transition in Asian academic publishing: Emerging trends and challenges. *Electronic Library, 38*(4), 657–671.

Kim, J., & Park, H. (2021). AI-driven academic advising: A case study in South Korean universities. *Asia Pacific Journal of Education, 41*(3), 435–450.

Kim, J., & Lee, W. (2021). The dark side of AI in education. *Journal of Educational Computing Research, 59*(2), 223–244.

Kim, H., & Lee, Y. (2022). The role of AI in enhancing student support in Korean higher education. *Asia Pacific Journal of Education, 42*(3), 345–360.

Kim, Y., & Park, H. (2013). Confucianism and the Korean education system: Examining the old to understand the new. *Asian Education and Development Studies, 2*(2), 8–21.

Kuleto, V., Ilić, M., Dumangiu, M., Ranković, M., Martins, O., Păun, D., & Mihoreanu, L. (2021). Exploring opportunities and challenges of artificial intelligence and machine learning in higher education institutions. *Sustainability, 13*(18), 10424.

Lee, M. (2000). The influence of Confucianism on the educational system of Korea. *Journal of Asian Studies, 45*(2), 245–260.

Lee, K., Kim, H., & Cho, Y. (2019). The role of robotics in enhancing laboratory learning in higher education. *Journal of Science Education and Technology, 28*(4), 396–407.

Lee, J., & Kim, M. (2021). Ethical considerations in AI-driven academic publishing. *Ethics in Science and Technology, 14*(4), 200–213.
Li, Q., & Xu, J. (2021). Squirrel AI and personalized learning in China's educational landscape. *Educational Technology & Society, 24*(4), 233–247.
Li, S., & Zhang, T. (2021). National strategies for AI development in Chinese higher education. *Journal of Educational Policy, 35*(1), 52–67.
Li, X., & Agha, L. (2019). The evolution of academic publishing in Asia: Challenges and opportunities. *Journal of Scholarly Publishing, 50*(3), 123–139.
Li, Y. (2019). The application of AI in education: The Squirrel AI case. *Educational Technology Research and Development, 67*, 1435–1449.
Liu, H., Zhang, J., & Yang, G. (2022a). The use of artificial intelligence in education and its implications for individualized learning. *Journal of Educational Technology & Society, 25*(1), 129–140.
Liu, J., Zhang, H., & Lee, K. (2022b). Open access movement in Asian academic publishing: A case study of Plan S. *Journal of Academic Librarianship, 48*(1), 102–110.
Lopez, G., & Hernandez, F. (2022). Overcoming language barriers in scientific writing using AI. *International Journal of Linguistic Studies, 12*(1), 22–36.
Manhiça, R., Santos, A., & Cravino, J. (2022, June). The use of artificial intelligence in learning management systems in the context of higher education: Systematic literature review. In *2022 17th Iberian conference on information systems and technologies (CISTI)* (pp. 1–6). IEEE.
Marginson, S. (2011). Higher education in East Asia and Singapore: Rise of the Confucian model. *Higher Education, 61*, 587–611.
Martinez, S., & Rodriguez, F. (2021). Ethical considerations in AI-assisted scientific research. *Journal of AI Ethics, 6*(2), 89–101.
Miller, R., & Davis, S. (2022). Combating plagiarism with AI tools in academic publishing. *Journal of Academic Integrity, 16*(1), 34–50.
Moore, R., & Patel, N. (2021). Personalized content delivery in academic publishing: The role of AI. *Journal of Information Technology in Education, 18*(2), 89–104.
Morrison, A., & Smith, D. (2019). The role of publications in academic identity formation. *Studies in Higher Education, 44*(4), 678–690.
Nguyen, T. (2015). Colonial influence on education in Vietnam. *Asian Journal of Education, 16*(3), 7–24.
Nguyen, T., Pham, T., & Nguyen, Q. (2021). AI in education: The Singapore experience. *Educational Technology Research and Development, 69*(1), 209–225.
Ong, M., & Lim, C. (2023). AI-enhanced academic publishing in Singapore: Balancing efficiency and transparency. *Singapore Journal of Publishing Studies, 15*(1), 89–103.
O'Neil, J., & Martin, P. (2018). Beyond publications: Rethinking the value of academic work. *Educational Philosophy and Theory, 50*(6), 597–605.
Pane, J. F., Baird, M. D., Steiner, E. D., Hamilton, L. S., & Pane, J. D. (2017). *How does personalized learning affect student achievement?* RAND Corporation.
Pappano, L. (2012). The year of the MOOC. *The New York Times.*
Park, S. (2021). Human-AI interaction in education: Challenges and opportunities. *Educational Technology Research and Development, 69*, 107–112.
Patel, K., & Patel, P. (2019). The role of augmented reality in enhancing education and training. *AI & Society, 34*(3), 567–581.
Patel, V., & Jackson, R. (2020). Publish or perish: Examining the implications for academic research and policy. *Education Policy Analysis Archives, 28*(17), 112–130.
Patel, V., & Smith, S. (2022). Ethical considerations for AI in higher education: A call for action. *Journal of Computing in Higher Education, 34*(3), 567–584.
Pedró, F. (2020). Applications of artificial intelligence to higher education: Possibilities, evidence, and challenges. *IUL Research, 1*(1), 61–76.
Peterson, G., & Green, L. (2021). Automated statistical review in scholarly publishing. *Journal of Statistical Software Review, 29*(1), 45–60.

Piaget, J. (1954). *The construction of reality in the child*. Basic Books.
Pinto, A. S., Abreu, A., Costa, E., and Paiva, J. (2023). How machine learning (ML) is transforming higher education: A systematic literature review. *Journal of Information Systems Engineering and Management, 8*(2), 21168. https://doi.org/10.55267/iadt.07.13227
Popenici, S. A., & Kerr, S. (2017). Exploring the impact of artificial intelligence on teaching and learning in higher education. *Research and Practice in Technology Enhanced Learning, 12*(1), 1–13.
Robinson, M., & Lee, A. (2022). The echo chamber effect in AI-driven academic recommendations. *Journal of Digital Scholarship, 6*(1), 54–69.
Saettler, P. (2004). *The evolution of American educational technology*. Information Age Publishing.
Saito, Y., Shoyama, Y., & Nakano, M. (2020). AI-powered language tutors: A revolution in language education in Japanese universities. *Journal of Language Learning Technologies, 54*(2), 58–75.
Shanahan, M. (2015). *The technological singularity*. MIT Press.
Shahjahan, R. A. (2020). On 'being for others': Time and shame in the neoliberal academy. *Journal of Education Policy, 35*(6), 785–811. https://doi.org/10.1080/02680939.2019.1629027
Sharma, P. (2017). The Gurukul system: Ancient education in India. *Journal of Ancient History, 5*(1), 112–117.
Shmueli, B., Ray, S., & Velasquez, N. (2021). Ethical implications of AI in higher education: Paradigms and paradoxes. *AI & Society, 36*(3), 875–885.
Siemens, G., & Gasevic, D. (2012). *Learning and knowledge analytics*. Routledge.
Singh, R., & Patel, N. (2022). The impact of AI on scientific manuscript writing. *Journal of AI in Academia, 10*(2), 117–130.
Singh, S., & Gupta, B. (2019). Global vs. regional: The dichotomy in academic publishing. *Asian Education and Development Studies, 8*(3), 312–327.
Smith, D. (2003). Critical thinking in Western education: A historical overview. *History of Education Quarterly, 43*(4), 515–543.
Smith, J., & Johnson, M. (2021). AI in academic publishing: An overview. *Journal of Technology in Higher Education, 19*(2), 112–127.
Smith, J., & Hamilton, D. (2021). Enhancing qualitative research with natural language processing: A case study in education research. *Journal of Educational and Behavioral Statistics, 46*(2), 204–225.
Syzdykbayeva, A., Baikulova, A., & Kerimbayeva, R. (2021). Introduction of artificial intelligence as the basis of modern online education on the example of higher education. In *2021 IEEE international conference on smart information systems and technologies (SIST)* (pp. 1–8). IEEE.
Tahiru, F. (2021). AI in education: A systematic literature review. *Journal of Cases on Information Technology (JCIT), 23*(1), 1–20.
Tan, J., & Ng, K. (2023). AI in Singaporean universities: Preparing graduates for the future. *Journal of Higher Education Policy and Management, 41*(2), 112–128.
Tanaka, K. (2021). AI in education: Japan's strategy and challenges. *Journal of Educational Computing Research, 59*(1), 4–22.
Tiwari, A., Kumar, A., Jain, S., Dhull, K. S., Sajjanar, A., Puthenkandathil, & Sajjanar, A. (2023). Implications of ChatGPT in public health dentistry: A systematic review. *Cureus, 15*(6). https://doi.org/10.7759/cureus.40367
Thompson, D., & Lee, C. (2021). Academic publishing and career advancement in higher education: An analysis of the 'publish or perish' culture. *Journal of Higher Education Policy, 33*(2), 210–225.
Thompson, M. (2023). The authenticity of AI-generated scientific writing. *Journal of Research Integrity, 17*(1), 33–47.
Tyack, D., & Cuban, L. (1995). *Tinkering toward utopia: A century of public school reform*. Harvard University Press.
Taylor, R. (2022). AI in selecting peer reviewers: Efficiency and accuracy. *Peer Review Quarterly, 8*(1), 22–37.

Vygotsky, L. S. (1978). *Mind in society: The development of higher psychological processes.* Harvard University Press.
Wang, F. (2018). Artificial intelligence in education: A review. *IEEE Access, 6*, 36154–36168.
Wang, Y., & Zhang, X. (2016). The first AI-authored paper in scientific literature. *Journal of AI History, 8*(4), 89–93.
Wang, T., Lund, B. D., Marengo, A., Pagano, A., Mannuru, N. R., Teel, Z. A., & Pange, J. (2023). Exploring the potential impact of artificial intelligence (AI) on international students in higher education: Generative AI, chatbots, analytics, and international student success. *Applied Sciences, 13*(11), 6716.
Wang, Y., & Zhang, Y. (2021). Peer review in Asian academic publishing: Tradition, tensions, and transformations. *Asian Studies Review, 45*(2), 301–317.
Watson, W. R., & Watson, S. L. (2007). An argument for clarity: What are learning management systems, what are they not, and what should they become? *TechTrends, 51*(2), 28–34.
Watters, A. (2020). The algorithms of discrimination: AI, equity, and early childhood education. *TechTrends, 64*(6), 806–815.
Welch, A. (2019). Higher education in Asia. In *The Oxford handbook of the history of education* (pp. 302–315).
White, K., & Zhao, L. (2022). Authorship and AI: Crediting AI contributions in scientific papers. *Journal of Scholarly Publishing, 53*(2), 102–118.
Williams, G. (2019). AI and the future of peer review. *Journal of Scholarly Publishing, 10*(4), 58–72.
Wong, L., & Chu, D. (2018). Balancing AI advancements with cultural values in Asian education. *Journal of Asian Educational Technology, 15*(2), 35–47.
Xia, P. (2020). Application scenario of artificial intelligence technology in higher education. In *International conference on applications and techniques in cyber intelligence ATCI 2019: Applications and techniques in cyber intelligence 7* (pp. 221–226). Springer International Publishing.
Xia, X., & Li, X. (2022). Artificial intelligence for higher education development and teaching skills. *Wireless Communications and Mobile Computing, 2*, 91–99.
Xie, H., & Zhu, F. (2020). Predicting student performance in higher education using machine learning algorithms. *International Journal of Higher Education, 9*(1), 291–303.
Yang, R. (2018). Foil to the West? Interrogating perspectives for observing East Asian higher education. In J. Jung, H. Horta, & A. Yonezawa (Eds.), *Researching higher education in Asia: History, development and future* (pp. 37–50). Springer. https://doi.org/10.1007/978-981-10-4989-7_3
Yang, R. (2019). Emulating or integrating? Modern transformations of Chinese higher education. *Journal of Asian Public Policy, 12*(3), 294–311.
Yonezawa, A., Kitamura, Y., Meerman, A., & Kuroda, K. (2014). The emergence of international dimensions in East Asian higher education: Pursuing regional and global development. In *Researching higher education in Asia: History, development and future* (pp. 1–13). Springer Netherlands.
Yun, S. (2015). Internationalization of East Asian higher education. *Journal of International Students, 5*(2), 204–206.
Zaman, B. U. (2023, October 5). *Transforming education through AI, benefits, risks, and ethical considerations.* TechRxiv. https://doi.org/10.48550/arXiv.2310.03478
Zhang, L., & Zhou, M. (2019). Personalized learning recommendation system based on big data analysis. *Journal of Educational Technology & Society, 22*(1), 273–286.
Zhang, X., Zhu, Y., & Liu, H. (2021). Using AI to analyze student engagement: A study in China. *Computers & Education, 160*, 104031.
Zhao, L., & Chen, W. (2022). The impact of AI on academic publishing efficiency in China. *Journal of Asian Academic Research, 28*(3), 75–90.
Zhou, M., Xu, Y., & Li, Z. (2020). Adaptive learning based on artificial intelligence: A survey. *Computer Science Review, 37*, 100274.

Ziguras, C. (2016). Globalization and the transformation of Asian higher education. In *The Palgrave handbook of Asia Pacific higher education* (pp. 73–88).

Márton Gosztonyi is an assistant professor (Ph.D.) at ELTE Faculty of Social Sciences, Institute of Empirical Studies. He specialises in social network analysis, data science and the social economy. He previously held an academic position at Universiti Malaya, as well as teaching and research roles at other universities in Europe. With over 15 years of teaching experience in higher education, his interdisciplinary work bridges sociology, artificial intelligence and participatory research. Márton has published widely in high-ranking academic journals and led projects across Europe and Southeast Asia. His research focuses on informal lending, marginalised communities and the application of computational methods in social science.

Chapter 7
Harnessing AI Innovations for Future-Ready Education in Malaysia Vis-à-Vis Thailand, Singapore and South Korea

Johari Abdullah, Kee Man Chuah, Erlinda Mikal, and Syahrul Nizam Junaini

Abstract This chapter explores the transformative role of Generative AI technologies in potentially redefining the educational landscape of Malaysia, with comparative insights drawn from Thailand, Singapore and South Korea. In the post-COVID-19 era marked by rapid technological change and evolving pedagogical needs, the chapter examines how these nations are integrating AI into their education systems. It critically analyses the impact of AI tools on educators, students and key stakeholders across these countries, highlighting differences in strategy, implementation and outcomes. Special focus is given to the integration of AI in teaching, learning and assessment practices, providing a regional perspective on innovation and inclusivity. The chapter also discusses opportunities and challenges faced by Malaysia and its neighbours in adopting AI in classrooms, offering strategic insights for effective and ethical integration. By comparing approaches and outcomes, this chapter provides a roadmap for educators and policymakers to collaboratively steer AI-driven educational reform in Malaysia and beyond. Ultimately, this chapter contributes to the broader discourse on future-ready education in Asia, positioning Malaysia at the forefront of educational innovation in the face of global technological advancements.

J. Abdullah (✉) · S. N. Junaini
Faculty of Computer Science and Information Technology, Universiti Malaysia Sarawak, Samarahan, Malaysia
e-mail: ajohari@unimas.my

S. N. Junaini
e-mail: syahruln@unimas.my

K. M. Chuah · E. Mikal
Faculty of Education, Language and Communication, Universiti Malaysia Sarawak, Samarahan, Malaysia
e-mail: kmchuah@unimas.my

E. Mikal
e-mail: merlinda@unimas.my

A. Pandian et al. (eds.), *Future of Education in Asia*,
https://doi.org/10.1007/978-3-032-07285-6_7

Keywords AI innovations · Generative AI · Teaching and learning · AI in assessment · Future-ready education · Educational technology

7.1 Introduction

The emergence of Generative Artificial Intelligence (AI) technologies such as ChatGPT, Claude 3 and Google Gemini marks a new age in the field of education (Labadze & Grigolia, 2023), with the capacity to profoundly revolutionise teaching and learning methods, particularly in the Malaysian setting. This chapter investigates how these technologies are reshaping the educational landscape in Malaysia, while drawing comparative insights from Thailand, Singapore and South Korea. The urgency of this transformation has been amplified by the COVID-19 pandemic, which accelerated digital adoption and highlighted the need for more flexible, inclusive and future-focused educational practices. In Malaysia, the incorporation of Generative AI technologies presents a promising opportunity to strengthen teaching methods, effectively involve learners and boost educational results in all areas.

The educational sector faces distinct difficulties and opportunities in the era following the COVID-19 pandemic (Aboagye et al., 2020). The rapid transition to digital learning platforms and subsequent recognition of technology's potential in education have emphasised the necessity for a methodical and deliberate use of AI tools in teaching and learning procedures. This chapter seeks to examine the intricacies of teaching and learning in the current period, which is defined by swift technology advancements and shifting educational requirements.

While Malaysia sees promise in integrating Generative AI to enhance teaching methodologies and student engagement, its progress is uneven. In contrast, countries like Singapore have made significant strides through robust policy frameworks, and South Korea and Thailand are advancing with strategic investments in AI-powered learning environments. This chapter offers a critical examination of these comparative trajectories, addressing the challenges, innovations and outcomes of AI implementation in each context.

Focusing on AI's integration in instruction, engagement and assessment, the chapter seeks to understand how regional best practices can inform Malaysia's strategy towards future-ready education. By analysing these varied approaches, it aims to support educators, policymakers and institutions in developing balanced, inclusive and ethically guided AI adoption models across Asia.

7.2 Key Concepts and Issues

7.2.1 Understanding Artificial Intelligence

Artificial Intelligence (AI) aims to replicate human-like intelligence, enabling systems to understand language, recognise patterns, make decisions and solve complex problems. The field's genesis in the mid-twentieth century was marked by Alan Turing's seminal question about machine reasoning, leading to pivotal developments. Among these, IBM's Deep Blue chess victory over Garry Kasparov in 1997 and Google's AlphaGo defeating a professional Go player in 2016 stand out, showcasing AI's capacity for analysing complex patterns and strategic decision-making (Granter et al., 2017).

Early AI research concentrated on theoretical problem-solving and symbolic methods, resulting in significant programmes like the Logic Theorist and ELIZA. These initiatives proved that machines could undertake tasks previously believed to require human intelligence. However, the initial optimism was tempered by technological limitations, leading to periods known as 'AI winters,' characterised by diminished funding and interest (Toosi et al., 2021). The landscape shifted with advancements in computing power, data availability and machine learning (ML) techniques, transitioning AI from a focus on rule-based processing to data-driven learning.

Machine Learning, as a vital component of AI, revolutionised the field by enabling systems to learn from data, thus identifying patterns without direct programming (Rebala et al., 2019). This shift laid the groundwork for further advancements in AI, notably through deep learning. Deep learning employs intricate, multi-layered neural networks to discern complex patterns, significantly enhancing AI's capabilities in fields like image and speech recognition, natural language processing and autonomous driving (Liu et al., 2021; Sarker, 2021).

Building on these advancements, Generative AI emerged as a cutting-edge frontier, capable of producing novel content, ranging from text to images and music. This development extends AI's capabilities beyond mere pattern recognition, enabling the creation of original, innovative outputs (Cao et al., 2023). The evolution from AI's theoretical underpinnings to its practical applications across various sectors highlights its transformative impact and underscores the field's continuous growth and potential for future innovation.

7.2.2 The Evolution of AI in Education

The integration of AI into various levels of education has witnessed a shift from automated systems for drills and practice to intelligent systems capable of promoting personalised learning experiences. As aptly pointed out by Roll and Wylie (2016), the evolution transcends mere introduction of new technologies into the classroom.

The field of AI in education has also achieved significant breakthroughs in theoretical development, which has contributed to making learning experiences more engaging and personalised in line with the progress to Education 4.0 and beyond (Bonfield et al., 2020). It is, in fact, through this expansion in theoretical understanding that AI technologies can be developed or integrated to be more relevant for educational purposes.

In the early days of AI implementation in education around the 70s, the main interest of the researchers and educators was on improving computer-assisted instruction (CAI). The role of AI in CAI was limited to providing basic drill and practice exercises (Dede, 1986). Carbonell (1970) claimed that these programmes were effective for rote memorisation and practice but lacked the sophistication to adapt to the diverse needs of learners or to provide in-depth instructional support. This restricted usage mirrors the educational paradigm of that era, which was largely characterised by behaviourist learning principles (Schunk, 2012). Even though these early systems were simple, they showed that computers could provide tailored learning experiences, which paved the way for more advanced AI applications.

As technology advanced, the 1980s and 1990s saw the emergence of intelligent tutoring systems (ITS), which represented a significant leap forward (Nwana, 1990; VanLehn, 2011). Further, Kulik and Fletcher (2016) in their meta-analytic review showed that these systems could provide more customised feedback and adapt to individual learners' needs. ITS utilised primarily rule-based systems to model the domain knowledge and the learner's understanding. This makes it possible to tailor instructions and offer targeted guidance to learners (Nwana, 1990). The core idea of ITS lies in the quest to create a personal tutor capable of addressing the inherent challenges of traditional one-size-fits-all educational models (Ferster, 2022). These systems aim to replicate the effectiveness of one-on-one tutoring by offering a highly personalised and adaptive learning experience. However, during the 80s and 90s, ITS were limited to institutions or schools that had the necessary technological infrastructure and financial resources to support such advanced systems.

The advent of the internet and the proliferation of digital data in the late 1990s and early 2000s ushered in a new era for AI in education. The availability of vast amounts of educational data and the development of machine learning algorithms enabled the creation of more advanced and adaptive learning environments (Cheung et al., 2021; Park et al., 2023). These technologies could analyse learner data in real-time, which eventually provides insights into learning patterns and identifies knowledge gaps encountered by the learner. In the last decade, the rise of big data, natural language processing and advanced analytics has further revolutionised AI's role in education (Cui et al., 2023). AI-powered educational platforms now offer a wide range of functionalities, from automated grading systems and plagiarism detectors to chatbots that provide instant support and feedback to learners (Chuah & Kabilan, 2021; Gillani et al., 2023). Furthermore, AI is being used to develop immersive learning experiences through virtual and augmented reality, making learning more engaging and accessible.

Building upon these advancements, the birth of a new generation of AI tools known as Generative AI in recent years has caught the attention of many educational

Table 7.1 Prominent generative AI tools in the market

Category	Generative AI tools
Texts	ChatGPT, Claude3, Google Gemini, Microsoft Copilot, Jasper
Images	DALL-E (also integrated in ChatGPT and Copilot), MidJourney, StableDiffusion (Stability AI), Imagen 3, Adobe Firefly
Video	Synthesia, HeyGen, D-ID
Audio	ElevenLabs, PlayHT, Descript
Coding	AlphaCode, GitHub Copilot

researchers and educators (Stokel-Walker & Van Noorden, 2023). Generative AI tools can produce original content ranging from text to images and even simulations. The key to generative AI tools is that users can generate various forms of content just by providing textual instructions or commonly known as prompts. They offer unprecedented opportunities for personalised and interactive learning experiences, as learners can ideally embrace self-regulated learning by engaging with these tools. Table 7.1 summarises some of the prominent Generative AI tools according to its categories.

Despite these advancements in AI, its adoption in the Malaysian education system remains slower than expected. Although the Ministry of Science, Technology and Innovation Malaysia (MOSTI) launched the Malaysia National Artificial Intelligence Roadmap, 2021–2025 (MOSTI, 2021), the implementation of AI in education remains at the conceptual level with minimal outcomes. The focus on education was also geared towards using machine learning for learning analytics and the prediction of educational outcomes. There is an apparent need for a more consolidated effort to integrate relevant AI technologies and tools for teaching and learning. However, a renewed effort can be seen in recent developments within the Malaysian education system. While the adoption of AI in education has been slower than anticipated, there are several promising initiatives and strategies in place to accelerate its integration and improve outcomes.

7.3 Discussion of Implications

7.3.1 Comparative Review on AI Integration in Education: Malaysia, Thailand, Singapore and South Korea

Malaysia, like many nations in Asia, is actively exploring the integration of artificial intelligence (AI) in education to improve teaching and learning outcomes. The Malaysia National Artificial Intelligence Roadmap (2021–2025) outlines key strategies, including personalised learning, STEM education and administrative automation, yet its implementation faces challenges. Limited infrastructure, particularly in rural areas and varying levels of teacher readiness have slowed AI adoption. Pilot projects such as 1BestariNet demonstrate the potential of AI integration, though these efforts have faced criticism for connectivity gaps and inconsistent implementation (MOSTI, 2021; Shanmugam et al., 2019).

In Thailand, the Digital Thailand and Thailand 4.0 policies prioritise AI to bridge educational inequities (Ajanapanya, 2024). Projects like the Smart Classroom Project aim to provide adaptive learning opportunities for rural students, leveraging partnerships with international tech companies to accelerate AI adoption. These programmes are aligned with Thailand's national goal of achieving educational equity through technology (Machmud et al., 2021).

Singapore, on the other hand, sets a regional benchmark with its Smart Nation 2.0 initiative, which fully integrates AI into education (Choudhury, 2024). Through tools such as predictive analytics, personalised learning systems and smart classrooms, Singapore emphasises scalability and equity. The EdTech Masterplan allocates substantial resources to ensure robust AI integration aligned with the national curriculum, making it an exemplary model of AI-driven education (EdTech Masterplan 2030, 2023).

The South Korean government has outlined three major policy directions for AI education, focusing on developing AI education content, advancing AI education policies and preparing for the new curriculum set to be implemented in 2025. These policies aim to integrate AI smoothly into classrooms and address emerging challenges (Lee & Jeong, 2023). AI curriculum design for K-12 education in Korea is still in its nascent stages, with efforts starting in 2020 to train teachers in AI education. The curriculum is designed by analysing AI education methods from advanced countries and adapting them to the Korean context (Park & Lim, 2023). The integration of AI into the Korean education system is progressing, with significant efforts in policy formulation, curriculum design and teacher training.

7.3.2 Impact of AI on Teaching and Learning

The impact of AI on teaching and learning is profoundly reshaping instructional methodologies across the globe, including in Malaysia. AI tools, such as adaptive learning software, AI-driven analytics and intelligent tutoring systems, are revolutionising traditional teaching methods (Gligorea et al., 2023). These technologies enable personalised learning experiences, allowing educators to tailor their instruction to meet the individual needs of each student (Pataranutaporn et al., 2021). For instance, adaptive learning platforms use AI algorithms to analyse a student's performance and learning style, adjusting the curriculum dynamically to suit their pace and understanding (Kabudi et al., 2021). In Malaysia, initiatives like the Malaysia Education Blueprint, 2013–2025 have emphasised the integration of such technologies into the curriculum, aiming to foster a more innovative and personalised learning environment (MOE, 2013).

The effect of AI technologies on student engagement and learning outcomes is significantly positive. AI-driven educational tools make learning more interactive and engaging, capturing the attention of students who might otherwise be disinterested in traditional teaching methods (Huang et al., 2023). These tools often incorporate gamification elements, interactive simulations and real-time feedback, making learning more appealing and relevant to students (Bachiri et al., 2023). This increased engagement often translates to improved learning outcomes, as students are more motivated and better equipped to understand complex concepts. In Malaysia, the use of AI in language learning apps has shown promising results in enhancing English proficiency among students, a key goal of the country's educational policy (Mohd Nasir et al., 2023). Such technologies also support educators in tracking student progress more accurately, enabling data-driven decisions to optimise teaching strategies and improve overall academic performance.

Case studies from Malaysian educational institutions highlight the transformative impact of AI on teaching and learning. For instance, in the study by Chew and Chua (2020), AI was utilised to enhance Chinese language learning by integrating the NAO robot into traditional classroom settings, which complemented human tutoring with interactive and engaging robot-led activities. This blend of human and robotic tutoring leveraged the robot's emotion recognition and body language capabilities to demonstrate Chinese words, making the learning process more interactive and memorable for students.

Also, the study by Rahim et al. (2022) addresses the transformative impact of AI in higher education by developing a model for adopting AI-based chatbots. This model, derived from a combination of Partial Least Squares Structural Equation Modelling (PLS-SEM) and Neural Network modelling, aims to enhance the understanding and effective integration of AI chatbots in educational environments. The research highlights the potential of AI to revolutionise educational practices, particularly through the use of advanced chatbot technologies that can facilitate more efficient, interactive and personalised learning experiences. These examples highlight the potential of AI to not only augment the role of educators but also to provide students with a more adaptive and immersive learning experience. As AI continues to evolve, its integration into the Malaysian education system promises to further enhance teaching methodologies and learning outcomes (Bachiri et al., 2023).

7.3.3 AI in Assessment and Evaluation

AI in educational assessment and evaluation presents opportunities to enhance efficiency and learning outcomes through personalised testing and automated grading. AI systems can tailor tests to individual students' needs, adjusting difficulty and question types based on past performance and learning preferences, ensuring engagement and knowledge retention (Chen et al., 2020; Minn, 2022). This personalised approach, supported by educational data mining, helps identify student challenges early, enabling targeted support (Romero & Ventura, 2020).

Automated grading through AI offers significant advantages, such as efficiency in grading essays and multiple-choice questions, reducing educator workload and providing timely feedback. Despite concerns about AI's current limitations in mimicking human judgement, advancements in deep learning are expected to address these challenges, improving grading consistency and bias reduction (Gardner et al., 2021; Hahn et al., 2021; Perrotta & Selwyn, 2020).

Furthermore, AI's analysis of assessment data can uncover trends and learning gaps, supporting data-driven decision-making in education. This approach helps identify at-risk learners and optimise resource allocation, although caution is advised due to potential biases in AI-driven insights (Sharma et al., 2019; Wang, 2021).

In summary, AI enhances educational assessments by offering personalised tests, automated grading and insightful data analysis. However, addressing challenges like bias and ensuring the irreplaceable role of human guidance is crucial for a responsible AI implementation in education.

7.3.4 Challenges and Opportunities

The integration of AI in education, while promising for enhancing learning experiences and administrative efficiency, encounters several significant challenges, including ethical concerns, resource allocation and the necessity for teacher training. Ethical issues, particularly regarding data privacy and security, necessitate stringent data protection measures to prevent breaches and misuse of student information (Stahl & Wright, 2018). Additionally, the substantial financial investment required for AI technologies' infrastructure poses a challenge, especially in economically disadvantaged regions, potentially exacerbating educational inequalities without strategic resource management (Schiff, 2021; Vrontis et al., 2022). Effective utilisation of AI tools in education also depends on teachers' proficiency with these technologies, highlighting the need for comprehensive training programmes to equip educators with the skills to integrate AI into teaching practices effectively (Celik et al., 2022; Jeon et al., 2020).

Despite these hurdles, AI presents opportunities for enhancing inclusivity and personalisation in Malaysian education. By accommodating the varied needs of students, including those with disabilities or from remote areas, AI can help overcome geographical and socioeconomic barriers. Tools offering real-time translation and subtitles expand access to non-native speakers and hearing-impaired students, while early identification of learning disabilities through AI enables tailored support (Hashim et al., 2022; Subashini & Krishnaveni, 2021).

7.3.5 Strategies for Effective AI Integration

Recent scholarly discourse presents a multifaceted view of the potential and challenges inherent in this technological evolution. A pivotal study underscores the importance of comprehending educators' perceptions and necessities vis-à-vis AI and other nascent educational technologies (Kizilcec, 2023). This foundational understanding is critical to ensure that implementing AI in classrooms aligns with pedagogical goals and teacher competencies.

Another intriguing aspect explored in the literature is the role of AI-driven tools, such as chatbots, in refining the educational process. While these technologies hold immense potential for enhancing student–teacher interactions and learning outcomes, they also bring forth significant ethical and practical considerations that necessitate careful deliberation (Adiguzel et al., 2023). Integrating AI in education is not merely a technical upgrade but also involves navigating complex moral landscapes.

Furthermore, the concept of 'smart classrooms', which amalgamate AI and advanced technological infrastructures, emerges as a revolutionary approach to education. Studies indicate these environments can substantially improve various facets of education, including classroom management, teaching methodologies and performance assessments (Dimitriadou & Lanitis, 2023). Smart classrooms represent a significant change, redefining the traditional educational space into a dynamic, responsive and interactive learning environment.

In addition, there is an increasing emphasis on the critical analysis of existing online AI educational resources. The current discourse advocates for these resources to be in alignment with effective pedagogical practices. There is a call for future AI curricula to adopt comprehensive designs that offer support, guidance and adaptability to cater to diverse learning needs and environments (Druga et al., 2022).

Lastly, the pioneering use of machine learning models, such as EDU-AI, in the educational sector represents a cutting-edge development. These models have shown potential in automating the design of classroom layouts, playing a vital role in the early stages of architectural projects (Karadag et al., 2023). This application of AI extends beyond the immediate realm of teaching and learning, illustrating its transformative impact on educational infrastructure.

7.3.6 Guidelines for Incorporating AI in Education

To effectively integrate AI into education while preserving its human elements, educators and policymakers must engage in collaborative efforts to enhance AI literacy through comprehensive training programmes. Olari and Romeike (2021) emphasise the importance of incorporating AI and data literacy into teacher education, preparing educators across all subjects for AI's expanding role. Such programmes should encompass technical, pedagogical, ethical and educational aspects of AI to facilitate its integration into teaching strategies, ultimately enriching student learning.

Moreover, the development of ethical guidelines for AI use in education is crucial, necessitating collaboration among policymakers, technologists, educators and ethicists. These guidelines should address critical issues, including data protection, consent, bias, fairness and algorithm transparency, ensuring AI technologies respect students' and staff's rights and dignity.

Addressing educational disparities is imperative, with regulations aimed at ensuring equitable AI resource access across different regions and socioeconomic backgrounds (Gilbert, 2021; Olari & Romeike, 2021; Xiang, 2022). Investments in infrastructure, such as high-speed internet and digital devices, are essential, particularly for schools in underserved areas, to bridge the digital divide. AI applications should prioritise augmenting rather than replacing human interactions (Bryson & Theodorou, 2019; Horvatić & Lipic, 2021; Steels, 2023), supporting educators, personalising learning and fostering interactive, cooperative learning environments. The focus should be on enhancing human capabilities, including critical thinking, creativity and empathy. Continuous research and evaluation are necessary to monitor AI's impact on educational outcomes, fairness and student welfare. Policymakers should implement systems for ongoing assessment of AI applications in educational settings, incorporating feedback from educators, students and other stakeholders to refine AI tools and strategies.

These policy recommendations aim to guide educators and policymakers in integrating AI into education thoughtfully, ensuring that technological advancements complement and enhance human-centred educational practices.

7.4 Future Directions

The integration of Artificial Intelligence (AI) into education represents a significant shift towards personalised learning and advanced pedagogical methods. AI technologies, particularly Natural Language Processing (NLP), are instrumental in analysing student feedback to enhance educational practices and infrastructures. Highlighting the importance of NLP, studies such as Shaik et al. (2022) emphasise its role in refining educational feedback mechanisms, indicating a growing reliance on AI for pedagogical improvement.

Building on the utility of AI in analysing feedback, its application extends to enhancing student performance, notably in STEM fields. The incorporation of AI technologies in classrooms has been associated with increased student engagement and academic achievement, as discussed by García-Martínez et al. (2023). This positive correlation underscores AI's potential to revolutionise education by making learning more accessible and engaging. Furthermore, the interaction between human intelligence and AI is transitioning from merely automating tasks to augmenting human capabilities, supporting informed decision-making in educational settings. Ifenthaler and Schumacher (2023) discuss how AI's ability to process large datasets in real-time can enhance educational strategies and outcomes. In the realm of online learning, AI-enabled tools like Smart Web-Based Interactive System Modelling (SWISM) are pioneering personalised education through predictive analytics, as described by Yang et al. (2023).

However, the deployment of AI, especially NLP in education, faces challenges, including the complexity of language and the need for advanced techniques for sentiment analysis and understanding domain-specific jargon (Shaik et al., 2022). The emergence of AI in education (AIEd) brings to light ethical considerations and the impact on stakeholders, necessitating ongoing research to address these issues effectively (Ojha et al., 2023).

The expanding research in AI for online learning, with significant contributions from global leaders such as China, India and the United States, focuses on enhancing online teaching and learning processes. This research, as Dogan et al. (2023) pointed out, includes developing adaptive learning systems and predicting student behaviours, indicating a broad interest in optimising online education through AI. The ethical dimensions of AI in education are paramount for fostering a trust-based learning environment. Yu and Lu (2021) advocate for ethical AI deployment, emphasising transparency, fairness and privacy. Furthermore, the post-pandemic potential of leveraging Extended Reality (XR) and AI in online higher education enhances learning accessibility and engagement, as suggested by Rangel-de Lázaro and Duart (2023).

In Malaysia, Thailand, Singapore and South Korea, AI's implementation in education has followed diverse trajectories shaped by national priorities, infrastructural readiness and policy ecosystems. While Malaysia is focusing on enhancing language education and supporting national STEM objectives, Thailand is leveraging AI to address rural–urban disparities through adaptive learning initiatives. Singapore, with its Smart Nation 2.0 framework, prioritises full-scale AI integration aligned with national curricula, and South Korea is actively developing AI-specific curricula and teacher training as part of its 2025 reform.

These regional developments emphasise the need for Malaysia to establish a robust ethical and operational framework that not only supports AI adoption domestically but also draws from successful practices in neighbouring countries. Comparative insights suggest that cross-border learning and collaboration can accelerate AI maturity in education systems.

Future research directions should include:

- Adapting AI tools to align with the diverse learning environments in Malaysia while benchmarking against regional leaders.
- Evaluating long-term impacts of AI on student learning outcomes and teacher effectiveness across Asian contexts.
- Developing inclusive, scalable AI policies grounded in ethical principles that resonate with Southeast and East Asian socio-cultural realities.

By navigating ethical, cultural and logistical challenges alongside Thailand, Singapore and South Korea, Malaysia can position itself as a regional leader in AI-driven educational innovation, shaping a collaborative and future-ready education ecosystem across Asia.

7.5 Conclusion

This chapter explores the transformative potential of AI and Generative AI in enhancing educational outcomes in Malaysia through personalised learning, while drawing comparative insights from Thailand, Singapore and South Korea. By tailoring instruction to individual needs, AI fosters greater student engagement, promotes educational equity and cultivates critical thinking and problem-solving skills essential for the modern workforce.

The comparative analysis highlights that while Malaysia is making strides, it can benefit from the strategic frameworks, infrastructure investments and policy innovations observed in its regional counterparts. Singapore's robust EdTech infrastructure, South Korea's structured AI curriculum development and Thailand's inclusive approaches to bridging digital divides offer valuable lessons for shaping a future-ready Malaysian education system.

However, the integration of AI in education across all four countries presents common challenges, including data privacy, algorithmic bias, infrastructural limitations and the need for comprehensive teacher training. Addressing these issues requires a collaborative and interdisciplinary approach involving educational institutions, AI specialists and policymakers across the region.

To harness AI's full potential, continuous research and cross-border knowledge exchange are essential to monitor its impact, adapt best practices and ensure inclusive and ethical implementation. Through these efforts, Malaysia, alongside its Asian peers, can build a resilient and innovative educational ecosystem that empowers learners and educators in an increasingly digital and interconnected global economy.

Highlights

- AI, especially NLP and Generative AI, is reshaping personalised learning and assessment practices across Asia, with Malaysia, Thailand, Singapore and South Korea at varying stages of implementation.
- Singapore's EdTech-driven Smart Nation policy and South Korea's curriculum reforms offer strategic models for AI integration that Malaysia can adapt and localise.
- Thailand's initiatives focus on bridging urban–rural education gaps through AI-enhanced learning environments, presenting key lessons in equity-driven implementation.

- Ethical considerations, including bias, privacy and teacher readiness, remain central to AI deployment across all four countries, underscoring the need for region-specific policy frameworks.
- Malaysia's strategic adoption of AI, informed by regional comparisons, positions it to accelerate innovation in STEM and language education while promoting educational equity.

References

Aboagye, E., Yawson, J. A., & Appiah, K. N. (2020). COVID-19 and e-learning: The challenges of students in tertiary institutions. *Social Education Research, 2*(1), 1–8. https://doi.org/10.37256/ser.212021422

Adiguzel, T., Kaya, M. H., & Cansu, F. K. (2023). Revolutionizing education with AI: Exploring the transformative potential of ChatGPT. *Contemporary Educational Technology, 15*(3), ep429. https://doi.org/10.30935/cedtech/13152

Ajanapanya, N. (2024, March 17). Thailand charts bold course in AI development, emphasizing ethics and innovation. *The Nation Thailand*. https://www.nationthailand.com/business/tech/40042660

Bachiri, Y.-A., Mouncif, H., & Bouikhalene, B. (2023). Artificial intelligence empowers gamification: Optimizing student engagement and learning outcomes in e-learning and MOOCs. *International Journal of Engineering Pedagogy (IJEP), 13*(8), 4–19. https://doi.org/10.3991/ijep.v13i8.40853

Bonfield, C. A., Salter, M., Longmuir, A., Benson, M., & Adachi, C. (2020). Transformation or evolution?: Education 4.0, teaching and learning in the digital age. *Higher Education Pedagogies, 5*(1), 223–246. https://doi.org/10.1080/23752696.2020.1816847

Bryson, J. J., & Theodorou, A. (2019). How society can maintain human-centric artificial intelligence. In *The ethics of artificial intelligence: Balancing opportunity and risk* (pp. 305–323). https://doi.org/10.1007/978-981-13-7725-9_16

Cao, Y., Li, S., Liu, Y., Yan, Z., Dai, Y., Yu, P. S., & Sun, L. (2023). *A comprehensive survey of AI-generated content (AIGC): A history of generative AI from GAN to ChatGPT*. arXiv. http://arxiv.org/abs/2303.04226

Carbonell, J. (1970). AI in CAI: An artificial-intelligence approach to computer-assisted instruction. *IEEE Transactions on Man-Machine Systems, 11*(4), 190–202. https://doi.org/10.1109/TMMS.1970.299942

Celik, I., Dindar, M., Muukkonen, H., & Järvelä, S. (2022). The promises and challenges of artificial intelligence for teachers: A systematic review of research. *TechTrends, 66*(4), 616–630. https://doi.org/10.1007/s11528-022-00715-y

Chen, L., Chen, L., Chen, G., Chen, P., Chen, P., Lin, Z., & Lin, Z. (2020). Artificial intelligence in education: A review. *IEEE Access*. https://doi.org/10.1109/access.2020.2988510

Cheung, S. K. S., Kwok, L. F., Phusavat, K., & Yang, H. H. (2021). Shaping the future learning environments with smart elements: Challenges and opportunities. *International Journal of Educational Technology in Higher Education, 18*(1), 16. https://doi.org/10.1186/s41239-021-00254-1

Chew, E., & Chua, X. N. (2020). Robotic Chinese language tutor: Personalising progress assessment and feedback or taking over your job? *On the Horizon, 28*(3), 113–124. https://doi.org/10.1108/OTH-04-2020-0015

Choudhury, A. R. (2024, March 17). *Singapore's Smart Nation 2.0 policy focuses on AI and building resilience.* GovInsider. https://govinsider.asia/intl-en/article/singapores-smart-nation-20-policy-focuses-on-ai-and-building-resilience
Chuah, K.-M., & Kabilan, M. K. (2021). Teachers' views on the use of chatbots to support English language teaching in a mobile environment. *International Journal of Emerging Technologies in Learning (IJET), 16*(20), 223. https://doi.org/10.3991/ijet.v16i20.24917
Cui, Y., Ma, Z., Wang, L., Yang, A., Liu, Q., Kong, S., & Wang, H. (2023). A survey on big data-enabled innovative online education systems during the COVID-19 pandemic. *Journal of Innovation and Knowledge, 8*(1), 100295. https://doi.org/10.1016/j.jik.2022.100295
Dede, C. (1986). A review and synthesis of recent research in intelligent computer-assisted instruction. *International Journal of Man-Machine Studies, 24*(4), 329–353. https://doi.org/10.1016/S0020-7373(86)80050-5
Dimitriadou, E., & Lanitis, A. (2023). A critical evaluation, challenges, and future perspectives of using artificial intelligence and emerging technologies in smart classrooms. *Smart Learning Environments, 10*(1), 12. https://doi.org/10.1186/s40561-023-00231-3
Dogan, M. E., Dogan, T. G., & Bozkurt, A. (2023). The use of artificial intelligence (AI) in online learning and distance education processes: A systematic review of empirical studies. *Applied Sciences, 13*(5), 3056. https://doi.org/10.3390/app13053056
Druga, S., Otero, N., & Ko, A. J. (2022). The landscape of teaching resources for AI education. In *Proceedings of the 27th ACM conference on innovation and technology in computer science education* (Vol. 1, pp. 96–102). ACM. https://doi.org/10.1145/3502718.3524782
Ferster, B. (2022). Intelligent tutoring systems. In *Intelligent tutoring systems* (pp. 1–20). Routledge. https://doi.org/10.4324/9781138609877-REE6-1
García-Martínez, I., Fernández-Batanero, J. M., Fernández-Cerero, J., & León, S. P. (2023). Analysing the impact of artificial intelligence and computational sciences on student performance: Systematic review and meta-analysis. *Journal of New Approaches in Educational Research, 12*(1), 171. https://doi.org/10.7821/naer.2023.1.1240
Gardner, J., O'Leary, M., & Yuan, L. (2021). Artificial intelligence in educational assessment: 'Breakthrough? or buncombe and ballyhoo?' *Journal of Computer Assisted Learning, 37*(5), 1207–1216. https://doi.org/10.1111/jcal.12577
Gilbert, E. J. (2021). Equitable AI. In *Extended abstracts of the 2021 CHI conference on human factors in computing systems* (pp. 1–2). ACM. https://doi.org/10.1145/3411763.3457780
Gillani, N., Eynon, R., Chiabaut, C., & Finkel, K. (2023). Unpacking the 'black box' of AI in education. *Educational Technology and Society, 26*(1), 99–111. https://www.jstor.org/stable/48707970
Gligorea, I., Cioca, M., Oancea, R., Gorski, A.-T., Gorski, H., & Tudorache, P. (2023). Adaptive learning using artificial intelligence in e-learning: A literature review. *Education Sciences, 13*(12), 1216. https://doi.org/10.3390/educsci13121216
Granter, S. R., Beck, A. H., & Papke, D. J. (2017). AlphaGo, deep learning, and the future of the human microscopist. *Archives of Pathology and Laboratory Medicine, 141*(5), 619–621. https://doi.org/10.5858/arpa.2016-0471-ED
Hahn, M. G., Baldiris Navarro, S. M., De La Fuente Valentin, L., & Burgos, D. (2021). A systematic review of the effects of automatic scoring and automatic feedback in educational settings. *IEEE Access, 9*, 108190–108198. https://doi.org/10.1109/ACCESS.2021.3100890
Hashim, S., Omar, M. K., Ab Jalil, H., & Mohd Sharef, N. (2022). Trends on technologies and artificial intelligence in education for personalized learning: Systematic literature review. *International Journal of Academic Research in Progressive Education and Development, 11*(1). https://doi.org/10.6007/IJARPED/v11-i1/12230
Horvatić, D., & Lipic, T. (2021). Human-centric AI: The symbiosis of human and artificial intelligence. *Entropy, 23*(3), 332. https://doi.org/10.3390/e23030332
Huang, A. Y. Q., Lu, O. H. T., & Yang, S. J. H. (2023). Effects of artificial intelligence-enabled personalized recommendations on learners' learning engagement, motivation, and outcomes in a

flipped classroom. *Computers and Education, 194*, 104684. https://doi.org/10.1016/j.compedu.2022.104684
Ifenthaler, D., & Schumacher, C. (2023). Reciprocal issues of artificial and human intelligence in education. *Journal of Research on Technology in Education, 55*(1), 1–6. https://doi.org/10.1080/15391523.2022.2154511
Jeon, I.-S., Jun, S., & Song, K.-S. (2020). *Teacher training program and analysis of teacher's demands to strengthen artificial intelligence education.* Semantic Scholar. https://api.semanticscholar.org/CorpusID:225328566
Kabudi, T., Pappas, I., & Olsen, D. H. (2021). AI-enabled adaptive learning systems: A systematic mapping of the literature. *Computers and Education: Artificial Intelligence, 2*, 100017. https://doi.org/10.1016/j.caeai.2021.100017
Karadag, I., Güzelci, O. Z., & Alaçam, S. (2023). EDU-AI: A twofold machine learning model to support classroom layout generation. *Construction Innovation, 23*(4), 898–914. https://doi.org/10.1108/CI-02-2022-0034
Kizilcec, R. F. (2023). To advance AI use in education, focus on understanding educators. *International Journal of Artificial Intelligence in Education.* https://doi.org/10.1007/s40593-023-00351-4
Kulik, J. A., & Fletcher, J. D. (2016). Effectiveness of intelligent tutoring systems. *Review of Educational Research, 86*(1), 42–78. https://doi.org/10.3102/0034654315581420
Labadze, L., & Grigolia, M. (2023). Role of AI chatbots in education: Systematic literature review. *International Journal of Educational Technology in Higher Education.* https://doi.org/10.1186/s41239-023-00426-1
Lee, J., & Jeong, H. (2023). Keyword analysis of artificial intelligence education policy in South Korea. *IEEE Access, 11*, 102408–102417. https://doi.org/10.1109/ACCESS.2023.3317261
Liu, Y., Sun, P., Wergeles, N., & Shang, Y. (2021). A survey and performance evaluation of deep learning methods for small object detection. *Expert Systems with Applications, 172*, 114602. https://doi.org/10.1016/j.eswa.2021.114602
Machmud, M. T., Widiyan, A. P., & Ramadhani, N. R. (2021). The development and policies of ICT supporting educational technology in Singapore, Thailand, Indonesia, and Myanmar. *International Journal of Evaluation and Research in Education (IJERE), 10*(1), 78. https://doi.org/10.11591/ijere.v10i1.20786
Malaysia Education Blueprint 2013–2025. (2013).
Malaysia National Artificial Intelligence Roadmap 2021–2025. (2021).
Ministry of Education Singapore. (2023). *EdTech masterplan 2030.* Ministry of Education Singapore. https://www.moe.gov.sg/education-in-sg/educational-technology-journey/edtech-masterplan
Minn, S. (2022). AI-assisted knowledge assessment techniques for adaptive learning environments. *Computers and Education: Artificial Intelligence, 3*, 100050. https://doi.org/10.1016/j.caeai.2022.100050
Mohd Rahim, N. I., A. Iahad, N., Yusof, A. F., & Al-Sharafi, M. A. (2022). AI-based chatbots adoption model for higher-education institutions: A hybrid PLS-SEM-Neural network modelling approach. *Sustainability, 14*(19), 12726. https://doi.org/10.3390/su141912726
Mohd Nasir, H., Brahin, N. M. A., Sani, F. E. M., Mispan, M. S., & Abd Wahab, N. H. (2023). AI educational mobile app using deep learning approach. *JOIV: International Journal on Informatics Visualization, 7*(3), 952–958. https://doi.org/10.30630/joiv.7.3.1247
Nwana, H. S. (1990). Intelligent tutoring systems: An overview. *Artificial Intelligence Review, 4*(4). https://doi.org/10.1007/BF00168958
Ojha, S., Narendra, A., Mohapatra, S., & Misra, I. (2023). *From robots to books: An introduction to smart applications of AI in education (AIEd).* arXiv. https://arxiv.org/abs/2301.10026v1
Olari, V., & Romeike, R. (2021). Addressing AI and data literacy in teacher education: A review of existing educational frameworks. In *The 16th workshop in primary and secondary computing education* (pp. 1–2). ACM. https://doi.org/10.1145/3481312.3481351

Park, M., & Lim, W. (2023). AI-integrated mathematics education: A review of literature in the Korean context. *The Korean Society of Educational Studies in Mathematics - Journal of Educational Research in Mathematics, 33*(3), 771–795. https://doi.org/10.29275/jerm.2023.33.3.771

Park, E., Ifenthaler, D., & Clariana, R. B. (2023). Adaptive or adapted to: Sequence and reflexive thematic analysis to understand learners' self-regulated learning in an adaptive learning analytics dashboard. *British Journal of Educational Technology, 54*(1), 98–125. https://doi.org/10.1111/bjet.13287

Pataranutaporn, P., Danry, V., Leong, J., Punpongsanon, P., Novy, D., Maes, P., & Sra, M. (2021). AI-generated characters for supporting personalized learning and well-being. *Nature Machine Intelligence, 3*(12), 1013–1022. https://doi.org/10.1038/s42256-021-00417-9

Perrotta, C., & Selwyn, N. (2020). Deep learning goes to school: Toward a relational understanding of AI in education. *Learning, Media and Technology, 45*(3), 251–269. https://doi.org/10.1080/17439884.2020.1686017

Rangel-de Lázaro, G., & Duart, J. M. (2023). You can handle, you can teach it: Systematic review on the use of extended reality and artificial intelligence technologies for online higher education. *Sustainability, 15*(4), 3507. https://doi.org/10.3390/su15043507

Rebala, G., Ravi, A., & Churiwala, S. (2019). Machine learning definition and basics. In *An introduction to machine learning* (pp. 1–17). Springer International Publishing. https://doi.org/10.1007/978-3-030-15729-6_1

Roll, I., & Wylie, R. (2016). Evolution and revolution in artificial intelligence in education. *International Journal of Artificial Intelligence in Education, 26*(2), 582–599. https://doi.org/10.1007/s40593-016-0110-3

Romero, C., & Ventura, S. (2020). Educational data mining and learning analytics: An updated survey. *WIREs Data Mining and Knowledge Discovery, 10*(3). https://doi.org/10.1002/widm.1355

Sarker, I. H. (2021). Deep learning: A comprehensive overview on techniques, taxonomy, applications and research directions. *SN Computer Science, 2*(6), 420. https://doi.org/10.1007/s42979-021-00815-1

Schiff, D. (2021). Out of the laboratory and into the classroom: The future of artificial intelligence in education. *AI and Society, 36*(1), 331–348. https://doi.org/10.1007/s00146-020-01033-8

Schunk, D. H. (2012). *Learning theories: An educational perspective*. Pearson.

Shaik, T., Tao, X., Li, Y., Dann, C., McDonald, J., Redmond, P., & Galligan, L. (2022). A review of the trends and challenges in adopting natural language processing methods for education feedback analysis. *IEEE Access, 10*, 56720–56739. https://doi.org/10.1109/ACCESS.2022.3177752

Shanmugam, K., Zainal, N. K., & Gnanasekaren, C. (2019). Technology foresight in the virtual learning environment in Malaysia. *Journal of Physics: Conference Series, 1228*(1), 012068. https://doi.org/10.1088/1742-6596/1228/1/012068

Sharma, R. C., Kawachi, P., & Bozkurt, A. (2019). The landscape of artificial intelligence in open, online and distance education: Promises and concerns. *Asian Journal of Distance Education, 14*, 1–2. https://api.semanticscholar.org/CorpusID:213415701

Stahl, B. C., & Wright, D. (2018). Ethics and privacy in AI and big data: Implementing responsible research and innovation. *IEEE Security and Privacy, 16*(3), 26–33. https://doi.org/10.1109/MSP.2018.2701164

Steels, L. (2023). Conceptual foundations of human-centric AI. In M. Chetouani, V. Dignum, P. Lukowicz, & C. Sierra (Eds.), *Human-centered artificial intelligence: Advanced lectures* (pp. 8–35). Springer International Publishing. https://doi.org/10.1007/978-3-031-24349-3_2

Stokel-Walker, C., & Van Noorden, R. (2023). What ChatGPT and generative AI mean for science. *Nature, 614*(7947), 214–216. https://doi.org/10.1038/d41586-023-00340-6

Subashini, P., & Krishnaveni, M. (2021). Artificial intelligence-based assistive technology. In *Artificial intelligence theory, models, and applications* (pp. 217–240). Auerbach Publications. https://doi.org/10.1201/9781003175865-13

Toosi, A., Bottino, A. G., Saboury, B., Siegel, E., & Rahmim, A. (2021). A brief history of AI: How to prevent another winter (a critical review). *PET Clinics, 16*(4), 449–469. https://doi.org/10.1016/j.cpet.2021.07.001

VanLehn, K. (2011). The relative effectiveness of human tutoring, intelligent tutoring systems, and other tutoring systems. *Educational Psychologist, 46*(4), 197–221. https://doi.org/10.1080/00461520.2011.611369

Vrontis, D., Christofi, M., Pereira, V., Tarba, S., Makrides, A., & Trichina, E. (2022). Artificial intelligence, robotics, advanced technologies and human resource management: A systematic review. *The International Journal of Human Resource Management, 33*(6), 1237–1266. https://doi.org/10.1080/09585192.2020.1871398

Wang, Y. (2021). When artificial intelligence meets educational leaders' data-informed decision-making: A cautionary tale. *Studies in Educational Evaluation, 69*(June), 100872. https://doi.org/10.1016/j.stueduc.2020.100872

Xiang, A. (2022). Diversity must be at the heart of equitable AI development. *Engineering and Technology, 17*(9), 21–21. https://doi.org/10.1049/et.2022.0904

Yang, H., Shankar, A., & Velliangiri, S. (2023). Artificial intelligence-enabled interactive system modeling for teaching and learning based on cognitive web services. *International Journal of E-Collaboration, 19*(2), 1–18. https://doi.org/10.4018/ijec.316655

Yu, S., & Lu, Y. (2021). Frontiers of AI in education. In *An introduction to artificial intelligence in education.* https://doi.org/10.1007/978-981-16-2770-5_8

Johari Abdullah is serving as an Associate Professor at the Faculty of Computer Science and IT, UNIMAS, Sarawak. He received his Ph.D. (Computing Science) from Newcastle University (UK), his Master of IT from Queensland University of Technology, Brisbane, Australia, and Bachelor of Computer Science (Networking) degree from Universiti Putra Malaysia. His interest in ICT is wide and ranging from Trusted System, Blockchain Technology, Web System Design and development, System Architecture, problem solving using tools such as TRIZ, ICT education for children and youth through Computational Thinking, Scratch and Computer Science Unplugged, and also open-source system and software.

Kee Man Chuah is a senior lecturer at the Faculty of Education, Language and Communication, with expertise in the fields of educational technology, computational linguistics, learning sciences and instructional design. He has won numerous awards at national and international levels for his innovative approaches to teaching and learning, as well as assistive technology.

Erlinda Mikal is currently serving as a senior lecturer at the Faculty of Education, Language, and Communication at UNIMAS, Sarawak. She received her Ph.D. in Education with a focus on Curriculum and Instruction from the University of Kansas (US), her Master of Science in Learning Sciences, and her Bachelor of Education (TESL) from Universiti Malaysia Sarawak (UNIMAS). Her research interests include curriculum and instruction, language and education, as well as language and identity.

Syahrul Nizam Junaini is a Professional Technologist and holds MSc in Electronic Imaging and Media Communications from the University of Bradford, UK. Previously he was the Deputy Director (Learning Technology) at the Centre for Applied Learning and Multimedia (CALM), Universiti Malaysia Sarawak. He is also a Senior Lecturer at the Faculty of Computer Science and Information Technology (FCSIT), UNIMAS. He is a fellow of UNIMAS Fellowship of Teaching and Learning Advancement (FoTL). He is also the recipient of Teaching Award (Applied Science) at the prestigious National Academic Award (AAN 2017). He also received Malaysia Digital Transformation Awards 2024.

Chapter 8
Building the Tomorrow: Technology's Transformative Role in ASEAN's Built Environment Education

N. M. Alwi, N. Z. Abidin, N. H. H. Lim, and L. Hasnan

Abstract The potential of transformative technologies in the built environment industry and continuing education across ASEAN has been widely acknowledged. AI applications, Virtual Reality and Geographic Information Systems are ways digital literacy is tested and examined as the keystones of innovation for future education in countries including Malaysia, Thailand and Singapore. This chapter presents practical applications of these technologies and digital literacy by discussing their real-world impact evident in built environment education and how it can inspire different modes of teaching and learning development. Subtopics explore how these elements converge to construct a future built environment that champions efficiency, cultural richness, sustainability and historic preservation in ASEAN's context. The key takeaways from each section offer insights into the seamless amalgamation of these technologies, forging a path towards cities that honour their heritage while boldly advancing into a technologically empowered future.

Keywords Transformative technologies · Digital literacy · Built environment innovation · Technological convergence · Artificial intelligence · Virtual reality

N. M. Alwi (✉) · N. Z. Abidin · L. Hasnan
Department of Architecture, Faculty of Built Environment, University of Malaya, Kuala Lumpur, Malaysia
e-mail: najah.alwi@um.edu.my

N. Z. Abidin
e-mail: nurdiyana@um.edu.my

L. Hasnan
e-mail: liyana_h@um.edu.my

N. H. H. Lim
Department of Urban and Regional Planning, Faculty of Built Environment, University of Malaya, Kuala Lumpur, Malaysia
e-mail: shimalim@um.edu.my

A. Pandian et al. (eds.), *Future of Education in Asia*,
https://doi.org/10.1007/978-3-032-07285-6_8

8.1 Introduction

The built environment sector in ASEAN is undergoing a rapid transformation driven by emerging digital technologies, built environment innovations for the growing need for sustainable and inclusive urban development. As cities become smarter and more connected, the education systems that train future architects, planners and designers must evolve accordingly. Technology plays a central role in this transformation, offering new opportunities for collaboration, spatial analysis, immersive learning and data-informed design.

This chapter explores how four key technologies, Digital Literacy, Virtual Reality (VR), Geographic Information Systems (GIS) and Artificial Intelligence (AI), are reshaping built environment education across ASEAN. It specifically examines Malaysia, Thailand and Singapore, highlighting innovations and ongoing challenges within their higher education systems. The aim is to identify how these technologies can serve as levers for educational reform, professional readiness and regional development.

The study is guided by four research questions:

1. How has digital literacy been integrated into the curriculum for design education in the built environment across ASEAN?
2. What are the advantages and disadvantages of using Virtual Reality in built environment education?
3. How do existing challenges in Geographic Information Systems impact planning education in Malaysia, Thailand and Singapore?
4. What are the potential implications and recommendations for integrating Artificial Intelligence into higher education and industry practices in the built environment?

In addressing these questions, the chapter uses a qualitative methodology based on selected case studies, document analysis and thematic comparison. The goal is not only to assess the current landscape but to propose strategic directions that can support a more future-ready, inclusive and digitally fluent built environment education ecosystem for ASEAN.

Understanding how technology has the potential to significantly improve the sustainability and efficiency of built environment education in by integrating sustainability education into curricula and utilising innovative teaching methods. The Sustainability Graduate Community (SGC) framework, designed for engineering programmes in Malaysia, serves as an exemplary case of how structured learning pathways can instil sustainability competencies in graduates, aligning with industry expectations and the Malaysian Education Blueprint 2025 (Johan & Turan, 2016). Similarly, the ProSPER.Net project emphasises the importance of embedding sustainability into engineering and built environment curricula through professional development for educators, highlighting the need for industry collaboration and a regional approach to sustainability education (Iyer-Raniga & Andamon, 2014).

The integration of Education for Sustainable Development (ESD) is also essential for achieving the Sustainable Development Goals (SDGs), particularly SDG 4, which focuses on quality education and lifelong learning opportunities. ESD equips students with the skills necessary to tackle global challenges such as climate change and resource management (Opoku & Guthrie, 2018; Opoku et al., 2024). Additionally, developing a modular framework for sustainability education within the built environment curriculum can act as a benchmarking tool, ensuring that graduates are well-prepared to engage with the green agenda and contribute to sustainable practices (Ekundayo et al., 2018). By embedding sustainability literacy at all educational levels, countries can cultivate professionals who are not only knowledgeable but also motivated to implement sustainable solutions in the built environment sector (Opoku et al., 2024). Therefore, when combined with comprehensive educational frameworks and industry collaboration, technology can play a crucial role in advancing sustainability and efficiency in ASEAN's built environment education.

8.2 Literature Review

8.2.1 Digital Literacy

Digital literacy is central to 21st-century education and is considered a foundational competency in architecture, planning and design. In Malaysia, digital literacy has evolved from being an optional skill to an essential requirement. The COVID-19 pandemic, through the MCO, revealed substantial weaknesses in both access to technology and institutional preparedness. Educators faced limited support, while students struggled with inconsistent internet access and a lack of devices (Tan & Lim, 2022).

Curriculum implementation of digital tools remains uneven. While some institutions use Building Information Modelling (BIM), digital drafting tools and simulation software, others still rely heavily on traditional methods. The American Library Association (2022) (Faruqe, Watkins, & Medsker, 2022). defines digital literacy not just as technical use but also as cognitive evaluation of digital content. This includes the ability to search, filter, evaluate and apply digital information effectively in academic and professional work.

Singapore offers a more structured integration of digital tools. Government initiatives such as SkillsFuture and Smart Nation Education encourage digital skill development from an early stage. Architecture schools have embedded digital proficiency rubrics and collaborative learning platforms. Thailand is advancing with hybrid design studios, though implementation remains regionally inconsistent due to infrastructural gaps.

Students with high digital literacy are better able to collaborate, innovate and conduct research. Tools like Autodesk Revit, Adobe Creative Suite, Rhino and Grasshopper are becoming standard, requiring educators and institutions to offer continuous training. The future of digital literacy lies not only in tools but in critical thinking, data interpretation and ethical engagement with technology.

8.2.2 Virtual Reality (VR)

VR is transforming how students interact with design environments. It enables immersive exploration of 3D models and virtual walkthroughs of unbuilt spaces, facilitating spatial reasoning and empathetic design (Chronaki & Donn, 2019; Wojtowicz & Butelski, 2017). In the architectural studio, VR promotes iterative design by allowing real-time changes in scale, materiality and lighting (Mahalingam & Surapathi, 2017).

Singapore has applied VR in urban planning consultations, enabling public feedback through virtual city simulations. VR-based design reviews improve communication between clients, students and stakeholders (Mak & Lu, 2019). Thailand's academic institutions have begun integrating VR into design pedagogy through collaborative projects. These allow learners to test structural feasibility and environmental impacts in virtual environments.

Despite its benefits, VR adoption is hampered by high initial costs, technical glitches and educator unfamiliarity. Students often face a learning curve in navigating VR software. The lack of dedicated VR labs, especially in Malaysia, limits its potential. Yet, as hardware becomes more affordable and software more intuitive, the role of VR in built environment education is expected to expand.

8.2.3 Geographic Information Systems (GIS)

Geographical Information Systems (GIS) was first conceptualised in the 1960s to quantify geographical information while facilitating better spatial analysis via locational visualisation. In other words, it links information that would otherwise be visualised as numbers and graphs to locations. It is a powerful tool, acknowledged by urban planners and geographers as well as other related professions alike, that promotes rational decision-making and efficient monitoring of spatial changes. It was introduced in Malaysia by one of the GIS pioneers in Malaysia, the late professor Ahris Yaakup in the early 1990s. (Abdullah, Mohmadisa, & Suhaily, 2006; Nordin, 2006). To this day, GIS has been embedded into all Urban and Regional Planning programmes offered in local universities across Malaysia, directed either as a core subject or a supplementary subject.

GIS is a core component of planning education, offering spatial analysis tools that support data-driven decision-making. Since the early 1990s, GIS has helped students in Malaysia develop spatial citizenship, an understanding of how space influences society (De Miguel González & De Lázaro Torres, 2020). Due to the nature of this field, students learn about obtaining geospatial data sources and conducting data mapping in the real world by using local data sets.

GIS is widely used to map land-use patterns, infrastructure needs and environmental risk zones. Students learn to layer population data, transportation systems and zoning maps to identify planning solutions. However, access to accurate local datasets remains a barrier, compounded by limited software availability and training (Masron & Samat, 2009; Singh & Singh, 2013).

Thailand's education system has introduced GIS at the secondary level, encouraging early spatial thinking. Singapore's smart nation framework incorporates GIS into urban development courses, preparing students for careers in data-intensive planning. Malaysian institutions could benefit future for Built Environment innovation from open data platforms and cross-sectoral partnerships to expand GIS learning.

8.2.4 Artificial Intelligence (AI)

Artificial Intelligence (AI) is becoming a critical element in the future of built environment education, particularly in architecture and planning design ideation. Recent developments show AI contributing significantly to generative design, pattern recognition and cross-disciplinary integration. In ASEAN contexts, the uptake is varied but growing, with Singapore and South Korea pioneering integration into university-level curricula (Sadek, 2023).

In design ideation, AI can assist students by generating multiple options based on performance or style inputs. Tools like Midjourney, DALL·E and parametric platforms powered by AI enable rapid prototyping and feedback cycles. These platforms analyse vast datasets and offer designers unprecedented speed and diversity in concept generation. In Malaysia, early studies highlight a lack of structured AI curriculum, with most applications being experimental or confined to tech-savvy institutions.

Beyond ideation, AI also supports simulation environments and personalised education platforms. Universities globally are exploring AI-assisted project management tools, code compliance checks and systems that analyse user behaviour to optimise design proposals. In the ASEAN region, collaborations with industry are essential to drive AI skill development.

However, the integration of AI also presents risks and ethical concerns. These include algorithmic bias, data security and over-reliance on machine-generated content. Educational institutions must introduce policies and frameworks that support critical engagement with AI while ensuring it complements, rather than replaces, human creativity and decision-making. Scholars such as Zawacki-Richter et al. (2019) and Dignum (2021) emphasise the need for ethical AI education, particularly regarding transparency, accountability and responsible design practices.

The literature reveals that AI's transformative potential will remain untapped without curriculum reform, faculty upskilling and industry-academic collaboration. Therefore, AI should be approached not only as a tool but as a pedagogical partner shaping the future of built environment education in ASEAN.

8.3 Methodology

This study employed a qualitative, exploratory research methodology grounded in four guiding research questions that explored the integration of transformative technologies, Digital Literacy, Virtual Reality (VR), Geographic Information Systems (GIS) and Artificial Intelligence (AI), within the built environment education systems of ASEAN countries, particularly Malaysia, Singapore and Thailand.

8.3.1 Research Design and Data Sources

This study employed an exploratory qualitative design, specifically inductive qualitative content analysis. The first step involved data preparation, where related documents or materials in the form of books, book chapters, journal articles and official government documents are gathered. Next, the content of those materials is analysed while taking into consideration both the manifest and latent content, guided by the research questions established earlier. This means that the interpretation of content by the researchers are used to derive meaningful categories and discussion. Then, abstraction is done so as to coagulate the categories. Sources were selected based on relevance to built environment education and the timeframe of 2015–2024. Cross-referencing academic research with national education frameworks provided a multi-layered understanding of each technology's implementation. Validity was enhanced through triangulation of multiple data types and perspectives.

8.3.2 Study Limitations

While this study offers valuable insights into the transformative role of technology in education across Malaysia, Thailand and Singapore, it is essential to acknowledge certain limitations that invite forward-thinking considerations. Firstly, the qualitative nature of the research, while rich and nuanced, may not fully capture the broader trends and patterns that quantitative methods could unveil. This opens opportunities for future research to adopt mixed-method approaches, integrating qualitative depth with quantitative breadth to provide a more comprehensive understanding of technology integration in education. Also, the focus on specific cultural contexts may limit the generalisability of the findings.

However, this presents a unique opportunity for scholars to explore similar themes in diverse educational landscapes, fostering a comparative analysis that can enhance our understanding of technology's impact across different regions. Moreover, the ethical considerations surrounding Artificial Intelligence and technology integration highlight an area ripe for further exploration. Future studies can probe deeper into the implications of these technologies, examining their effects on equity, access and student agency in various educational settings. In recognising these limitations, we not only validate the current research but also pave the way for subsequent investigations that can build on these insights, ultimately contributing to a more holistic understanding of the evolving landscape of educational technologies.

8.4 Findings

8.4.1 Digital Literacy

Digital literacy remains a cornerstone of innovation in built environment education, yet its implementation across ASEAN is marked by uneven progress. In Malaysia, the Movement Control Order (MCO) during the COVID-19 pandemic highlighted significant disparities in digital infrastructure, particularly between urban and rural institutions (Tan & Lim, 2022). While PricewaterhouseCoopers Malaysia (2021) noted increased digital adoption among students and professionals, the digital divide persists. In Singapore, structured policies and institutional investment ensure digital literacy is embedded from early stages in architectural curricula. In contrast, Thailand's advancements have largely centred in urban universities, with rural areas still catching up.

Educators across the region report a gap between students' ability to use digital tools and their understanding of deeper competencies such as data management, visual communication and digital research (American Library Association, 2022). These skills are often acquired informally, leading to inconsistency in student outcomes (Abdullah & Lim, 2015).

Lim and Tan (2018) argue that digital proficiency should be systematically taught, not assumed. The lack of teacher training further exacerbates disparities, impacting the quality of digital engagement and innovation in design education. A more cohesive integration of digital literacy, including ethical awareness and critical evaluation, is necessary to ensure that students are equipped not just with tools but with adaptive, future-ready mindsets.

Aligning curriculum with national strategies, such as the Malaysian Education Blueprint 2025, can bridge these gaps. Embedding digital competencies at all educational levels and across disciplines, supported by infrastructure, faculty training and institutional policy, will better prepare students to lead in a digital built environment landscape. Embedding digital literacy across all levels of education, supported by continuous training and infrastructure, is critical for preparing future professionals in the built environment sector.

8.4.2 *Virtual Reality (VR)*

Virtual Reality (VR) has emerged as a pedagogical tool for immersive learning in design education. In Singapore, VR is used in architectural design reviews, stakeholder consultations and public engagement exercises. Thailand has piloted VR for participatory urban planning. In Malaysia, institutions such as Universiti Malaya have begun incorporating VR in architectural studios.

Students benefit from enhanced spatial understanding, immediate design feedback and the ability to iterate in real time. VR simulations allow them to walk through their digital designs and evaluate proportions, light and circulation in a way traditional 2D models cannot provide (Chen et al., 2018; Mahalingam & Surapathi, 2017). These environments foster intuitive engagement with space and scale, encouraging students to make faster, more informed design decisions.

VR also strengthens collaboration. Students and instructors can meet in shared virtual spaces to discuss ongoing work, improve communication and reduce misunderstandings. In Singapore, collaborative VR-based design review systems have been tested for urban projects where planners, architects and community stakeholders interact in virtual forums to evaluate and modify designs (Zhang et al., 2020).

Several universities in ASEAN are exploring VR in construction management, historical preservation and disaster preparedness scenarios. In Thailand, VR has been used to simulate flood-risk areas and allow planning students to test urban design interventions. In Malaysia, smaller institutions have adopted low-cost alternatives like mobile-based VR tools for teaching architectural history and cultural landscape interpretation (Radford, 2018).

However, VR integration remains limited by high costs of headsets, software licensing and the need for trained technical staff. Learning curves associated with new interfaces can slow adoption among both students and educators (Sung et al., 2019). Further, realism in VR environments still lacks sensory fidelity, such as soundscape simulation or tactile feedback, which can impact the perceived accuracy of user experiences (Mak & Lu, 2019).

Moreover, ensuring VR's alignment with learning objectives is essential. Without clear pedagogical frameworks, institutions risk using VR as a novelty rather than a transformative tool. Curriculum designers must embed VR thoughtfully into studio-based learning, guided by measurable outcomes and supported by research in spatial cognition and immersive learning (Maher et al., 2019).

In conclusion, while VR holds substantial promise for architectural and urban planning education in ASEAN, its full potential will only be realised through sustained investment, strategic curriculum integration and educator training.

8.4.3 Geographic Information Systems (GIS)

GIS education in Malaysia is widespread at the university level, particularly in urban planning programmes. Students use GIS for analysing land-use patterns, infrastructure coverage and environmental vulnerabilities. However, difficulties accessing reliable datasets and software licensing remain major limitations. According to Masron and Samat (2009), a lack of experienced trainers and inconsistent data access policies have hindered effective GIS adoption.

Singapore offers a robust example of GIS integration through its Smart Nation initiative and urban simulation labs. Planners and students alike use GIS tools to visualise transportation systems, demographic shifts and infrastructure plans. In Thailand, the Ministry of Education has supported GIS in secondary education, helping build spatial literacy at a younger age. Students in urban planning and geography departments engage with real-world projects, using GIS to map flood-prone areas and propose mitigation strategies (De Miguel González & De Lázaro Torres, 2020; Mzuza & Van Der Westhuizen, 2019).

In Malaysia, GIS instruction often emphasises technical skills but lacks integration with critical thinking or real-world applications. Students may learn how to use the software but struggle to apply it creatively to complex urban problems. According to Singh and Singh (2013), few institutions extend GIS training beyond basic mapping. Furthermore, data access remains a persistent problem due to limited government transparency and outdated information systems.

To realise the full potential of GIS in education, students must be exposed to dynamic data sources and taught to layer and interpret multi-dimensional datasets. Open-data platforms and stronger collaboration with planning authorities could improve learning outcomes. The development of user-friendly GIS plug-ins and curriculum-based learning tasks would further encourage critical engagement with spatial data. By incorporating GIS across multiple disciplines, ASEAN universities can foster a new generation of planners and designers who are both technically competent and contextually aware.

8.4.4 Artificial Intelligence (AI)

AI adoption in built environment education is still in early stages across ASEAN. In Malaysia, AI tools are used experimentally for generative design, performance analysis and iterative modelling. Students are exploring platforms like Grasshopper with AI extensions, but formal curriculum integration remains limited. Research by Sadek (2023) and Chen et al. (2022) highlights the value of AI tools in transforming architectural ideation and design workflows, yet also emphasises the importance of structured guidance for educational use.

Singaporean and South Korean institutions lead in incorporating AI into studio-based learning. Students are exposed to AI-driven design iterations, smart performance feedback and hybrid learning environments. These technologies allow real-time assessment of design solutions and facilitate design decision-making based on data-driven insights (Zawacki-Richter et al., 2019).

Nonetheless, the ethical implications of AI, including data privacy, algorithmic bias and intellectual ownership, are not yet widely addressed in ASEAN curricula. Educators must also confront the pedagogical implications of AI. Faculty members have expressed concern about the complexity of integrating AI into lesson plans due to a lack of familiarity with algorithms and system functionalities (Lee & Perret, 2022). Further, few universities have institutional frameworks to manage or monitor AI-enhanced assessments, raising concerns about transparency and accountability (Kim et al., 2022).

Malaysia's rich cultural and environmental diversity could offer unique datasets to train context-sensitive AI models in the built environment field. Cross-disciplinary collaborations with computer science and environmental engineering departments could foster more innovative applications and research outputs.

To maximise AI's potential in built environment education, institutions should prioritise faculty development, ethical literacy and industry collaboration. With these supports in place, AI can help shift built environment education from a reactive to a proactive model, equipping students to explore, simulate and iterate their ideas with precision, creativity and accountability.

8.5 Discussion of Implications

8.5.1 Pedagogical Challenges

The shift towards digital pedagogy has been uneven across ASEAN. In Malaysia, many educators acknowledge the value of digital tools, but actual implementation is hindered by limited access to training, hardware and consistent institutional support (Lim & Tan, 2018). The COVID-19 pandemic accelerated digital tool adoption, but many faculty members remain underprepared to fully leverage these technologies in their teaching.

Teacher beliefs also significantly influence the adoption of digital technologies. Some educators, especially in rural or resource-limited settings, prefer traditional teaching models and are hesitant to integrate advanced technologies (Snoeyink & Ertmer, 2001; Liu, Lin, & Zhang, 2017). Addressing this requires investment not only in hardware but also in long-term professional development and pedagogical reorientation.

8.5.2 Policy and Institutional Gaps

While Malaysia's Education Blueprint 2025 encourages digital integration, yet actual implementation varies across institutions. While Singapore benefits from strong centralised policy and infrastructure, countries like Thailand and Malaysia struggle with fragmented implementation and resource disparities. Institutions often lack coherent strategies to support the integration of technologies such as VR and AI into curricula.

Sustained collaborations between government, academia and industry are needed to support curriculum redesign, faculty training and infrastructure investment. Without this alignment, the gap between policy aspirations and classroom reality will persist.

8.5.3 Curriculum Development

Curricular reform is needed to embed transformative technologies as core components rather than elective or optional tools. Many built environment programmes still treat AI, VR and GIS as supplementary skills. Instead, design studios, planning labs and research modules should integrate these tools into the core curriculum.

Problem-based learning, design thinking and scenario simulation using these technologies can enrich learning outcomes. Institutions must also align their offerings with evolving industry expectations, ensuring that graduates are workplace-ready and capable of critical, creative and data-informed decision-making.

8.5.4 Regional Collaboration Opportunities

There is substantial potential for ASEAN-wide collaboration to share resources, practices and technologies. Initiatives like regional virtual exchanges, open-access GIS datasets and shared VR platforms could help bridge existing disparities. Countries like Singapore and South Korea can play a mentorship role, while Malaysia and Thailand can contribute cultural and contextual insights.

In encapsulating the principal concepts articulated within each segment examined in this chapter, thematic elements can be further categorised into distinct divisions about pedagogical and educational instruments tailored to the domain of built environment design and its functionalities:

1. Integrate digital literacy training into design curricula: Educators can incorporate digital literacy training into design courses to equip students with the skills necessary to navigate the digital landscape adeptly. This can include training on software and tools used in the industry and best data management and analysis practices.

2. Use Virtual Reality (VR) for design evaluation: VR can be a valuable tool for evaluating the real-world implications of designs. Educators can provide students with VR technology, allowing them to create immersive simulations of their designs and evaluate them from multiple perspectives.
3. Integrate GIS into urban planning courses: GIS can play a significant role in optimising resource allocation and promoting sustainable development in urban planning. Educators can incorporate GIS training into urban planning courses, allowing students to familiarise themselves with spatial data and learn how to use it to make informed decisions.
4. Use AI for design ideation: AI is rapidly reshaping design processes and educators can encourage students to explore the full potential of AI in the ideation process. This can include training on AI-powered design tools and opportunities to experiment with AI-generated design options. By incorporating these technologies and concepts into design education, educators can equip students with the skills and knowledge necessary to thrive in a technology-driven built environment industry.

A regional alliance for digital built environment education would strengthen collective capacity, promote standardised competencies and foster innovation. The future of ASEAN's built environment education relies on such partnerships to balance local relevance with global competitiveness.

8.6 Future Directions

ASEAN's built environment education must evolve into a digitally fluent, inclusive and ethically conscious ecosystem that equips students for both local relevance and global competitiveness. The following directions propose actionable strategies aligned with technological advancement and regional realities.

8.6.1 Digital Literacy Futures

For ASEAN to lead in future-ready education, digital literacy must move beyond basic competency to become a foundational and assessable outcome across all education levels. Built environment programmes need to embed digital skills such as data sourcing, visualisation and collaborative platforms within studio work and project-based learning. Institutions in Malaysia could draw on Singapore's structured digital competency frameworks to standardise student capabilities.

Teachers must receive ongoing digital pedagogy training. This includes not just technical tools but also ethical awareness, online research skills and critical evaluation of digital sources. The Malaysian Education Blueprint 2025 can serve as a guiding framework, but institutions must localise its objectives and translate them into course-level outcomes. Further, digital access equity, through infrastructure, devices and support, remains crucial in ensuring no student is left behind in ASEAN's digital transition.

8.6.2 *Advancing VR Integration*

Virtual Reality holds immense promise in reshaping experiential learning for architectural and planning disciplines. Future efforts must focus on making VR tools more accessible through scalable infrastructure, ranging from high-end labs in research universities to affordable VR kits in smaller institutions. Initiatives in Singapore that integrate VR into community design and planning consultation can inspire regional adaptations.

To further be inclusive in the teachings and learning of VR in the Built Environment Design, several strategies can be incorporated into the education system or syllabus, which can be separated into integration, exploration, collaboration, immersion, simulation and interactive visualisation.

These integrations include:

i. Integration into Architectural teaching Studios
ii. Urban Planning Courses with VR Exploration
iii. VR-Based Collaborative Design Reviews
iv. Immersive Learning Environments for Design Education
v. VR Simulations for Architectural Assessments
vi. Interactive VR Visualisations for User Engagement.

These approaches showcase the diverse ways virtual reality can be incorporated into built environment design, spanning from educational settings to collaborative design reviews and architectural assessments. Implementing these strategies can leverage the benefits of virtual reality, enhance the design process and improve the overall quality of built environments.

A roadmap for VR integration could include the development of open-source VR templates, cross-border collaborative VR studios and research grants for immersive environment pedagogy. Partnerships with industry can offset hardware costs while embedding real-world relevance. To maximise impact, VR use should be aligned with student learning outcomes, enabling iterative design, empathy-building and scenario testing.

8.6.3 *Expanding GIS Access*

GIS adoption in education should not remain confined to mapping and technical coursework. Future directions point to integrating GIS into participatory planning, climate risk analysis and socio-economic impact assessments. ASEAN countries can model from Thailand's use of GIS at the school level and Singapore's data-driven urban planning ecosystem.

In Malaysia, increasing institutional access to spatial datasets and simplifying licensing terms for educational use are necessary. Universities should also partner with local authorities to co-develop live GIS-based projects. Expanding GIS education to include plug-ins for real-time simulations (e.g., sea level rise, urban heat zones) will build students' capacity for complex spatial decision-making and climate-resilient design.

Ethical considerations pertaining to all technological domains necessitate that GIS be accorded precedence to guarantee the responsible and equitable utilisation of data, thereby addressing potential biases and fostering transparency within decision-making frameworks. The incorporation of these methodologies will not only augment the educational experience but will also equip emerging professionals with the requisite skills to address urgent environmental issues through innovative solutions.

8.6.4 Strengthening AI Pedagogy

Artificial Intelligence must be reframed from a technological frontier to a pedagogical partner. As students increasingly engage with generative tools and predictive modelling, built environment programmes must train educators to guide the ethical and effective use of AI. South Korean and Singaporean examples show that AI integration benefits from policy support and interdepartmental collaboration.

Malaysia can create AI-focused electives, capstone projects or even micro-credentials to familiarise students with current tools like Midjourney, Revit AI or neural network-based planning platforms. However, with increasing reliance on AI, ethical frameworks must be introduced early, covering algorithmic bias, data privacy and the limitations of automation.

A future-ready ASEAN must also invest in open research platforms and industry-academic partnerships to explore how AI can inform culturally grounded, socially responsible and ecologically resilient design. The shift towards AI in education should not only mirror industry trends but push innovation in how we teach, learn and imagine the built environment.

8.7 Conclusion

This chapter has explored the transformative role of digital technologies, specifically digital literacy, Virtual Reality (VR), Geographic Information Systems (GIS) and Artificial Intelligence (AI), in shaping the future of built environment education across ASEAN, particularly in Malaysia, Singapore and Thailand. Through a multi-method qualitative framework and thematic analysis of case studies and policy documents, the findings underscore a growing commitment to educational innovation across the region, albeit with differing levels of infrastructural readiness and pedagogical integration.

Digital literacy remains a foundational yet unevenly distributed skill, influenced by access to infrastructure, teacher training and institutional support. VR has emerged as a powerful design and learning tool, though its cost and complexity limit widespread adoption. GIS is well-established within urban planning curricula but requires enhanced data accessibility and institutional support. AI holds significant promise for design ideation and personalised education, yet it introduces ethical considerations and requires further curricular development.

The discussion highlighted the need for more cohesive policy implementation, faculty capacity-building, curriculum reform and cross-sector collaboration. Regional partnerships and shared digital infrastructures could accelerate progress and support more equitable outcomes across ASEAN. In addition to what has already been mentioned, it would be highly beneficial and advantageous to incorporate a variety of more comprehensive and detailed strategies that focus specifically on the ongoing professional development for educators, which would enable them to effectively and consistently keep pace with the rapid and ever-evolving technological advancements that are continuously emerging within the field of education.

Looking ahead, built environment education must evolve into a digitally fluent, ethically grounded and contextually relevant system that prepares future professionals to respond to sustainability, technological disruption and regional development goals. By embracing transformative technologies through structured, inclusive and future-facing educational strategies, ASEAN institutions can cultivate a generation of built environment professionals who are prepared not only to design the future, but to shape it responsibly.

Highlights

- AI is transforming the design ideation process in the built environment by generating numerous design alternatives, offering new perspectives and enabling students to rapidly explore different creative possibilities.
- AI facilitates the integration of cultural nuances and regional design preferences, ensuring that projects are not only technologically advanced but also contextually relevant and culturally sensitive.
- AI is reshaping teaching methods in architecture and urban planning education by introducing innovative, AI-powered tools and platforms that promote continuous learning and expose students to cutting-edge technologies.
- While AI enhances educational outcomes, its integration raises ethical issues such as data privacy, algorithmic bias and the need for comprehensive understanding, requiring educators to address these concerns thoughtfully.

- VR creates immersive and collaborative learning environments that improve design comprehension.
- GIS facilitates spatial thinking and data-driven planning from the secondary to university level.
- Digital literacy is foundational but must be expanded beyond basic software skills to include critical evaluation, research fluency and ethical reasoning.

References

Abdullah, A. H., Mohmadisa, M. F. D., & Suhaily, M. (2006). *Using GIS as a teaching tool in geography education: Geographic education issues in Malaysia.* UPSI Publisher.

Abdullah, A. B., & Lim, S. Y. (2015). The relevancy of digital literacy for Malaysian students for learning with Web 2.0 technology. *System, 98*, 22360–22364. https://doi.org/10.1016/j.system.2015.08.001

Abgaryan, H., Asatryan, S., & Matevosyan, A. (2023). Revolutionary changes in higher education with artificial intelligence. *Main Issues of Pedagogy and Psychology, 10*(1), 76–86. https://doi.org/10.24234/miopap.v10i1.454

Altomonte, S., Logan, B., Feisst, M., Rutherford, P., & Wilson, R. (2016). Interactive and situated learning in education for sustainability. *International Journal of Sustainability in Higher Education, 17*(3), 417–443. https://doi.org/10.1108/IJSHE-01-2015-0003

Chen, Z., Song, Z., Yuan, S., & Chen, W. (2022). Influence analysis of education policy on migrant children's education integration using artificial intelligence and deep learning. *Frontiers in Psychology, 13*, Article 910692. https://doi.org/10.3389/fpsyg.2022.910692

Chen, S., Luo, D., Gu, N., Li, Y., Liu, L., & Wang, X. (2018). Enhancing user engagement in architecture through VR visualizations with natural human–computer interaction. *Automation in Construction, 94*, 363–377. https://doi.org/10.1016/j.autcon.2018.06.022

Chronaki, C., & Donn, M. (2019). Enabling landscape architecture and urban planning students to design spaces in virtual reality. *Journal of Urban Technology, 26*(4), 33–48. https://doi.org/10.1080/10630732.2019.1575970

Darken, R. P., & Darken, C. J. (1996, April 10). *VR + AI = intelligent environments: A synergistic approach to engineering design support.* https://doi.org/10.1117/12.237450

De Miguel González, R., & De Lázaro Torres, M. L. (2020). WebGIS implementation and effectiveness in secondary education using the digital atlas for schools. *Journal of Geography, 119*(2), 74–85. https://doi.org/10.1080/00221341.2019.1630536

Dignum, V. (2021). The role and challenges of education for responsible AI. *London Review of Education, 19*(1), Article 1. https://doi.org/10.14324/lre.19.1.01

Ekundayo, D., Udeaja, C., Gyau, K., & Higham, A. (2018). Towards the development of a framework for incorporating sustainability education in the built environment curriculum.

Ertmer, P. A., Conklin, D., & Lewandowski, J. (2001). Increasing Preservice Teachers' Capacity for Technology Integration through Use of Electronic Models. https://eric.ed.gov/?id=ED470081

Faruqe, F., Watkins, R., & Medsker, L. (2022). Competency model approach to AI literacy: Research-based path from initial framework to model. *Advances in Artificial Intelligence and Machine Learning, 02*(04), 580–587. https://doi.org/10.54364/aaiml.2022.1140

Iyer-Raniga, U., & Andamon, M. (2014). Embedding sustainability education in a built environment curriculum, ProSPER.Net: Transforming higher education and creating sustainable societies (pp. 56–87).

Johan, K., & Turan, F. M. (2016). The development of Sustainability Graduate Community (SGC) as a learning pathway for sustainability education-a framework for engineering programmes in Malaysia Technical Universities Network (MTUN). In *IOP conference series: materials science and engineering 160*(1), 012074. IOP Publishing. https://doi.org/10.1088/1757-899X/160/1/012074

Kim, J., Lee, H., & Cho, Y. (2022). Learning design to support student-AI collaboration: Perspectives of leading teachers for AI in education. *Education and Information Technologies, 27*(5), 6069–6104. https://doi.org/10.1007/s10639-021-10831-6

Lee, I., & Perret, B. (2022). Preparing high school teachers to integrate AI methods into STEM classrooms. *Proceedings of the AAAI Conference on Artificial Intelligence, 36*(11), 12783–12791. https://doi.org/10.1609/aaai.v36i11.21557

Li, Y., & Yang, P. (2023). Higher education worries and responses in the era of artificial intelligence. *Education.* https://doi.org/10.54647/education880398

Lim, M. Y., & Tan, K. H. (2018). *A comparative study of digital literacy skills of ESL/EFL students in Malaysia and China.* Atlantis Press.

Liu, H., Lin, C.-H., & Zhang, D. (2017). Pedagogical beliefs and attitudes toward information and communication technology: A survey of teachers of English as a foreign language in China. *Computer Assisted Language Learning, 30*(8), 745–765. https://doi.org/10.1080/09588221.2017.1347572

Mahalingam, A., & Surapathi, A. (2017). Virtual reality in architecture and design: Potential and challenges. In *Handbook of research on recent developments in architecture and the built environment* (pp. 227–241). IGI Global.

Maher, M. L., Simoff, S., Cicognani, A., & Lau, K. H. (2019). VR in urban design and planning: Assessing the impact of visual realism on user's satisfaction and design intuition. *Journal of Urban Technology, 26*(4), 65–86. https://doi.org/10.1080/10630732.2019.1575983

Mak, C. M., & Lu, W. S. (2019). Exploring the use of virtual reality (VR) for architectural design communication. *Automation in Construction, 102*, 82–93. https://doi.org/10.1016/j.autcon.2019.02.015

Masron, T., & Samat, N. (2009). *GIS in school.* Prentice Hall.

Mzuza, M. K., & Van Der Westhuizen, C. P. (2019). Review on the state of GIS application in secondary schools in the southern African region. *South African Geographical, 101*(2), 175–191. https://doi.org/10.1080/03736245.2019.1570321

Nordin, S. (2006). Implementation of GIS in geography education in Sabah. *Environmental Geography and Education*, 109–122.

Opoku, A., & Guthrie, P. (2018). Education for sustainable development in the built environment*International Journal of Construction Education and Research, 14*(1), 1–3. https://doi.org/10.1080/15578771.2018.1418614

Opoku, A., Ekung, S., Kugblenu, G., & Mushtaha, E. S. (2024). Education for sustainable development, the built environment, and the sustainable development goals. In *The Elgar Companion to the Built Environment and the Sustainable Development Goals* (pp. 178–194). Edward Elgar Publishing. https://doi.org/10.4337/9781035300037.00019

PricewaterhouseCoopers Malaysia. (2021). *Digital skills improved since the pandemic, but more needed.* PwC Malaysia. https://www.pwc.com/my/digital-skills-improved

Radford, A. D. (2018). Virtual reality in design education: Defining desirable levels of interactivity. In *Proceedings of the 23rd international conference of the association for computer-aided architectural design research in Asia (CAADRIA)* (Vol. 1, pp. 167–176).

Sadek, M. (2023). Artificial intelligence as a pedagogical tool for architectural education: What does the empirical evidence tell us? *MSA Engineering Journal*, *2*(2), 133–148. https://doi.org/10.21608/msaeng.2023.291867

Samuel, J., Kashyap, R., Samuel, Y., & Pelaez, A. (2022). Adaptive cognitive fit: Artificial intelligence augmented management of information facets and representations. *International Journal of Information Management, 65*, 102505. https://doi.org/10.1016/j.ijinfomgt.2022.102505

Singh, S. S. B., & Singh, B. (2013). Integrating geography information systems in teaching geography in Malaysian secondary smart schools. *Education Journal, 2*(4), 149–154.
Sung, Y. H., Lee, S., & Kim, H. (2019). A study on the effect of VR design communication on user participation: Focused on the moderating effects of design complexity. *Automation in Construction, 101*, 74–86. https://doi.org/10.1016/j.autcon.2019.02.016
Tan, K. H., & Lim, S. Y. (2022). The impact of lockdown measures on the transformation of Malaysian education. *Journal of Education Transformation, 10*(3), 125–140.
Wojtowicz, J., & Butelski, K. (2017). Virtual reality in architectural design education: Defining possibilities. In *Proceedings of the 22nd international conference of the association for computer-aided architectural design research in Asia (CAADRIA)* (Vol. 1, pp. 661–670).
Yao, R. (2013). Sustainability in the Built Environment (pp. 1–22). Springer. https://doi.org/10.1007/978-1-4471-4781-7_1
Zawacki-Richter, O., Marín, V., Bond, M., & Gouverneur, F. (2019). Systematic review of research on artificial intelligence applications in higher education—Where are the educators? *International Journal of Educational Technology in Higher Education, 16*(1), Article 39. https://doi.org/10.1186/s41239-019-0171-0
Zhang, Y., Zhang, H., & Liu, Y. (2020). Collaborative design review in virtual reality for architectural and urban projects. *Frontiers of Architectural Research, 9*(3), 380–392. https://doi.org/10.1016/j.foar.2020.02.003

Najah Md. Alwi is a Senior Lecturer in the Faculty of Built Environment (FABE) at the Universiti Malaya. She has pursued academic and professional endeavours across the UK, New Zealand and Malaysia as a Project Architect and Researcher in Sustainable Design. Her expertise lies in studying social spaces and integrating technology in design while prioritising the needs of individuals and communities.

Nurdiyana Zainal Abidin obtained her Master of Architecture and Doctorate in Architectural Heritage and Conservation from Universiti Teknologi Malaysia. Previously, she was attached to the Center for the Study of Built Environment in the Malay World (KALAM), in UTM and also to the Department of Architecture, Kulliyyah of Architecture, International Islamic University Malaysia. Her research background focuses on architectural heritage, vernacular architecture, culture, and conservation. Her past publications included studies on Asian history, traditional architectural influences, sustainable vernacular timber construction, and Malay cultural studies.

Noor Hashimah Hashim Lim obtained her Ph.D. of Urban and Regional Planning from Universiti Teknologi Malaysia and her MSc Project Management from the National University of Singapore. Her research interest revolves around statistical modelling, examining the intricate interplay between environmental factors and human behaviours, emphasising population health, climate-community resilience, participatory planning, and digital urbanism. Her recent works include developing a virtual community mapping system that captures people-place attachment and engaging underprivileged B40 children in participatory design initiatives.

Liyana Hasnan is a researcher whose work investigates the cultural, historical, and political rhetoric within the discourses that have defined the architecture discipline. She earned her doctorate from Universiti Putra Malaysia. Her recent research examines architectural pedagogy modes, spatial politics, and architecture as an interface. Between 2013 and 2015, she embarked on her master's degree at Columbia University, GSAPP, in the programme Critical, Curatorial, and Conceptual Practices in Architecture (MSc. CCCP).

Chapter 9
Future Directions in Filipino Education: Translanguaging Practices in Online Mathematics Learning Environments

Cluster(s) of Future Studies: Post-Normality and Complexity

Karizza P. Bravo-Sotelo

Abstract Translanguaging is a growing research area that focuses on using several linguistic and semiotic resources for communication. However, translanguaging in online classes remains sparse, and little is known about the affordances and constraints of the technology to students' meaning-making practices in an online environment. This research aims to identify how the online setup facilitates or impedes students' translanguaging practices. Using purposive sampling, data were gathered from class video recordings, chat responses of students, video-stimulated recall, interviews with four mathematics teachers, and focus group discussion responses of 20 grade 8 students from four classes, and were analysed using thematic analysis to identify the affordances and constraints of online classes to students' translanguaging practices. Results show that the online modality facilitates the sharing of answers, communicates and co-solves class-related technical issues and conflates linguistic expressions with other semiotic resources, like emojis and key smashes. Conversely, the online modality presents challenges to students' translanguaging practices. Academically, students' opportunities for recitation, elaboration and problem-solving are limited, while socially, students are unable to do group work and interact with classmates on non-academic related tasks. Recommendations are in place to leverage the benefits and mitigate the challenges of online learning in mathematics to students' translanguaging practices.

Keywords Basic education · Cebuano · Mathematics discourse · Online learning environments · Philippines · Translanguaging practices

Present Address:
K. P. Bravo-Sotelo (✉)
Ateneo de Manila University, Quezon City, the Philippines
e-mail: kbravo-sotelo@ateneo.edu

A. Pandian et al. (eds.), *Future of Education in Asia*,
https://doi.org/10.1007/978-3-032-07285-6_9

9.1 Introduction

At a time when local and global situations are becoming highly volatile, uncertain, complex and ambiguous, having safety nets to ensure learning continuity is a welcome move. In the Philippines, for instance, online and flexible learning modes provide a viable alternative after the pandemic, particularly when onsite learning is disrupted by the aftermath of natural disasters, such as typhoons and earthquakes; extreme weather conditions, like high heat indexes; significant social issues, such as civil unrest, transportation strikes and labour disputes affecting educational institutions, among others. With the increasing use of online synchronous learning, whether as an immediate response to a pandemic or as part of harnessing technological tools for learning, understanding how students' translanguaging practices are impacted in technology-mediated mathematics discourse can be a point of interest for language and applied linguistics researchers as these modes have different dynamics as compared to face-to-face communication, which may compromise communication clarity. Translanguaging is a growing area of research which includes 'the use of multiple linguistic and semiotic resources to make meaning' (Mendoza & Parba, 2018, p. 2). This means that linguistic and non-linguistic resources form part of a speaker's communication and knowledge-generation repertoire and are utilised for meaning-making.

The choice of mathematics for this study stems from the long-standing concern of the Philippine education sector regarding the low performance of students, as evidenced by the results of high-stakes assessments, such as the Trends in International Mathematics and Science Study (TIMSS), TIMSS-Advanced and the *Programme for International Student Assessment* (PISA). As English is the language of assessment in the aforementioned international tests and the medium of instruction in mathematics in the Philippines from grade 4 onwards, this can cause difficulty among students whose mother tongues (MTs) are different from the languages of instruction and assessment (Commission and on Education II (EDCOM II), 2024; Canilao & De Los Reyes, 2024; Bravo, 2017).

The use of technology as a mode to deliver instructional content and how translanguaging is utilised in synchronous online environments is a research area worth exploring. Although the global health crisis has been contained at the time of writing, this study merits attention due to the continuous flux of technology and the pivotal role of technology in education (Gonzalez-Lloret, 2023). Given the unique context of online learning and the limited number of studies on technology-mediated translanguaging practices of students, this study endeavours to identify the mathematics students' communication constraints and affordances in an online context, a topic worthy of scholarly attention and investigation beyond the pandemic. To this end, the research aims to answer the following research question: How does the online environment (a) facilitate and (b) impede the translanguaging practices of mathematics students?

9.2 Literature Review

9.2.1 What is Translanguaging?

Translanguaging is a blanket term that refers to one's integrated linguistic repertoire and other semiotic resources for communication (Cenoz & Gorter, 2020; Mendoza, 2020). It originated from the Welsh word, *trawsieithu,* coined by Cen Williams in the 1980s to describe a pedagogical practice of bilinguals, switching the language of input and the output systematically and deliberately (Lewis et al., 2012). This practice capitalises on the stronger language of the students by providing a more understandable input in order to support the weaker language and create a balanced learning ecology. The English term, as it is known today, was adopted upon the agreement between Williams and Baker (Lewis et al., 2012).

Translanguaging has moved beyond education to include bilinguals' spontaneous language practices (Cenoz & Gorter, 2017) in diverse settings outside the academic context, such as, but not limited to, a karate club, shopping area and advertisements (see Turnbull, 2019; Zhu et al., 2017, 2019), showing its relevance and use in everyday life. Furthermore, current literature on translanguaging (e.g. Donley, 2022; Ho & Tai, 2021; Mendoza & Parba, 2018; Williams, 2022) shows how the definitions have departed from the original intent of dual language model (see García & Li, 2014; Lewis et al., 2012; Baker, 2011; Creese & Blackledge, 2011; García & Sylvan, 2011) to capture the multifaceted nature of communication. While early definitions of translanguaging centred around bilingual speakers' linguistic practices and experiences by shuttling between two languages to mediate meaning, the field has since expanded to include the nuances of multilingual practices that go beyond the use of one language as an input and another language as an output.

Translanguaging involves the use of the entire linguistic resources at one's disposal (García & Li, 2014) and recognises the fluid use of linguistic and non-linguistic cues for communication (Li, 2018; Williams, 2022). As a communication mode that decentres the focus from the use of linguistic resources, translanguaging leverages one's communicative resources to facilitate the interplay of linguistic and semiotic modes to convey messages. Thinking along these lines, translanguaging is defined in this study as a fluid communicative practice using one's semiotic resources for meaning-making. As speakers draw on their entire linguistic repertoire for knowledge construction and negotiation, other semiotic resources may overlay with language resources to complement, supplement or compensate for utterances.

9.2.2 Translanguaging as Multilingual, Multimodal and Multisemiotic Practices

Translanguaging practices in the classroom refer to 'any instance that involved integrative use of two languages, together with different modalities' (Fang & Liu, 2020,

p. 5). The multilingual view of translanguaging refers to an understanding of language as a dynamic and adaptable resource that speakers employ to navigate their complex social and cultural settings. This acknowledges the idea that speakers are not bound by a single linguistic code when expressing ideas but work around their multilingual assets, along with the rich context that goes with them, when they engage in a conversation.

Studies on translanguaging using multilingual resources of students discuss its potential to facilitate content knowledge, bridge language gaps or level the playing field for students (Canilao & De Los Reyes, 2024; De Los Reyes & Bagona, 2022; Perfecto, 2020; Tai & Li, 2020). For students, translanguaging was used to construct explanations, share experiences related to the lesson and interact with classmates during group activities (De Los Reyes & Bagona, 2022). Perfecto (2020) notes that translanguaging practices in multilingual classroom settings encourage the use of language resources for a more in-depth thinking process and meaning negotiation among students. Tai and Li (2020) offer insights into how translanguaging created a co-learning space between the teacher and students, allowing students to use their languages in co-constructing meanings. These studies show the affordances of multilingual resources in employing translanguaging practices.

The role of multimodality in translanguaging is especially highlighted in digital spaces where communication is expressed not only in speech or writing, but also in different modes like visuals, sounds and gestures which simultaneously interact with one another. Adinolfi and Astruc (2017) examined translanguaging in a virtual foreign classroom done in synchronous online English as a Foreign Language spaces, discussing how teachers move between communicative modes to prompt students for non-verbal responses which include the use of technological functions. Similarly, Canals (2021) examined multimodality and translanguaging in language-related episodes in a college-level online classroom, showing that students used English, Spanish and other languages in their interaction, while multimodal elements like gestures are used as complementary practices that enhanced and supplemented online interactions and communication.

The concepts of modality and semiosis are closely intertwined, and while multimodality can explain what modes or resources were used, modes in and of themselves are devoid of meaning as they can only be derived from the context in which the modes operate. Thus, the way to interpret modes is through social semiosis, which is 'the active making of signs in social (inter)actions' (Kress, 2010, p. 54). Individuals with different backgrounds, cultures, beliefs and ideas, who come into a social interaction, actively bring in these distinct characteristics and contribute to shaping communication.

Semiosis intersects with translanguaging as both constructs are part of a social process that involves interaction with other interlocutors in order to shape and reshape meaning. Lin (2018) introduces the concept of trans-semiotising, an analytical view of language that recognises its intersection with other semiotic resources, such as visuals, gestures and body movements, which complement translanguaging. He et al. (2016) explored the translanguaging and trans-semiotising (i.e. use of mathematical symbols, visuals, displays, etc.) practices in a multimodal mathematics presentation using PowerPoint slides. Findings show that language and visual resources complemented seamlessly and contributed to a better understanding of the lecture, especially for members of the audience whose languages differed from that of the professor (He et al., 2016).

This subsection has outlined the multilingual, multimodal and multisemiotic aspects of translanguaging, underscoring the relevance of these constructs in meaning-making and how they add depth to knowledge construction that emphasises the interplay of both linguistic and non-linguistic resources. But while these studies provide insights on either one or two translanguaging practices, no study to date has identified the affordances and constraints of the online environment in the translanguaging practices of grade 8 students in mathematics whose context, to the best of my knowledge, is underexplored.

9.2.3 Online Learning and Communication

Online learning is a learning delivery mode powered by technology, allowing the use of audio and video for synchronous or asynchronous instruction. Originally intended for business communication (Sondak & Sondak, 1995), this modality has been adopted outside business to facilitate communication between people who are connected to the Internet and not in the same physical or geographical space.

In education, the surge in the use of computer-mediated technology was seen during the COVID-19 pandemic, carrying out classes online for learning continuity. Applications like Zoom and Google Meet are among the commonly used software platforms as they have free versions, are user-friendly and are straightforward to use. Moreover, these platforms have learning-related features for synchronous online classes, like screen and video sharing, file transfer, chat, annotation tools, breakout rooms, virtual hand raising, reactions through emojis and captioning that both students and teachers can leverage for extended communication (Correia et al., 2020; Riordan, 2017).

Learning in an online environment provides unique opportunities for students to use multimodal resources in written and visual forms, such as emojis and key smashes, both of which are non-word entities, to support or amplify linguistic or emotive expressions. Offering a perspective on the roles of emojis in the translanguaging practices of online users, Zhang (2018) notes that emojis serve various translanguaging functions to manage communication, express emotions or project identities. Keyboard smash or key smash is another form of non-word expression born out of online communication and interaction. Park (2022), in her study of the linguistic behaviour of key smashes, provides empirical evidence on how key smashes project specific emotions, attitudes or sarcasm.

However, as Gonzalez-Lloret (2023) argues, technology in education, language learning in particular, has yet to reach its full integration and normalisation. While benefits are present, studies on the use of synchronous online classes present challenges of using technology as an alternative to the face-to-face modality (e.g. Ho et al., 2023; Trust & Goodman, 2023; Gherheş et al., 2021; Castelli & Sarvary, 2021). Some of the common challenges identified in the literature are the lack of interaction in online synchronous classes and the limited visual presence of students, both of which pose concerns for communication.

Castelli and Sarvary (2021) interrogated students' reasons for not turning on their cameras during online synchronous classes. This was a concern worth exploring, as the lack of students' facial expressions can influence students' communication and interaction in class. Similarly, Trust and Goodman (2023) studied the students' camera use in online classes and noted the multifaceted reasons for students' reluctance to turn on cameras in online classes, such as 'mirror anxiety' (p. 11), which is the idea of self- consciousness, and 'Zoom gaze' (p. 11), which is the awareness of being watched, preventing them from participating (Trust & Goodman, 2023).

Using a translanguaging view, it is worth considering how students' translanguaging practices in mathematics classes are harnessed or deterred by the online modality, given the unique and dynamic setup it presents.

9.3 Methodology

9.3.1 School Context

In Mindanao, the Philippines, the language in the classroom is guided by the policies set at the national level. Thus, schools in Mindanao adhere to the mother tongue-based multilingual education framework, with kindergarten up to grade 3 pupils using the mother tongues (MTs) (Department of Education, 2012). Filipino and English languages are gradually introduced as subjects in early grades, then become the medium of instruction from grades 4 to 10 in all subjects (Department of Education, 2019).

In other words, the school adopts English as the main medium of instruction in grade 8 mathematics classes (Department of Education, 2012). Considering that many students from Mindanao likely encounter Filipino and English only in an academic setup and use them sparingly in their day-to-day communication (Care & Pavlovic, 2016), it is assumed that mathematics instruction and assessment delivered in English could pose some challenges for grade 8 students.

9.3.2 Research Design and Sampling

This study is part of a larger work on translanguaging practices in the mathematics classroom. Following a qualitative design that allows for an in-depth analysis of specific contexts and cases (Creswell & Creswell, 2018), the researcher used a case study approach (Duff, 2011) to select four classes from a public junior high school in

Cagayan de Oro, Philippines, which affords government-subsidised basic education from grades 7 to 10 and caters to low- to middle-income families. The selection of the school for this study was guided by purposive sampling, which is the targeted and intentional selection of the locale based on identified criteria (Creswell & Creswell, 2018; Miyahara, 2020).

The location was selected due to the multilingual background of the area and its access to the internet, which enabled some of the students to join online synchronous classes during the research period. This is an important consideration as the researcher needed to observe the translanguaging practices in mathematics classrooms. As for the mathematics teachers, they were licensed professional teachers who held bachelor's degrees relevant to mathematics. They were locals from Mindanao speaking two or more languages and were working at the same public school and the same grade level in Cagayan de Oro at the time of the study.

As for the students, they were grade 8 multilingual students from the classes of the participating mathematics teachers. For the FGDs, five to six students were selected from each class based on the following criteria: Students must (1) express willingness to be interviewed and (2) show a high participation level in class throughout the observation period as this would show their ability to express themselves academically.

9.3.3 Data Gathering and Analysis Procedures

In School Year 2021–22, the researcher gathered data from four mathematics teachers and their respective students through 16 class video recordings, recorded chat responses in online classes, four individual semi-structured interview sessions with the teachers, and four focus group discussions (FGDs) with five to six students selected based on certain criteria. Collecting data from multiple sources is important for evidential corroboration, avoiding the potential misinterpretation of data from a singular perspective (Curdt-Christiansen, 2020; Paltridge, 2020). The online data gathering was in step with the context at the time the study was conducted, which limited face-to-face interactions during a pandemic.

For teachers, a semi-structured interview was preferred over other interview methods as it allowed the participant to express their views freely about the topic, but still within the context of the interview questions (Creswell & Creswell, 2018). For students, FGD is the selected interview method to promote a deeper sharing of insights, making them feel more comfortable in expressing ideas and less inhibited from conveying their opinions, unlike in a one-on-one interview (Ho, 2006). The researcher developed the interview protocols and FGD questions, which were checked for content validity and piloted to a comparable group prior to the implementation. The questions were in English, but the researcher used Filipino for follow-up questions. Interviewees and FGD participants were encouraged to speak in their language for ease of expression.

The researcher listened to the recordings twice for transcribing and for checking for accuracy. Transcription codes for class observations were derived from Barnard and McLellan (2014), while interview data were transcribed orthographically. After the classroom observations and interview data had been transcribed, the researcher sent them back to the mathematics teachers and students who participated in the FGDs for member checking and/or correction of transcripts. There was no correction from the research participants, and they agreed that the transcripts were accurate.

Data were analysed thematically using Braun and Clarke's (2006) thematic analysis framework. Thematic analysis is 'the search for and display of relationships in the data' (Noor Azam Haji-Othman, 2005, p. 129). This analysis method was chosen as it places primacy on experience, which is aligned with the study (Ho et al., 2023). Similar studies on translanguaging (e.g. Fang & Liu, 2020; Mendoza, 2020; Nagashima & Lawrence, 2022) have employed thematic analysis to code, organise and analyse salient points drawn out from the data.

9.4 Results and Discussion

Data culled from interviews, classroom observation and chats were organised thematically to present the affordances and constraints of the online platform to students' translanguaging practices.

For interview and FGD extracts, participants' responses were translated, and the English version was presented for brevity. For extracts from class observations, Filipino and Cebuano utterances are in italics, while the English translations, if any, are boldfaced in parentheses. *TMx* refers to the math teacher interviewee, followed by the number coding, and *S* for student.

The mathematics students in the study did not learn in the ideal conditions, but still, the study offers takeaway for applied linguistics researchers, yielding considerable and interesting results on the potential of translanguaging in online classes. The findings offer insights into the instructional and learning challenges of enacting translanguaging in online mathematics classes in view of the inherent spatial and proximal limitations native to virtual environments. These findings are therefore significant and relevant as they advance understanding of the nuanced and multifaceted nature of translanguaging in online classes. However, without being intentional on the translanguaging practices in the classroom, as there are no clear macro- or micro-level policy provisions, or there is just an implicit message that students can use their local language in the classroom, the hegemony of dominant languages may persist. The call is to 'open up spaces for translanguaging in schools' (Rosén, 2017, p. 52) and be intentional in encouraging and using multilingual repertoire in mathematics communication (Sahan & Rose, 2021). Otherwise, vexing views and practices may continue to perpetuate, which can reflect in the imbalanced language use and representation, as well as the lack of co-learning opportunities in the virtual classroom.

9.4.1 *Affordances of the Online Environment to Students' Translanguaging Practices*

The technological affordances of chat provide an alternative space for classroom communication, and data show how students use the chat feature to (1) communicate and co-solve class-related technical issues and (2) conflate linguistic expressions with other semiotic resources. These affordances, however, come with potential tradeoffs if used without clear parameters. The following discussions provide evidence from gathered data from the online classes of grade 8 students.

Communicate and co-solve class-related technical issues

The chat mode allows students to communicate problems to the teacher and help each other solve concerns on their own. Figure 9.1 shows an instance when the teacher was writing a solution on the board, but some students were unable to see it because the camera was misaligned, thereby not completely capturing what the teacher was writing on the board.

In this chat extract, TM1 was solving a math problem on the board, but her camera was not properly adjusted. Hence, lines 1 to 4 are chat messages from four different students indicating that they could not see what TM1 was computing. Line 1 was from a student who chatted in Filipino and English who said, '*Hindi po kita yung* left side' (The left side [of the computation] cannot be seen). Line 2 was from another student who also wrote in Filipino, '*Mula po sa -4 kita*' (Only the one from -4 [negative four] can be seen], telling about the part of the frame that can be seen on the student's screen. Line 3, which was written in Cebuano and English, says, '*Ma'am dili nako makita ang* full solution' (Ma'am, I cannot see the full solution), while line 4, echoing the same sentiment, says, '*Maam dili makita sa* solution' (Ma'am, the solution cannot be seen). All these chat messages confirmed that there was a viewing issue on some students' screens.

Line 1

Line 2

Line 3

Line 4

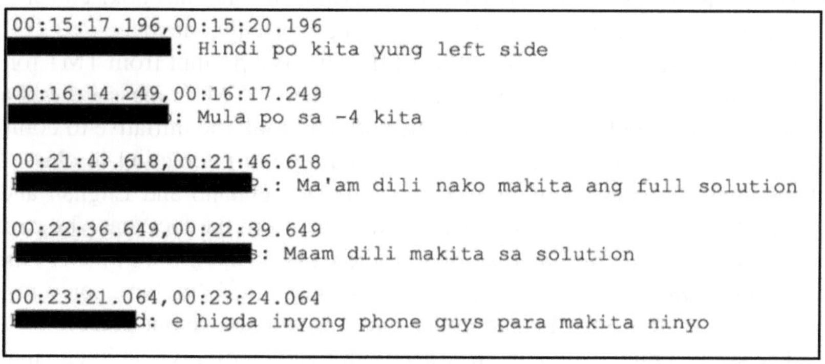
00:15:17.196,00:15:20.196
[redacted]: Hindi po kita yung left side

00:16:14.249,00:16:17.249
[redacted]: Mula po sa -4 kita

00:21:43.618,00:21:46.618
[redacted].: Ma'am dili nako makita ang full solution

00:22:36.649,00:22:39.649
[redacted]: Maam dili makita sa solution

00:23:21.064,00:23:24.064
[redacted]: e higda inyong phone guys para makita ninyo

Fig. 9.1 Students communicate via chat about a technical issue

```
00:37:51.765,00:37:54.765
[redacted]: write your answers here

00:40:59.140,00:41:02.140
[redacted] Ahh hi maa'm [redacted] ah maa'm sorry man hinay ako connection tas na late pa gyud maa'm saba kay sa amoa☹
```

Fig. 9.2 Sad emoji of a student

As TM1 was continuously solving on the board and did not immediately see the chat of the students, one of the students who noticed the classmates' dilemma provided an instruction in the chat. In Line 5, the student chatted using Cebuano, English and Filipino, '*e higda iyong phone guys para makita ninyo*,' (You adjust your phone to landscape orientation, guys, for you to see it), allowing the student to assist classmates who seemed to be having difficulties seeing the entire board work.

Figure 9.1 shows that the chat feature afforded immediate communication of problems and solutions without disrupting the class. In the absence of TM1's response, students were able to co-solve technical issues in languages they can all understand using a feature of the online modality accessible to them. Regardless of the use of Cebuano, Filipino or English, the students were able to communicate and self-manage through translanguaging.

Conflate linguistic expressions with multisemiotic resources

The online environment provides a unique opportunity for students to use multisemiotic resources, such as emojis or key smash, both of which are non-word entities, to support or amplify linguistic expressions. Using a translanguaging view, it is worth considering the added value of these semiotic resources as they intersect with linguistic resources. In doing so, samples from the chat and the interviews are presented to provide support for the conflation of text and non-text elements in online classroom discourse.

Figure 9.2 shows the use of a sad emoji (☹) at the end of a text-based explanation, which could be indicative of the student's attempt to support the written message. TM1 was asking students to type their answers in the chat box, but one student, instead of writing an answer, explained his/her current context.

Figure 9.2 shows that the first line of the chat is a prompt from TM1 for students to type in their answers: 'write your answers here'. TM1's message was a general statement intended for all students, but a student took the initiative to communicate the context that deterred them from responding and interacting in class. In line 2 of the chat, the student apologised in a mix of Cebuano and English and offered information that his/her internet was slow, apart from the fact that s/he came in late in their online class and it was noisy in their place: '*maa'm (sic) sorry man hinay ako connection tas na late pa gyud saba kay amoa*☹' (ma'am, sorry my internet connection is slow, then I came in late and it is so noisy in our place).

The student punctuated the message with a sad emoji (☹), an image that denotes sadness, which could appear as a way to support the message that the student is remorseful, and by extension, an implied expression of the student's inability to respond to TM1's question. This corroborates previous studies (Cramer et al., 2016;

Fig. 9.3 Thumbs up in support of 'okay'

```
00:06:45.614,00:06:48.614
[redacted]s: Ok ma'am

00:08:40.407,00:08:43.407
[redacted]s: 👍👍👍
```

Hsieh & Tseng, 2017), noting that emojis, which can represent human emotions, are used before and after a textual message to complement a message.

In the next example, Fig. 9.3 shows that a student used a thumbs-up emoji to complement a linguistic expression. A student responded to TM3's verbal message using the written mode, 'okay ma'am', then after a few seconds, the same student added three thumbs-up emojis, which is a symbolic way of saying 'okay'.

Considering the context, the emojis appear to be the equivalent of the conventionalised emblem gesture (McNeill, 2005), and the additional symbols appear to reinforce the student's written message in the earlier part of the chat. In the Philippines, thumbs up, whether as a physical gesture or an emoji, is a contextually acceptable way of expressing agreement.

Data from teachers' and students' interviews and FGD recognise that the use of emojis is a way to express emotions, show affinity or maintain formality. For a student, she said that she uses emoji in their online class 'to communicate [her] feelings'. Many students do not turn on their videos during online classes to avoid excessive mobile data consumption, resulting in scarce non-verbal expressions. As a proxy for facial expressions, emojis are used as an add-on to written messages to support or complement a statement. On the other hand, TM1 said that she avoids using emojis as it makes her look like their '*barkada*', a Filipino term that means 'friend': 'I do not use emojis because it feels like I'm their *barkada* (friend), so I'm not using it.'

The message implies that not using emojis is a way to maintain a formal stance in online interaction, sending a message regarding the interpersonal boundaries to be observed between the teacher and the students. Despite holding classes online, TM1's response reflects a view of the classroom as a social context in which role relationships are often clearly defined (Course, 2014). Hence, TM1 does not want students to misconstrue her message by using emojis and to make them interact with her casually. What is common in both statements is that emojis are meaning carriers that convey one's emotions, complement a message or amplify the meaning of expressions. By implication, if emojis are used indiscriminately, these can cross the formality-informality divide in academic discourse and may send mixed messages to online recipients.

Another example of the conflation of word and non-word expressions to send a message is using a key smash. Figure 9.4 shows a case of a student who used the Google Meet chat to make incoherent expressions and even write expletives in Cebuano, which, presumably, the teacher could have missed due to the absence of the teacher's spoken and written feedback. A key smash was used to preface a student's seeming frustration, which was eventually supported by relevant linguistic expressions.

```
        01:04:35.153,01:04:38.153
Line 1  [redacted]: Hgcuisiaihsjjjzz sa ka hqbjaiinwkvkakjkijrjajfieifiit

        01:04:35.170,01:04:38.170
        [redacted]: Bwhcjaioenbeajgiirjaiohvjaokhkakj was

Line 2  01:04:35.840,01:04:38.840
        [redacted]: Bhchisic

        01:05:12.759,01:05:15.759
Line 3  [redacted]: Answer nako kay ajdiwidiwijdsjjdjriedifi

        01:06:22.301,01:06:25.301
        [redacted]: Animal yawa

Line 4  01:08:21.335,01:08:24.335
        [redacted]: Wla ko kasabot hahah animal
```

Fig. 9.4 'Wla ko kasabot…'

In Fig. 9.4, the student did not use conventional words to express his thoughts and emotions, but typed random letters on the keyboard within split-second intervals, creating 'noise' in the online class, which could be equivalent to the noise from chatters in a face-to-face synchronous class. Lines 1 to 3 show that the student did a key smash in the chat. These were some random letter combinations that did not make any sense. In line 4, some understandable linguistic expressions were created, the first three of which, 'Answer *nako kay…*' which means in English, 'My answer is…', followed by a non-word keysmash, 'ajdiwidiwijdsjjdjriedifi', implying that the student had no sensible answer to the teacher's question. Almost a minute later, in line 5, the student typed in English and Cebuano, '*animal yawa* (devil or demon)' (free translation in Filipino: *Hayop na demonyo ito*), an expletive with no direct translation in English which roughly means, 'What an animal this devil is.'

Line 6 supports the assumption that the key smash could be indicative of the student's frustration when s/he expressed that the ongoing math discourse was vague by saying in Cebuano, '*Wla ko kasabot hahah animal*', which means, 'I do not understand anything hahah animal'. The 'hahah', an expression of laughter in technology-mediated communication, could be a way to soften his/her expression of frustration for being unable to follow the lesson, something akin to awkward laughter. Lastly, punctuating the line with 'animal,' a subtle cuss expression in Cebuano, could be a way of ending his discourse on a 'cool kid' note.

This analysis acknowledges that, without adequate data from the student, there is insufficient evidence to conclude that the Cebuano expletives '*animal yawa*' were directed to any particular person in their online class. Profane words are linguistic expressions that manifest one's creativity to convey emotions through language (Horan, 2013). In this case, it could be a general expression of frustration, as supported by the keysmashes in the first three lines and the expression, '*Wla ko kasabot*' (I do not understand anything). From the translanguaging perspective, what this dataset shows is that in an online discourse, linguistic and non-linguistic features can intersect and contribute to meaning-making, like an expression of difficulty or frustration, and by implication, teachers can use them as warning signs for students who might be having difficulties in class.

Consider, however, that teachers and students in a regular classroom setup abide by formal and informal classroom norms to maintain order and manage students. In an online class, students might take advantage of the unconventional setup and available tools to make unrestricted expressions. Students may unfortunately deviate from classroom norms, and expressions deemed unacceptable in a regular face-to-face class may slip through the cracks, allowing students to get away with it inadvertently. Thus, a note of caution on the findings is that while the chat provides an alternative space for students, this does not warrant any deviant expressions or demonstration of aggressive behaviours that go against acceptable academic codes and norms. Emojis and key smashes must be treated with caution as these can send mixed messages that confound any communicative endeavour.

Students are able to answer and express their ideas through the conflation of words with non-word resources to amplify a message. Similarly, emojis and key smashes give the idea that these non-linguistic symbols unique to online classes can communicate positive and negative affect, whether students are sad, happy, anxious or confident in a math class, which teachers can take as additional cues to navigate online classroom dynamics.

These findings suggest that the chat function in mathematics online classes provides an avenue to democratise recitation and affords a unique opportunity for students to use translanguaging through the interplay of multilingual, multimodal and multisemiotic resources, contributing to the growing understanding of translanguaging that goes beyond the use of linguistic resources. Though the findings may have just scratched the surface, these are significant as they shed light to some communicative practices native to online learning. However, the results raise awareness of the potential tradeoffs of these online affordances if used in the classroom without any restrictions, leading to the indiscriminate use of online tools and the compromise of the clarity of messages.

9.4.2 Constraints of the Online Environment on Students' Translanguaging Practices

The online environment, while beneficial to an extent, can be a deterrent to fully expressing grade 8 students' ideas in mathematics classes. Excerpts from the classes, interviews and FGDs show that students' translanguaging practices are not fully optimised in an academic setup. The online setup (1) limits opportunities for recitation, (2) limits opportunities to elaborate answers and (3) provides fewer opportunities for problem-solving.

Limited opportunities for recitation

Most of the students expressed that the online classes limited their opportunity to recite, and this was due to the following reasons: recitation (1) requires students to play an active role in classroom discourse, (2) induces fear and anxiety among students and (3) risks losing face.

In the FGDs, though some students said that they had more chances to speak in online classes, some students were unable to do so as this required an active role of muting and unmuting oneself, or of 'raising hand', a technological affordance of online platforms akin to the physical action of hand raising. Unlike in conventional face-to-face classes where students have to simply raise their hands and teachers will immediately see them, the online setup requires students to click on the 'raise hand' button, or any similarly labelled function and unmute themselves to be heard. Being called to recite was not immediate as this would depend on whether the teacher would see on the screen the raised hand or not.

This was echoed in the FGD responses of TM2, TM3 and TM4's students, and an extract from TM4's student elaborated on this concern, explaining that students do not get 'picked easily' by the teacher because their virtual hands might not have been seen on the screen;

> S: When you answer, ma'am, [in a face-to-face class] you raise your hand, but in an online class, if you want to answer, you won't be picked easily because you are not seen.

Thus, students might find it convenient not to recite anymore and simply listen to the online class instead of taking an active part in the discussion. This becomes a concern for the mathematics teachers because they would be left doing most of the talking. TM2 commented on this perceived imbalanced discourse dynamics, noting that it is as though the teacher is doing a 'monologue'.

Another reason why students recite less in a mathematics class, especially in an online environment, is because it induces fear and anxiety. Although language speaking anxiety phenomenon is not new in education, communication and applied linguistics research (see He, 2013, 2017; Ho et al., 2023), anxiety in mathematics learning in a virtual environment is a learning context worth exploring due to the interplay of language, content and technology.

The prescribed language to be used in mathematics classes in the Philippines is English. In the context of the study, as most of the students speak Cebuano as their mother tongue, some of them were anxious to express their ideas in English. Others expressed their fear of committing grammatical mistakes when they recite.

This was also echoed by TM2, observing that students would make calculated moves when reciting due to their fear of committing grammar mistakes;

> TM2: You know, in class, they are competing a bit. They are afraid to make [a] grammar mistake.

Although all the interviewed mathematics teachers said that while EMI is the prescription of the Department of Education in delivering instruction, there is no strict imposition of the use of English for recitation. However, students have an implicit understanding that English should be used to respond during recitations. Two students from two different classes express the same sentiments regarding the non-use of their local language in the class as they would appropriate the language of their reply with the language used by their mathematics teacher for questioning.

> S3: Because English is the language used by the teacher to ask, it is weird to answer in Bisaya. It is also possible that the meaning of your response would be different if expressed in Bisaya. (TM1 student)
>
> S2: Usually, ma'am, the teachers talk in English, so we appropriate our response and use English. It is embarrassing to respond in Bisaya when the teacher speaks in English. (TM4 student)

As students bring in various linguistic resources in class, English becomes the lingua franca of the observed multilingual mathematics classrooms. However, all the interviewed mathematics teachers understand this linguistic barrier in the online classroom, so they allow their students to recite using their mother tongue or any linguistic resource. Both mathematics teachers and students recognise this unwritten agreement on language choice in the classroom.

The observed mathematics teachers provide a translanguaging space for students, albeit unofficially, to use their available linguistic resources. This is similar to Fang and Liu's (2020) observation regarding the implicit translanguaging provisions in EFL classroom towards the achievement of pedagogical goals. All interviewed mathematics teachers agree that their classroom practices consider and recognise the rich multilingual background of their students, and while classroom language provisions are not overtly articulated, mathematics teachers allow students to express using their mother tongue or any available language, for that matter.

The allowance for the students' multilingual resources comes with a caveat that if what is used is not a shared language between the teacher and the student, the student or someone from the class may provide translation assistance, as TM2 expressed: '*Tapos* during sa class discussion, *ina*-accept *ko din po* but I need them to translate *dahil hindi po ako naka-* intindi' (Then during the class discussion, I accept [any language] too, but I need them to translate because I do not understand their language). Similar points were raised by TM4. If a student from a Tagalog-speaking area cannot use Cebuano, one is free to speak in Tagalog since this is a common language understood by teachers and students alike: 'I allow my students to use their language because math *ang tinuturo ko* (is what I'm teaching), so I'm not strict on the language.'

Alternatively, if the mother tongue of the student can neither be understood by the teacher or any of the students, they need to find a common ground for linguistic expressions, like Tagalog or Filipino, as TM3 mentioned: 'If I cannot understand what you are talking about, I will just tell you that 'maybe we can find another language which we can use that both of us could understand, like Tagalog or English'.

However, since there is no classroom norm or expression that invites students to use their language resources in the class, students' idea expression may be impeded by misconceptions about what is a 'valid' or 'invalid' language in mathematics classroom discourse. Hence, students may find recitation challenging and may choose not to participate in class discussions. This is similar to Heller's (1996) observation that social and institutional factors shape the perspective of a legitimate language in classroom interaction. Without explicit identification and a clear understanding of linguistic roles in the classroom, some languages may be favoured while other linguistic resources may not be explicitly fully optimised, which may result in limited interaction in classroom discourse.

As an implication, mathematics teachers can consider providing translanguaging spaces and subspaces that promote linguistic inclusivity where clear and specific provisions on language use in mathematics classroom are explicitly mentioned. Certain classroom agreements can be made between content teachers and students in order to negotiate classroom discourse norms. While the observed mathematics teachers show consideration regarding the use of students' linguistic resources and are not rigidly conforming to the prescribed EMI policy in consideration of the language ecology of the classroom where the majority of the students do not have a strong command of the English language, it will benefit the students if there is a mutual understanding on harnessing multilingual resources in the classroom. This is an opportunity to communicate the mathematics teacher's flexibility regarding classroom discourse, and to make students aware of the linguistic allowance given to them so that language may not be a major impediment or threat to meaning expression.

Apart from the language of communication in the classroom, not having a strong grasp of mathematical concepts can trigger anxiety among students during recitation. Mathematics, being a language on its own (O'Halloran, 1998; Schleppegrell, 2007), can be challenging for students as it often requires dealing with abstract ideas. TM2 observed that her students' ability to speak well in English does not automatically translate to the ability to explain mathematical concepts. Thus, while some students might be articulate in the English language, which is used in grade 8 mathematics, they might have second thoughts answering the teacher's question due to one's unfamiliarity with the mathematical language.

TM2 gave 'transpose' as an example of a commonly misused term in mathematics, explaining that the acceptable mathematical term is 'add to the additive inverse of negative one.' This example is expressed in O'Halloran's (2015) description of mathematical language as a specialised language that requires mathematical background knowledge. Thus, students' unfamiliarity with the mathematical expressions might be a deterrent for meaning expression and expanded discourse, and might be an added reason for anxiety. This corroborates the findings of Bravo (2017), noting that tertiary students' anxiety brought about by their inadequate linguistic and mathematical competencies prevents them from participating in mathematics discourses.

Another trigger for students' recitation anxiety is the online class setup itself. Some FGD participants from different classes said that they were not comfortable reciting as everyone would hear their voices loud and clear through a headset. These concerns are communicated by FGD participants from different classes.

S1 (TM1): Then when answering, I am shy to answer.

S2 (TM2): Everyone is on mute and listening.

S4 (TM2): Because one speaks and the others are muted.

Alternatively, if students choose not to speak, they would resort to using chat, as expressed by TM4's student.

S3: If you don't want to open your mic, you can chat [your answer]. (TM4 student).

Talking in an online environment inadvertently invites other online participants, who become a captive audience, to focus on the interlocutor's message, and so speakers may see this unwanted attention as a threat. As TM3 expressed, 'Students are afraid to answer when there are a lot of students listening'. Thus, in order to not invite undue attention to themselves, students in online classes would choose to speak only when necessary (i.e. when the teacher calls them for recitation), or would choose to be passive attendees in an online class, especially those students who lack the confidence either to explain mathematical concepts or to express their ideas in English fluidly.

Lastly, one reason that limits students from reciting is the risk of losing face. Students resort to silence, which is a face-saving mechanism for those who might be unable to answer the teachers' questions. This is consistent with the findings of Ho et al. (2023), noting that giving the wrong answer can put students at risk of losing face. Other students thought that they would be silently judged by their classmates, and so they would rather not recite voluntarily. Thus, assuming that students heard the teacher call their names, silence can be a fallback option to avoid recitation. In this way, the online setup can conveniently allow students to not answer by ignoring the teacher's call or by saying that their internet connection is unstable. The teacher, meanwhile, can only do so much and can only call the student once or twice before deciding to make the next discourse move.

This subsection provided empirical evidence on how the online setup limits opportunities for students' recitation in mathematics classes. These findings suggest that internal and external barriers add to the complexity of online learning and contribute to the inability of students to fully utilise their translanguaging and meaning-making capacities. Students are hindered from reciting as it requires students to play an active role in classroom discourse, induces fear and anxiety or risks losing face. Having an understanding of these concerns can help mathematics teachers identify and mitigate linguistic and socio-cognitive burdens, and foster an environment that encourages students to participate in classroom discourse using their available multilingual, multimodal and multisemiotic resources.

Limited opportunities to elaborate answers

Mathematics teachers in the observed classes made some conscious effort to engage students in classroom discourse and ask them questions just as how they would do it in face-to-face classes. In the observations, however, it was noticeable that some students who managed to recite had short and limited responses, thus language use was not elaborate. As close-ended questions were used by teachers in exchanges, students' responses were restricted to yes–no answers.

This is similar with the findings of Bravo (2017), where tertiary students from math classes would simply respond with one-line answers instead of elaborating since most of the questions did not require them to provide in-depth answers and engage in extended classroom discourse. This is evident in online classes where most of the talking was done by the teachers. Hence, students' translanguaging practices in online spaces were not fully optimised.

From the logistic viewpoint, asking close-ended questions makes the online interaction more manageable and efficient, allowing teachers to manage online class time well and to ensure brevity of responses. Unlike in regular face-to-face classes where some classes can go overtime, the online classes had to be facilitated efficiently as some students might get disconnected due to mobile data issues and limitations.

> TM2: I am conscious of the time. I'll be in a hurry to finish the lesson so that the student's mobile load won't run out. (TM2 interview)

However, the downside of asking closed-ended questions is the limited opportunity to elaborate answers, hence, students' translanguaging practices were not maximised. This can be seen in the number of close-ended questions as opposed to open-ended ones asked by four mathematics teachers. Close-ended questions are the ones that can be answered by predetermined options (i.e. yes/no, agree/disagree or multiple-choice items) or that can generate one- to two-word responses (Course, 2014). Using the IRF pattern (Sinclair & Coulthard, 2002) to analyse classroom discourse and identify the initiated questions that the mathematics teachers made in all of their classes, it can be seen that the cumulative number of closed-ended questions (n = 580) of all four math teachers in all four classes was more than the open-ended ones (n = 143) (see Table 9.1).

Some questions do not require elaboration because only numerical values are needed. In Table 9.1, extract from TM3's class shows an exchange between the teacher and students where students are asked to carry out basic arithmetic and analytical procedures and supply information. See Extract from TM3's class (Class 1, 11:25–12:26).

1 TM3: So in here, the sum of the lengths of any two sides is greater than the length of the third side. So example, determine the following measures of the sides of the triangle. Okay number 1, the sides are 10 cm—let us assume that is centimetres—8 < cm > and 10 cm. So *atong buhaton ana* (what we'll do there) is that we will add the first and second. So *unsa man ang* (what are the) first and second sides?
2 Ss: 10 *ug* (and) 8
3 TM3: Okay. So 10 plus 8, *pila* (how many)?

Table 9.1 Number of open-ended and closed-ended questions in all observed classes

Math teacher	Open-ended questions	Closed-ended questions	Total
TM1	7	63	70
TM2	87	176	263
TM3	18	217	235
TM4	31	124	155
Total	143	580	723

4 Ss: 18
5 TM3: Okay, 18. 18 is greater than 10, yes or no?
6 S4: Yes
7 TM3: Okay so wala pa na siya ha kay ato pa na siya i-try kay tulo mani sila kabuok (it does not end there because there are three sides). So correct. How about si 8 plus 10?
8 S5: Mao ra gihapon ma'am. (It's the same, ma'am.)
9 TM3: Parehas ra gihapon (It's the same). Tapos (Then) 18 is greater than 10, correct?
10 Ss: Yes
11 TM3: How about kaning ika-tulo (the third one), 10 plus 10?
12 Ss: 20
13 TM3: Is it < 20 > greater than 8?
14 S6: Yes ma'am
15 Therefore, since check siya lahat dahil greater than (all of them satisfy the greater-than rule), we can now make a triangle. Kung (If) greater than ma'am, makacreate siya ug (it can create a) triangle, pero (but) if less than ma'am, we cannot create triangle. Nasabtan ra sya? (Do you understand?) Meaning to say, dili sya makahimo ug triangle (it cannot create a triangle). Okay, so mao na sya *ang* explanation (that is the explanation). Yes, the given numbers are the measures of the sides of a triangle since it satisfies the inequalities.

In discussing the triangle inequality theorem, which states that 'the sum of any two sides of a triangle is greater than or equal to the third side' (Hosch, 2023), TM3 engaged the students in proving that the sum of any two sides was greater or equal to the third side by making them add the values of the triangle's sides. Notice, however, that the synthesis (line 15) came from the teacher, and the students' discourse responses (lines 2, 4, 6, 8, 10, 12, 14) were limited to providing the sum of added values. Opportunities for elaboration and critical thinking were scarce as questions were close-ended.

The inclination of all four mathematics teachers towards asking more closed-ended questions resonates with the interview response of TM2 regarding the need to fast-track the lesson as the students might not have sufficient mobile data to stay throughout the online class. As a result, students did not have ample opportunities to use their linguistic resources to explain how they came up with their answers.

Fewer opportunities for problem-solving

Face-to-face mathematics classes normally involve board work, allowing teachers and students to present their solutions on the board and explain the process that led to their answers. All interviewed math teachers expressed the importance of board work for mathematics teachers and students, with TM2 and TM4 elaborating on the importance of board work in class as 'it can help explain well' and encourage idea exploration through 'trial and error'.

Consequently, the absence of a physical board made students rely on teachers' problem-solving, making them passive learners in mathematics classes. Since problem-solving on a physical board was impossible, students just watched how the teacher would solve the math problem, which was unnatural for a process-oriented subject that requires the computational and visual representation of solutions. As O'Halloran (1998) notes, mathematics discourse requires the interplay of semiotic resources, such as mathematical symbols (i.e. equations), linguistic resources (i.e. verbal descriptions) and visual cues (i.e. charts and diagrams).

Furthermore, the lack of board work was restricting for students as they did not allow real-time computations and there were limited opportunities to verbally communicate the step-by-step process of solving a mathematical problem. Given that visual cues, such as writing the solution, complement verbal explanations when solving mathematics problems and reinforce what is being said (O'Halloran, 1998), the lack of board computations restricts students' learning and communicative opportunities in mathematics classes.

Kress (2010) argues that 'learning is best seen in the frame of a learner's principled transformative engagement with the characteristics of a prior prompt in terms of the learner's interest'. Engagement and interaction are pivotal to meaningful learning. As Ho et al. (2023) present in their findings, both written and oral communication in the classroom are indicators of meaningful learning. Considering that students take no active role in problem-solving in mathematics classes, this begs the question if any learning is taking place.

9.5 Future Directions

Understanding the affordances and constraints of the online modality to mathematics learning can provide opportunities to harness the translanguaging practices of students and mitigate factors that can restrict students from fully optimising multilingual and multimodal resources. This study is clearly significant as it provides empirical data that adds to the existing knowledge on the complexities and intricacies of mathematics discourses in online classes.

The translanguaging practices presented in this study show how technology mediates or inhibits meaning-creation, particularly how these semiotic resources and new technologies may support or impinge on content instruction and pedagogy. It also presents how the lack of use of the local language in online classes can disrupt students' learning, especially since online classes are becoming a viable alternative to face-to-face classroom instruction in the Philippines and in the Asian context.

This paper highlights that the online environment provides students with unique opportunities to translanguage as they share their answers, communicate and co-solve class-related technical issues and conflate linguistic expressions with other semiotic resources. In addition to verbally expressing their answers and using their multilingual resources, students utilise the chat feature to communicate linguistically and non-linguistically, as in the case of using emojis and key smashes. While these expressions have inherent value in showing one's emotions and supporting messages, they have to be nuanced and tempered in classroom discourse as they can send mixed messages if used indiscriminately.

Nonetheless, the use of emojis and key smashes as part of the students' translanguaging practices is a research area that has not been fully explored yet. This study provides insights into how these semiotic resources prevalently utilised in the online realm are part and parcel of meaning negotiation and expression in digital spaces, adding to existing studies (Murniati, 2023; Sindoni, 2023) that identify emojis as meaning and emotion carriers that transcend linguistic boundaries (Hopkyns & Dovchin, 2024).

On the other hand, the online environment can deter students from accomplishing certain academic tasks or engaging in social activities, which, to an extent, limits them from using their multilingual and multimodal resources. The learning ecology native to the online environment can inhibit students from reciting, elaborating their answers and doing board work. Similarly, collaborative activities done in a face-to-face setup cannot be fully enacted in online mathematics classes due to teachers' and students' technical limitations, thus depriving students of the chance to co-create knowledge or negotiate meaning with their peers. Also, the lack of use of the local language in online classrooms impedes learning, thus disrupting learning, as seen in the classroom discourses and interview responses which validate the dominance of English in mathematics instruction (Dovchin & Wang, 2024).

This case study, while it has its own benefits, has limitations in terms of the generalisability of data. The findings are unique to the context in which the study was done. Nonetheless, the case study makes it relatable to other case studies that have similar contexts. Employing an ethnographic approach can yield more insights through immersion into the classroom culture.

Furthermore, as this study is qualitative in orientation, future research can look into the quantitative aspect of translanguaging. The descriptive nature of the study can only go as far as identifying the translanguaging practices of grade 8 mathematics teachers and students and the influence of online platforms on teachers' and students' communicative practices. Future studies can do a quantitative or a mixed-methods approach that investigates teachers' and students' attitudes towards translanguaging, the correlation of translanguaging with other learning variables or its effects on students' mathematics performance.

There is also a potential to make comparative cross-country studies of the translanguaging practices in online contexts in different countries in Asia or around the world, which can allow researchers and educators to identify the intersectionality and variances of translanguaging practices in different contexts.

9.6 Conclusion

While the world has reeled from the aftermath of the pandemic and classes have returned to in-person meetings as of late, online learning remains a strong and feasible alternative to face-to-face classes globally. With the integration of digital tools and resources in the classroom, technology use to support pedagogical goals is ever-relevant. As technology-driven digital learning has become a mainstay in the education sector, especially with the adoption of hybrid classes, there is a call to leverage technologies, especially for low-income households, to continue learning amidst emergencies.

This can ensure that learning remains uninterrupted and context-sensitive. Hence, these new learning tools and spaces, even extending to potential sophisticated technological tools that will be invented in the future, invite researchers and educators to reimagine how co-construction of knowledge can be promoted in a virtual environment, as well as to optimise communicative practices that will be beneficial for learners.

The findings recognise that the online environment offers advantages and poses disadvantages to grade 8 students' translanguaging in mathematics classes, and understanding these affordances and constraints can allow mathematics teachers who use the online modality to reflect on their teaching practices, leverage the technological affordances and create classroom provisions that can promote inclusivity in meaning expression among students through translanguaging.

Highlights

- The online learning environment facilitates the use of multiple linguistic and semiotic resources, such as emojis and chat responses, to enhance communication and collaborative problem-solving in mathematics classes.
- Online mathematics classes pose challenges to students' translanguaging practices, particularly due to limited opportunities for recitation and elaboration of answers, hindering active engagement.
- The absence of face-to-face interaction in online classes restricts students' ability to work collaboratively and engage in spontaneous, non-academic communication, limiting social and academic discourse.
- Future research should focus on optimising the integration of translanguaging practices in online mathematics classrooms, addressing both technological affordances and constraints to promote more inclusive learning environments.

9.7 Highlights

References

Adinolfi, L., & Astruc, L. (2017). An exploratory study of translanguaging practices in an online beginner-level foreign language classroom. *Language Learning in Higher Education, 7*(1). https://doi.org/10.1515/cercles-2017-0008

Baker, C. (2011). *Foundations of bilingual education and bilingualism* (5th ed.). Multilingual Matters.
Barnard, R., & McLellan, J. (Eds.). (2014). *Codeswitching in university English-medium classes: Asian perspectives*. Multilingual Matters.
Braun, V., & Clarke, V. (2006). Using thematic analysis in psychology. *Qualitative Research in Psychology, 3*(2), 77–101. https://doi.org/10.1191/1478088706qp063oa
Bravo, K. (2017). *Codeswitching use of tertiary level math teachers and students' class participation* [Unpublished master's thesis]. University of the Philippines Diliman.
Canals, L. (2021). Multimodality and translanguaging in negotiation of meaning. *Foreign Language Annals, 54*, 647–670. https://doi.org/10.1111/flan.12547
Canilao MLE, De Los Reyes R (2024) *Language practices in Philippine education and other public spaces*. Springer
Care, E., & Pavlovic, M. (2016). *Longitudinal study of learning achievement of students in the Autonomous Region in Muslim Mindanao* (Policy Forum Report 2). Assessment Curriculum and Technology Research Centre. https://actrc.files.wordpress.com/2017/01/executive-summary_policy-forum-report-2-oct20161.pdf
Castelli, F., & Sarvary, M. (2021). Why students do not turn on their video cameras during online classes and an equitable and inclusive plan to encourage them to do so. *Academic Practice in Ecology and Evolution, 11*, 3565–3576. https://doi.org/10.1002/ece3.7123
Cenoz, J., & Gorter, D. (2017). Minority languages and sustainable translanguaging: Threat or opportunity? *Journal of Multilingual and Multicultural Development, 38*(10), 901–912. https://doi.org/10.1080/01434632.2017.1284855
Cenoz, J., & Gorter, D. (2020). Teaching English through pedagogical translanguaging. *World Englishes, 39*(2), 1–12. https://doi.org/10.1111/weng.12462
Congressional Commission on Education II (EDCOM II). (2024). *Miseducation: The failed system of Philippine education, EDCOM II year one report*. Second Congressional Commission on Education.
Correia, A.-P., Liu, C., & Xu, F. (2020). Evaluating videoconferencing systems for the quality of the educational experience. *Distance Education*, 1–24. https://doi.org/10.1080/01587919.2020.1821
Course, S. (2014). ELT students' use of teacher questions in peer teaching. *Procedia - Social and Behavioral Sciences, 158*, 331–336. https://doi.org/10.1016/j.sbspro.2014.12.096
Cramer, H., de Juan, P., & Tetreault, J. (2016). Sender-intended functions of emojis in US messaging.In *Proceedings of the 18th international conference on human-computer interaction with mobile devices and services—mobile HCI'16*. https://doi.org/10.1145/2935334.2935370
Creese, A., & Blackledge, A. (2011). Separate and flexible bilingualism in complementary schools: Multiple language practices in interrelationship. *Journal of Pragmatics, 43*, 1196–1208.
Creswell, J. W., & Creswell, J. D. (2018). *Research design: Qualitative, quantitative, and mixed methods approaches* (5th ed.). SAGE Publications, Inc.
Curdt-Christiansen, X. L. (2020). Observations and field notes: Recording lived experiences. In J. McKinley & H. Rose (Eds.), *The Routledge handbook of research methods in applied linguistics* (pp. 336–347). Routledge.
De Los Reyes, R. A., & Bagona, E. (2022). Translanguaging in teaching and learning science in selected multilingual elementary classrooms in the Philippines. *Journal of Language, Identity & Education*, 1–15. https://doi.org/10.1080/15348458.2022.2070848
Department of Education. (2012). *Policy guidelines on the implementation of Grades 1–10 of the K to 12 basic education curriculum (BEC) effective school year 2012–2013*. https://www.deped.gov.ph/wp-content/uploads/2012/04/DO_s2012_31.pdf
Department of Education. (2019). *Policy guidelines on the K to 12 basic education program*. https://www.deped.gov.ph/wp-content/uploads/2019/08/DO_s2019_021.pdf

Donley, K. (2022). Translanguaging as a theory, pedagogy, and qualitative research methodology. *NABE Journal of Research and Practice, 12*(6), 1–16. https://doi.org/10.1080/26390043.2022.2079391

Dovchin, S., & Wang, M. (2024). The resistance to translanguaging, spontaneous translanguagers and native speaker saviorism. *Critical Inquiry in Language Studies, 21*, 429–446. https://doi.org/10.1080/15427587.2024.2336469

Duff, P. (2011). How to carry out case study research. In A. Mackey & S. Gass (Eds.), *Research methods in second language acquisition: A practical guide* (1st ed., pp. 95–116). Wiley-Blackwell.

Fang, F., & Liu, Y. (2020). 'Using all English is not always meaningful': Stakeholders' perspectives on the use of and attitudes towards translanguaging at a Chinese university. *Lingua, 247*, 102959. https://doi.org/10.1016/j.lingua.2020.102959

García, O., & Li, W. (2014). *Translanguaging: Language, bilingualism and education.* Palgrave Macmillan.

García, O., & Sylvan, C. E. (2011). Pedagogies and practices in multilingual classrooms: Singularities in pluralities. *The Modern Language Journal, 95*, 385–400.

Gherheş, V., Şimon, S., & Para, I. (2021). Analysing students' reasons for keeping their webcams on or off during online classes. *Sustainability, 13*(6), 1–13. https://doi.org/10.3390/su13063203

Gonzalez-Lloret, M. (2023). The road system travelled: Five decades of technology in language education. *System.* https://doi.org/10.1016/j.system.2023.103124

He, D. (2013). What makes learners anxious while speaking English: A comparative study of the perceptions held by university students and teachers in China. *Educational Studies, 39*(3), 338–350.

He, D. (2017). How to cope with foreign language speaking anxiety effectively? The case of university students in China. *Electronic Journal of Foreign Language Teaching, 14*(2), 159–174.

He, P., Lai, H., & Lin, A. M. Y. (2016). Translanguaging in a multimodal mathematics presentation. In C. M. Mazak & K. S. Carroll (Eds.), *Translanguaging in higher education: Beyond monolingual ideologies* (pp. 91–120). Multilingual Matters. https://doi.org/10.21832/9781783096657-008

Heller, M. (1996). Legitimate language in a multilingual school. *Linguistics in Education, 8*(2), 139–157. https://doi.org/10.1016/s0898-5898(96)90011-x

Ho, J., & Tai, K. W. H. (2021). Translanguaging in digital learning: The making of translanguaging spaces in online English teaching videos. *International Journal of Bilingual Education and Bilingualism*, 1–22. https://doi.org/10.1080/13670050.2021.2001427

Ho, D. G. E., Sa'adi, M., He, D., & Hoon, C. Y. (2023). Silence over the wire: Student verbal participation and the virtual classroom in the digital era. *Asia Pacific Education Review.* https://doi.org/10.1007/s12564-023-09834-4

Ho, D. G. E. (2006). The focus group interview: Rising to the challenge in qualitative research methodology. *Australian Review of Applied Linguistics, 29*(1), 5.1–5.19. https://doi.org/10.2104/aral0605

Hopkyns, S., & Dovchin, S. (2024). Translanguaging and emotionality of English as a second language (ESL) teachers. *International Review of Applied Linguistics in Language Teaching, 62*(3), 1257–1278. https://doi.org/10.1515/iral-2024-0094

Horan, G. (2013). 'You taught me language; and my profit on't/Is, I know how to curse': Cursing and swearing in foreign language learning. *Language and Intercultural Communication, 13*(3), 283–297. https://doi.org/10.1080/14708477.2013.804533

Hosch, W. L. (2023, June 7). Triangle inequality. *Encyclopedia Britannica.* https://www.britannica.com/science/triangle-inequality

Hsieh, S. H., & Tseng, T. H. (2017). Playfulness in mobile instant messaging: Examining the influence of emoticons and text messaging on social interaction. *Computers in Human Behavior, 69*, 405–414. https://doi.org/10.1016/j.chb.2016.12.052

Kress, G. (2010). *Multimodality: A social semiotic approach to contemporary communication.* Routledge.

Lewis, G., Jones, B., & Baker, C. (2012). Translanguaging: Origins and development from school to street and beyond. *Educational Research and Evaluation, 18*(7), 641–654. https://doi.org/10.1080/13803611.2012.718488

Li, W. (2018). Translanguaging as a practical theory of language. *Applied Linguistics, 39*(1), 9–30. https://doi.org/10.1093/applin/amx039

Lin, A. M. Y. (2018). Theories of trans/languaging and trans-semiotizing: Implications for content-based education classrooms. *International Journal of Bilingual Education and Bilingualism.* https://doi.org/10.1080/13670050.2018.1515175

McNeill, D. (2005). *Gesture and thought.* University of Chicago Press.

Mendoza, A., & Parba, J. (2018). Thwarted: Relinquishing educator beliefs to understand translanguaging from learners' point of view. *International Journal of Multilingualism, 16*(3), 270–285. https://doi.org/10.1080/14790718.2018.1441843

Mendoza, A. (2020). *A linguistic ethnography of laissez faire translanguaging in two high school English classes* [Doctoral dissertation]. University of Hawai'i at Mānoa]. http://hdl.handle.net/10125/69070

Miyahara, M. (2020). Sampling: Problematizing the issue. In J. McKinley & H. Rose (Eds.), *The Routledge handbook of research methods in applied linguistics* (pp. 52–62). Routledge.

Murniati. (2023). The reasons for using emojis on WhatsApp chats: A case study of Indonesian students' translingual practice. *Journal of English Language and Culture, 13*(2), 105–113.

Nagashima, Y., & Lawrence, L. (2022). To translanguage or not to translanguage: Ideology, practice, and intersectional identities. *Applied Linguistics Review.* https://doi.org/10.1515/applirev-2019-0040

Noor Azam Haji-Othman. (2005). *Linguistic diversity in Negara Brunei Darussalam: An ecological perspective* [Doctoral dissertation]. University of Leicester. https://leicester.figshare.com/account/articles/10131284

O'Halloran, K. L. (1998). Classroom discourse in mathematics: A multisemiotic analysis. *Linguistics and Education, 10*(3), 359–388. https://doi.org/10.1016/S0898-5898(99)00013-3

O'Halloran, K. L. (2015). The language of learning mathematics: A multimodal perspective. *The Journal of Mathematical Behavior, 40*(Part A), 63–74. https://doi.org/10.1016/j.jmathb.2014.09.002

Paltridge, B. (2020). Multi-perspective research. In J. McKinley & H. Rose (Eds.), *The Routledge handbook of research methods in applied linguistics* (pp. 29–88). Routledge.

Park, A. (2022). *On the linguistic behavior of keysmashes.* UC San Diego: San Diego Linguistic Papers. https://escholarship.org/uc/item/8cc2z9nx

Perfecto, M. (2020). English language teaching and bridging in mother tongue-based multilingual education. *International Journal of Multilingualism*, 1–17. https://doi.org/10.1080/14790718.2020.1716771

Riordan, M. (2017). Emojis as tools for emotion work: Communicating affect in text messages. *Journal of Language and Social Psychology, 36*(5), 549–567. https://doi.org/10.1177/0261927x17704238

Rosén, J. (2017). Spaces for translanguaging in Swedish education policy. In B. Paulsrud, J. Rosén, B. Straszer, & A. Wedin (Eds.), *New perspectives on translanguaging and education* (pp. 38–55). Multilingual Matters.

Sahan, K., & Rose, H. (2021). Problematising the E in EMI: Translanguaging as a pedagogic alternative to English-only hegemony in university contexts. In B. Paulsrud, Z. Tian, & J. Toth (Eds.), *English-medium instruction and translanguaging* (pp. 1–14). Multilingual Matters. https://doi.org/10.21832/PAULSR7321

Schleppegrell, M. (2007). The linguistic challenges of mathematics teaching and learning: A research review. *Reading and Writing Quarterly, 23*, 139–159.

Sinclair, J., & Coulthard, M. (2002). Towards an analysis of discourse. In M. Coulthard (Ed.), *Advances in spoken discourse* (pp. 1–34). Routledge.

Sindoni, M. G. (2023). *Multimodality and translanguaging in video interactions.* Cambridge University Press.

Sondak, N., & Sondak, E. (1995). Video conferencing: The next wave for international business communication. In *Proceedings of the annual conference on languages and communication for world business and the professions* (pp. 1–10). https://files.eric.ed.gov/fulltext/ED386781.pdf

Tai, K. W. H., & Li, W. (2020). Co-learning in Hong Kong English medium instruction mathematics secondary classrooms: A translanguaging perspective. *Language and Education*, 1–27. https://doi.org/10.1080/09500782.2020.1837860

Trust, T., & Goodman, L. (2023). Cameras optional? Examining student camera use from a learner-centered perspective. *TechTrends*. https://doi.org/10.1007/s11528-023-00855-9

Turnbull, B. (2019). Beyond bilingualism in Japan: Examining the translingual trends of a "monolingual" nation. *International Journal of Bilingualism, 24*(4), 634–650. https://doi.org/10.1177/1367006919873428

Williams, M. (2022). *A case study of emergent bilinguals meaning-making during multimodal science lessons in a bilingual primary school* [Doctoral dissertation]. Curtin University

Zhang, L. (2018). *The relationship between online translanguaging practices and Chinese teenagers' self-identities*. [Doctoral dissertation], University of Leeds. https://etheses.whiterose.ac.uk/22538/1/Thesis_LZ.pdf

Zhu, H., Li, W., & Lyons, A. (2017). Polish shop(ping) as translanguaging space. *Social Semiotics, 27*(4), 411–433. https://doi.org/10.1080/10350330.2017.1334390

Zhu, H., Li, W., & Jankowicz-Pytel, D. (2019). Translanguaging and embodied teaching and learning: Lessons from a multilingual karate club in London. *International Journal of Bilingual Education and Bilingualism*. https://doi.org/10.1080/13670050.2019.1599811

Karizza Bravo-Sotelo holds a PhD in Applied Linguistics from Universiti Brunei Darussalam. Her scholarly works are published in peer-reviewed journals like Asian Englishes and the International Journal of Multilingualism. Currently, she is an Assistant Professor at Ateneo de Manila University, the Philippines. Her current research interests include code-switching, translanguaging, and multilingual education. Karizza is a research fellow under the Tagpuan: Ateneo Center for Dialogue, Research, and Collaboration.

Chapter 10
Shaping the Future of Higher Education in Asia: Impact of Hybrid Learning Models on Student Outcomes in Vietnam, Thailand and China

Cluster(s) of Future Studies: Technological Trends, Past and Futures

Mai Phuc Thinh

Abstract This chapter presents an empirical investigation into the impact of hybrid courses on student learning in Vietnamese higher education, examining how the integration of digital and traditional learning environments influences educational outcomes. It further contextualises Vietnam's experience through comparative insights from Thailand and China, offering a broader regional perspective on the implementation of hybrid learning models across Asia. The primary aim is to assess the effectiveness, scalability and challenges of blended learning within diverse policy and infrastructural landscapes. Utilising qualitative methods, semi-structured interviews, focus groups and document analysis captures rich narratives from both students and faculty, providing a nuanced understanding of hybrid learning's influence on educational access, student engagement and digital competency development. Key findings indicate that hybrid models enhance flexibility, support learner autonomy and build critical digital skills, though persistent challenges remain, including infrastructural disparities and uneven levels of digital literacy. By comparing Vietnam's centralised digital transformation strategy with Thailand's decentralised model and China's high-tech, surveillance-oriented approach, this chapter identifies context-sensitive pathways for scaling hybrid learning. It concludes by proposing policy and pedagogical recommendations focused on equitable digital access, curricular innovation and cross-institutional collaboration. These insights contribute to the development of more resilient, inclusive and adaptive higher education systems across Asia.

Keywords Digital transformation · Hybrid learning models · Educational technology integration · Higher education · Blended learning

M. P. Thinh (✉)
College of Education, Can Tho University, Can Tho, Vietnam
e-mail: phucthinhpq@gmail.com

A. Pandian et al. (eds.), *Future of Education in Asia*,
https://doi.org/10.1007/978-3-032-07285-6_10

10.1 Introduction

The adoption of hybrid courses in Vietnamese higher education signifies a pivotal evolution towards a more integrative learning experience, blending traditional pedagogical methods with digital advancements. This transition aligns with global educational trends, particularly in Asia, where there is increasing acceptance and implementation of hybrid learning models due to their flexibility and accessibility. For instance, countries such as China, India and South Korea have significantly invested in hybrid learning infrastructures, recognising the need to cater to diverse learner needs and the growing demand for lifelong learning opportunities (Bordoloi et al., 2021; Wang et al., 2021; Zaugg et al., 2021).

Globally, the educational landscape is undergoing rapid transformation driven by technological advancements and changing pedagogical paradigms. Hybrid learning, which combines face-to-face instruction with online components, has emerged as a significant trend in higher education. This model leverages the strengths of both traditional and digital learning environments, aiming to enhance student engagement, improve learning outcomes and provide greater accessibility to education (Anthonysamy et al., 2020; Müller & Mildenberger, 2021; Nayar & Koul, 2020). The COVID-19 pandemic has further accelerated the adoption of hybrid learning, highlighting its potential to ensure continuity in education during disruptions (Hill & Reimer, 2022; Sato et al., 2023).

In the context of Asia, the integration of hybrid learning is particularly noteworthy. The region is home to a diverse and rapidly growing student population, necessitating innovative educational solutions that can address varying needs and preferences. Countries such as Singapore and Malaysia have made significant strides in adopting hybrid learning, supported by strong governmental policies and investments in educational technology (Mustapha et al., 2022; Zaugg et al., 2021). These initiatives reflect a broader regional commitment to modernising education systems and preparing students for the demands of the twenty-first century.

Despite these advancements, the implementation and impact of hybrid courses on student learning and engagement remain underexplored within the Vietnamese context, presenting a critical gap in both scholarly literature and practical application. Vietnam, with its unique socio-economic dynamics and rapid technological adoption, presents an ideal context for investigating the efficacy of hybrid learning models. The country is undergoing significant educational reforms aimed at improving access to and the quality of higher education, making it a pertinent case study for examining the integration of hybrid learning approaches (Tang & Tien, 2020).

Vietnam's higher education system is in the midst of a strategic digital transformation—a comprehensive shift that involves integrating digital technologies into all aspects of teaching, learning, administration and governance—as the government recognises the critical role of digital tools in fostering a competitive, knowledge-driven economy. With a growing population and a rapidly expanding economy, Vietnam is well-positioned to leverage digital transformation for broader socio-economic impact. Policies such as the *National Digital Transformation Programme*

(Decision No. 749/QD-TTg, 2020) and the *Strengthening the Application of Information Technology and Digital Transformation in Education and Training Project* (Decision No. 131/QD-TTg, 2022) (Prime Minister of Vietnam, 2020, 2022) reflect a deliberate approach to embedding digital skills and infrastructure across educational institutions. These policies aim to establish digital learning environments as a routine part of teaching and learning, ensuring that a minimum of 50 per cent of students and educators have access to and engage with digital platforms by 2025. Furthermore, the government has set ambitious targets to achieve full integration of digital administration systems by the same year, with all educational institutions managing resources, records and student data digitally on a nationwide scale.

Despite these initiatives, the digital transformation of higher education in Vietnam faces significant challenges. First, there is a pressing need for a cohesive legal and regulatory framework to support digital practices across institutions, particularly as universities increasingly adopt online and blended learning models, a pedagogical approach that combines face-to-face classroom instruction with online learning components to create more flexible, engaging and personalised educational experiences. Policies must be flexible enough to incorporate emerging technologies, yet robust enough to uphold quality standards and protect data security. Second, the shift to digital education requires a profound change in mindset and an increase in management capacity among academic and administrative staff. School leaders are tasked with driving this transformation, which demands innovative management practices and an openness to new pedagogical models (Pitt et al., 2022; Tri & Hoang, 2023).

Infrastructure remains another critical hurdle, especially in remote and under-resourced regions. Stable and high-speed internet access is essential for digital learning, yet connectivity issues persist in many areas, limiting equitable access to online resources. Alongside internet infrastructure, institutions require compatible and integrated digital platforms to support all aspects of university operations, from classroom management to administrative tasks. Furthermore, the success of digital transformation hinges on the skills and adaptability of educators, who must acquire new competencies to effectively engage students in a digital setting. Teachers need support to develop digital teaching skills that enable them to create interactive and inclusive online learning experiences.

The Vietnamese government's commitment to digital transformation in higher education reflects a forward-looking strategy aimed at enhancing educational quality and fostering socio-economic development. However, for these ambitious goals to be realised, there must be a sustained investment in infrastructure, legal frameworks and human capital development, especially in rural and underserved areas. This research aims to fill this void by evaluating the effectiveness of hybrid learning models in enhancing student outcomes in Vietnamese higher education institutions, while also drawing comparative insights from Thailand and China. By examining how these three countries approach digital transformation in education through centralised, decentralised and technology-intensive models, respectively, this chapter highlights both shared and context-specific challenges and innovations.

Through a qualitative inquiry involving interviews, focus groups and document analysis, the study uncovers the perceptions, challenges and opportunities presented by hybrid courses from the perspectives of both educators and students in Vietnam. By situating this investigation within the broader discourse on educational innovation and technology-enhanced learning, and by comparing policy and implementation approaches in Thailand and China, the study not only contributes to the theoretical understanding of blended learning's impact but also offers practical insights for policymakers and institutional stakeholders across the region.

The necessity of this research is underscored by the transformative potential of hybrid courses to redefine educational paradigms, fostering a more engaged, personalised and effective learning experience. Hybrid learning environments facilitate active engagement with the material, allowing students to build understanding through exploration, experimentation and reflection (Islam et al., 2022; Raes, 2022). Additionally, hybrid models can bridge gaps in educational access, particularly for students in remote or underserved areas, by providing flexible learning opportunities that are not bound by geographical constraints (Raes, 2022; Sato et al., 2023). Comparative discussions with Thailand and China further contextualise Vietnam's progress, offering cross-national lessons for inclusive and scalable educational transformation in Asia.

10.2 Literature Review

10.2.1 Theoretical Foundations of Blended Learning

Blended learning represents a pedagogical strategy that integrates traditional face-to-face classroom methods with online learning activities, aiming to provide a more flexible, engaging and effective educational experience (Armellini & Rodriguez, 2021; Demillo, 2019; Garrison & Vaughan, 2008). At its core, the theoretical foundations of blended learning are anchored in several key educational theories that support its pedagogical strategies.

Constructivism, a theory that posits learners construct knowledge through their experiences rather than passively receiving information, plays a pivotal role in blended learning environments. This approach encourages active engagement with the material, allowing students to build understanding through exploration, experimentation and reflection. Blended learning environments facilitate these constructivist principles by providing diverse resources and activities online, alongside traditional classroom interactions, enabling learners to control their learning pace and path (Tarnopolsky, 2012).

Social Learning Theory, proposed by Bandura and Walters (1977), emphasises the importance of observational learning, imitation and modelling. Blended learning leverages these concepts through collaborative activities, discussion forums and peer reviews, promoting knowledge acquisition through social interaction. This theory underscores the significance of social presence in the learning environment, which blended learning fosters by enabling interactions both in-person and online.

The Community of Inquiry (CoI) framework further supports blended learning by highlighting the critical dimensions of social presence, cognitive presence and teaching presence in creating meaningful online learning experiences (Garrison & Arbaugh, 2007). Blended learning environments strive to balance these elements, ensuring that learners engage deeply with the content, participate actively in the community and benefit from effective guidance and facilitation (De Vriendt, 2015).

Active learning, a key principle in blended learning, aligns with these theories by prioritising student engagement, critical thinking and problem-solving activities over passive content consumption. Blended learning environments often incorporate simulations, case studies and interactive assignments to promote active learning, enhancing student outcomes by making learning more relevant and engaging (Armellini & Rodriguez, 2021).

10.2.2 Global Practices and Trends in Blended Learning

Globally, blended learning has been adopted across various educational settings, demonstrating its adaptability to different contexts and disciplines. In countries around the world, institutions have implemented blended learning to cater to diverse learner needs, leveraging technology to complement traditional teaching methods.

Studies and reports have illustrated the global landscape of blended learning, revealing trends in its application and highlighting its flexibility. For instance, in higher education, blended learning has been used to enrich coursework with digital resources, foster collaboration through online forums and accommodate flexible scheduling (Apandi & Raman, 2020; Castro, 2019; Zaugg et al., 2021). In K-12 settings, it has supported personalised learning paths and provided supplementary resources to enhance understanding (Li & Wang, 2022).

The variability in implementation strategies reflects the adaptability of blended learning, accommodating local educational cultures, technological infrastructure and pedagogical goals. This diversity underscores the model's potential to transform education globally, offering insights into effective practices and challenges encountered in different regions.

10.2.3 Opportunities and Impact of Blended Learning Implementation

Implementing blended learning presents a range of logistical, technological and pedagogical challenges. Institutions must navigate issues related to infrastructure, access to technology and digital literacy, ensuring that both instructors and students are equipped to engage effectively with blended formats. Pedagogically, designing coherent and integrated learning experiences that leverage the strengths of both online and face-to-face components requires careful planning and expertise (Boelens et al., 2017; Rasheed et al., 2020).

However, blended learning also offers significant opportunities to overcome traditional educational barriers. By facilitating accessibility through online components, it can reach diverse learner populations, including those with scheduling constraints or geographical limitations. Personalisation and scalability are further advantages, with technology enabling tailored learning experiences and the efficient delivery of education to large numbers of students (Al-Samarraie & Saeed, 2018; Anthonysamy et al., 2020; Castro, 2019).

Empirical studies on blended learning have yielded valuable insights into its effects on student outcomes, including academic achievement, engagement and satisfaction. Research generally indicates positive outcomes, with blended learning associated with higher levels of student engagement and satisfaction compared to traditional formats (Fisher et al., 2021).

Academic achievement under blended learning has also shown promising results, with some studies reporting improvements in grades and understanding (Kazu & Demirkol, 2014; Müller & Mildenberger, 2021). However, the evidence is not uniformly positive, with variations across contexts and disciplines (Xu et al., 2023). These inconsistencies suggest the importance of careful design and implementation, highlighting factors such as the quality of online materials, the integration of online and face-to-face components and the role of instructor guidance.

Despite these promising trends, gaps in the literature exist, particularly regarding long-term outcomes and the comparative effectiveness of different blended learning models. Further research is needed to understand the conditions under which blended learning most effectively enhances student outcomes.

10.3 Methodology

This study adopted a qualitative research design to explore the efficacy and impact of hybrid courses in Vietnamese higher education. The qualitative approach was chosen for its strength in uncovering deep insights into participants' perceptions, experiences and the contextual factors influencing the implementation and outcomes of hybrid learning. Through semi-structured interviews, focus groups and document analysis, the research aimed to capture a comprehensive understanding of the complexities and nuances associated with hybrid learning environments.

10.3.1 Participants and Sampling

The study engaged a purposive sample of participants from four universities in Vietnam, including three public universities and one private university. A total of eighteen students were selected for semi-structured interviews, ensuring a balanced representation in terms of gender, academic disciplines and year of study. Additionally, two focus groups were conducted with twelve lecturers who had experience designing and teaching hybrid courses across the four universities. This sample size provides a comprehensive representation and strengthens the validity of the findings.

Given the qualitative nature of this study, the chosen sample size is appropriate for generating in-depth insights and understanding the contextual dynamics of hybrid learning within Vietnamese higher education. This approach allows for the exploration of rich, detailed narratives that can uncover underlying themes and patterns relevant to the local context. Although the study is situated within Vietnam, the methodological framework and insights gained have broader implications that can inform similar educational contexts across Asia. By addressing specific local challenges and opportunities, this research provides a foundational case study that can be adapted and expanded in future studies within the region.

The purposive sampling strategy aimed to select participants who could provide rich, relevant and diverse perspectives on the research questions. This approach ensures that the study captures a wide range of experiences and insights, which are crucial for understanding the multifaceted nature of hybrid learning.

10.3.2 Data Collection Methods

Data collection involved three primary methods: semi-structured interviews, focus groups and document analysis. The semi-structured interviews with students were designed to explore their experiences, perceptions and outcomes related to participating in hybrid courses. The interview guide included open-ended questions to allow for in-depth responses and follow-up queries tailored to each participant's experiences. This method ensures that the data collected is both comprehensive and detailed, providing valuable insights into the student perspective.

The focus groups with lecturers were conducted to gather collective insights into the pedagogical strategies, challenges and perceived benefits of hybrid learning. The discussions were structured around themes such as course design, student engagement, assessment methods and technological integration. Conducting focus groups with a larger number of lecturers allowed for a richer and more diverse set of data, reflecting a broader range of experiences and viewpoints.

Document analysis supplemented these methods, involving the review of course syllabi, teaching materials and institutional policies related to hybrid learning. This triangulation of data sources enriched the study's findings by providing multiple perspectives on the implementation and impact of hybrid courses. The use of document analysis helps to contextualise the interview and focus group data, adding depth to the overall understanding of the hybrid learning environment.

10.3.3 Data Analysis

Thematic analysis (Clarke et al., 2015) was employed to analyse the data collected from interviews, focus groups and documents. This involved a systematic process of coding the data inductively to identify patterns, themes and categories related to the research questions. The analysis was iterative, moving back and forth between the dataset and the developing themes to ensure a comprehensive and nuanced understanding of the data.

Thematic analysis is particularly well-suited for qualitative research as it allows for flexibility in identifying themes and patterns within the data. By employing this method, the study was able to uncover deep insights into the participants' experiences and perceptions, providing a robust foundation for the research findings.

10.3.4 Ethical Considerations

Ethical approval for the study was obtained from the research ethics boards of the participating universities, ensuring that the study adhered to ethical guidelines throughout its execution. Informed consent was sought from all participants, ensuring they were fully aware of the study's purpose, the confidentiality of their responses, and their right to withdraw at any time without penalty. The study adhered to ethical guidelines by anonymising participant information and securely storing data to protect privacy and confidentiality. Additionally, care was taken to address potential power dynamics, especially in the focus group discussions, to ensure a respectful and inclusive environment for all participants. By maintaining these ethical standards, the study ensured the integrity and reliability of its findings.

10.4 Findings

10.4.1 Perceptions of Hybrid Learning

The hybrid learning model, which merges online and traditional classroom teaching, has generated a broad spectrum of insights from both students and faculty, reflecting its impact on the educational landscape. Students have highlighted the flexibility and accessibility of hybrid learning as one of its most valued features, allowing them to tailor their educational pursuits around personal and professional commitments. One student remarked, 'Being able to access lectures and resources at any time has revolutionised my approach to learning. It makes it more adaptable and less constrained by traditional schedules'.

The use of digital tools and platforms in hybrid learning has been praised for enhancing the educational experience, introducing students to innovative methods of understanding complex subjects. This integration of technology into learning processes enriches students' academic journeys, offering them new perspectives and methodologies that were previously inaccessible in a purely traditional setting. A student noted, 'Before COVID-19, we hardly had any online classes. Now, with 30% of our courses online, I've learned to use various digital tools that make studying more engaging'.

Additionally, the hybrid model fosters diverse interaction opportunities, expanding communication and collaboration beyond the confines of the classroom. Students appreciate this broader engagement with peers and instructors, which is facilitated through a combination of online forums and face-to-face discussions, thus enriching the educational experience by fostering a more inclusive and interactive learning environment. 'The online discussions have been great for getting to know classmates I wouldn't have interacted with otherwise', commented another student.

10.4.1.1 Student Engagement and Responsibility

Hybrid learning combines various modes of learning, including in-class, online and individual self-study, which requires students to be proactive and responsible for completing their learning tasks. If students do not complete their assignments, the online management system records this and marks them as not meeting the requirements. The ability to adjust the pace of self-study to fit personal needs is a significant advantage. A student mentioned, 'The online learning system allows me to study at my own pace, fitting my study schedule around my personal commitments, which is highly beneficial'.

Assignments in hybrid learning can be shared publicly for peer review and collaborative learning, enhancing accountability and encouraging serious efforts in individual and group assignments. Through hybrid learning, students also develop essential soft skills such as communication, critical thinking, collaboration, time management, planning and listening. Compared to fully online or face-to-face courses, hybrid learning provides a more comprehensive learning experience by allowing students to discuss lessons in class, participate in online forums and practice what they have learned. This approach helps students grasp knowledge more firmly and receive consistent support from instructors and peers. Studies in Vietnam have shown that students in hybrid learning are more engaged compared to traditional learning models, due to the diverse activities and personalised attention they receive.

10.4.1.2 Faculty Perspectives and Challenges

From the perspective of faculty members, hybrid learning presents both opportunities and challenges. Educators have embraced the model as a chance to innovate teaching methods, incorporating more interactive and engaging content to captivate students' interests. This shift has led to an observed increase in student engagement, notably among those who typically remain less vocal in a conventional classroom setting. 'I've noticed that students who were usually quiet in class are more active in online discussions', a lecturer observed.

However, the transition to hybrid teaching has not been without its difficulties. Adapting teaching strategies to suit both online and offline environments requires significant effort and adaptation, a process that some faculty members have found initially daunting. 'Balancing online and offline teaching methods has been a steep learning curve', another lecturer mentioned. Hybrid learning demands a significant change in lesson planning, often requiring faculty to overhaul their traditional teaching plans to achieve the best results. The use of technology helps faculty manage student activities more effectively, with automated assignment delivery and submission tracking.

Hybrid learning also enhances the faculty's technological proficiency, driving them to continuously improve their teaching methods and support for students. For example, online tools can track how often students view materials, the time spent on videos and can provide questions for self-study. Initially, some faculty may hesitate due to their technology skills, but hybrid learning challenges them to seek training and learn from experts, ultimately increasing their technological competency and commitment to innovative teaching.

10.4.1.3 Technology and Infrastructure

Hybrid learning offers several advantages, such as meeting credit-based training requirements by reducing in-class hours, objective assessment through online quizzes and peer evaluation and cost savings (less need for classroom space and in-person attendance). However, effective implementation requires proper technological tools like personal computers, internet access and online learning software.

In Vietnam, many university students are equipped with personal computers for their studies, but first-year students might find it challenging to meet these technological requirements. Furthermore, internet quality can vary, and not all classroom areas may be equipped with the necessary infrastructure. Both faculty and students might experience psychological pressure compared to traditional face-to-face courses, with faculty spending more time and effort designing blended teaching plans.

10.4.2 Practical Applications and Innovations

The integration of practical applications and innovations in education through blended learning models represents a significant evolution in how universities approach teaching and learning. This transformation, particularly evident in the context of Vietnamese higher education, showcases the adaptability and forward-thinking of institutions in response to the digital age's demands. The following discussion highlights key examples of such innovative practices, illustrating their impact on both teaching methodologies and student engagement.

Blended learning, which combines traditional face-to-face instruction with online learning, has been successfully implemented across several universities. This model not only accommodates the diverse learning preferences of students but also prepares them for the digital workforce. For instance, one university introduced a blended learning model with a 70% focus on traditional methods and 30% on e-learning for specific subjects like computer science and foreign languages. This approach enabled a flexible learning environment, allowing students to access learning materials and participate in online discussions outside the traditional classroom setting.

10.4.2.1 Face-to-Face Model

The Face-to-Face model involves traditional classroom settings where students receive direct instruction from their teachers and engage with digital tools. This model is particularly effective in classes with diverse levels of student ability, providing a structured and interactive learning environment.

10.4.2.2 Rotation Model

Another variant of blended learning, the Rotation model, was adopted for subjects such as finance, banking, accounting, auditing, management and marketing. This model facilitated a seamless integration of online and in-person sessions, which were determined by the instructor based on the specific needs of each course. Such flexibility not only catered to different learning styles but also encouraged students to engage with the material on a deeper level. 'The rotation model has allowed us to balance the best of both worlds', commented a lecturer.

10.4.2.3 Flex Model

The Flex model predominantly delivers content through digital platforms, allowing students to engage in online learning with direct guidance and discussion from instructors. This model is ideal for students with time constraints or those balancing education with work commitments.

10.4.2.4 Labs Model

In the Labs model, students participate in online programmes while being physically present in specialised computer labs. This model optimises physical resources and teaching personnel, as assistants manage the labs and address student queries.

10.4.2.5 Self-Blend Model

The Self-Blend model allows students to complement their traditional classroom learning with online courses that are outside the standard curriculum. This model encourages students to independently explore digital resources, enhancing their knowledge and skills. It is particularly suited for university-level education, promoting self-directed learning and professional development.

10.4.2.6 Online Driver Model

The Online Driver model enables students to complete their courses entirely online, with interactions between students and instructors occurring through digital platforms. This model provides maximum flexibility, accommodating students facing geographical or logistical challenges.

10.4.3 Evaluating Practical Effectiveness of the Blended Learning Models

To evaluate the practical effectiveness of the Blended Learning model in universities, the teaching and learning levels need to be considered. This includes interaction skills, discussion assignments and learning outcomes. For example, the time spent on online platforms like learning management systems (LMS) indicates active student engagement, often peaking in the evenings as students complete assignments.

- Time Flexibility: This can be fixed or flexible depending on the course requirements, allowing students to manage their schedules effectively.
- Role of Learners: This encompasses participation levels, self-study, group work and guided learning. Effective blended learning models encourage active participation and self-directed learning, enhancing the overall educational experience.
- Role of Instructors: This involves guiding or not guiding learners, designing online lessons and managing online classrooms. Instructors play a crucial role in creating engaging learning environments and supporting student learning through well-designed online resources.
- Student Support Activities: These include the diversity of learning materials, availability of academic advisors, technological infrastructure and software tools.

Robust support systems ensure that students have access to necessary resources and assistance to succeed in a blended learning environment.

The practical benefits of these innovative educational models became particularly evident during the COVID-19 pandemic. Universities with established online learning platforms and digital resources were able to quickly adapt to fully online teaching, minimising disruptions to students' education. For example, one university with over a decade of experience in online education was able to transition 20% of its courses to a blended format, combining direct classroom instruction with online learning. This proactive approach ensured continuity in education during lockdowns, demonstrating the resilience and flexibility of blended learning models.

Moreover, the adoption of blended learning has facilitated access to a wealth of digital resources, including electronic library systems and global databases such as Elsevier, Emerald and the World Bank. This access has significantly enriched the learning experience, enabling students to explore a broader range of academic materials and engage in self-directed learning. 'Having access to a wide range of digital resources has been incredibly beneficial', noted a student.

The shift towards blended learning models has also had a positive impact on teaching staff. With the creation of digital course materials, instructors have been able to reduce redundancy in their teaching, focusing their efforts on guiding discussions and facilitating hands-on activities that reinforce core concepts. This shift has not only improved the efficiency of teaching but also enhanced the quality of student learning outcomes. 'Creating digital course materials has allowed me to spend more time on interactive activities in class', shared a lecturer.

By integrating these six models of blended learning, Vietnamese higher education institutions have demonstrated a comprehensive approach to modernising their teaching practices, making education more flexible, engaging and accessible for all students.

10.5 Discussion of Implications

The advent of hybrid learning models, which combine online and traditional classroom instruction, marks a pivotal transformation in the educational sphere, driven by technological advancements and shifting pedagogical paradigms. Hybrid learning's flexibility offers unprecedented access to education, allowing students to adapt their learning experiences to personal and professional commitments. In Asia, where educational needs are diverse and access to resources varies significantly, hybrid learning has emerged as a critical tool for addressing these challenges.

For instance, in Vietnam, hybrid models gained momentum during the COVID-19 pandemic, with many universities shifting to a 70 per cent offline and 30 per cent online model, and some courses being offered fully online. This adaptability not only preserved educational continuity but also demonstrated the resilience of hybrid learning systems in crisis situations, a trend also observed in countries like Singapore and Malaysia (Mustapha et al., 2022; Zaugg et al., 2021).

Vietnam, Thailand and China each provide unique perspectives on how digital transformation policies shape and support hybrid learning in higher education, reflecting different national priorities and challenges. In Vietnam, the National Digital Transformation Programme (Decision No. 749/QD-TTg, 2020) and the Strengthening the Application of Information Technology and Digital Transformation in Education (Decision No. 131/QD-TTg, 2022) (Prime Minister of Vietnam, 2020, 2022) represent a government-led, centralised framework that emphasises improving digital literacy and addressing the urban–rural digital divide. This framework has contributed to making hybrid learning models more inclusive by increasing access to digital resources, especially in underserved regions. However, challenges remain in terms of infrastructure and digital literacy, particularly in rural areas, where consistent internet access and familiarity with digital platforms are still limited (Pitt et al., 2022; Tri & Hoang, 2023).

Thailand's approach, under the Thailand 4.0 agenda, takes a more decentralised stance, granting autonomy to universities to adapt digital strategies based on local needs. However, Thailand's digital transformation in education reflects the country's political dynamics, where Bangkok's centrality in policymaking has resulted in resource disparities (Stokols, 2024). Urban universities, particularly those in Bangkok, are well-equipped with advanced infrastructure, while institutions in rural areas face limitations in digital capacity. This decentralised approach fosters flexibility but risks deepening regional inequalities, suggesting that a more balanced policy intervention could be beneficial. Hybrid learning in Thailand shows promise in urban regions but faces challenges in achieving equitable access across the country, a concern that is also relevant in the broader context of Thailand's educational policies (Laorach & Tuamsuk, 2022).

China's experience underscores a high-investment, technology-driven model that extends beyond basic digital literacy to the integration of artificial intelligence, big data and cloud computing under the Education Informatisation 2.0 Action Plan (Ministry of Education of the People's Republic of China, 2018) (Yan & Yang, 2021). This centralised model facilitates rapid deployment of smart campuses and digital classrooms across the nation, making China one of the most digitally advanced education systems. However, China's strategy also raises concerns about data privacy and the role of surveillance in educational spaces, given the extensive use of monitoring technologies (Stokols, 2024). China's model, while effective in creating uniform digital access, is resource-intensive and presents ethical challenges (Munna & Shaikh, 2023) that may not align with the financial and social landscapes of countries like Vietnam and Thailand.

With reference to Table 10.1, the hybrid learning models in these three countries reveal valuable insights into the scalability and sustainability of digital transformation in higher education across Asia. Vietnam's centralised approach, which prioritises inclusivity, may benefit from drawing lessons from Thailand's flexible model, which encourages institutions to tailor digital solutions to their contexts. Meanwhile,

Table 10.1 Digital transformation in higher education in Vietnam, Thailand and China

Country	Digital transformation approach	Challenges	Key outcomes
Vietnam	Centralised; aims for inclusivity with targets for digital engagement and administration	Digital divide between urban and rural areas; limited digital literacy among educators	Progress in urban centres; ongoing rural access challenges; commitment to equitable digital access
Thailand	Decentralised; flexibility under Thailand 4.0 agenda; Bangkok-dominant policy structure)	Urban–rural disparities; political centralisation that favours urban areas	Innovation in urban universities; need for policy to address regional inequalities
China	Highly centralised, technology-intensive model under Education Informatisation 2.0	Surveillance and data privacy concerns; high costs and limited flexibility	High-level technology adoption; model serves both educational and national security goals

China's model exemplifies the power of substantial government investment, though it also illustrates the need for careful consideration of privacy and ethical implications.

As Vietnam continues to refine its digital transformation efforts, a hybrid model that combines centralised policies with institutional autonomy may offer a balanced pathway to achieving digital inclusivity while addressing the unique challenges of rural and underserved communities. This approach could enhance the resilience and adaptability of Vietnam's hybrid learning system, allowing it to keep pace with rapid technological advancements while remaining sensitive to socio-economic disparities.

10.6 Future Directions

To enhance the effectiveness and scalability of blended learning in Vietnam, a comprehensive approach involving both policy and practice is essential. Vietnam's commitment to digital transformation, as exemplified by initiatives like the National Digital Transformation Programme and ICT Vision 2025, underscores the importance of investing in digital infrastructure. Expanding reliable internet access and ensuring the availability of digital devices across educational institutions, particularly in underserved rural areas, are crucial steps. Policymakers should continue to prioritise these infrastructural investments, which form the foundation for sustainable hybrid learning models.

Another essential component is the professional development of educators. The transition to blended learning requires that teachers are proficient not only in traditional pedagogical methods but also in digital tools and platforms. The Ministry of Education and Training (MOET) has already taken steps to enhance teachers' digital competencies through professional development programmes. Expanding these initiatives can ensure that educators are equipped to facilitate a seamless blend of online and face-to-face instruction, enhancing educational quality and accessibility. Additionally, curriculum adaptation is needed to incorporate digital literacy and online learning skills, as seen in programmes at institutions like Hanoi University of Science and Technology. Such curricular reforms ensure that students are prepared for the demands of a digital economy.

Providing comprehensive support services is also critical to maximising the benefits of hybrid learning. Institutions like Can Tho University have established support centres offering counselling, tutoring and technical assistance to help students navigate hybrid environments. Expanding these support structures across Vietnam will ensure that students from diverse backgrounds, particularly those facing barriers to digital access, can fully engage with and benefit from blended learning.

Vietnam's experience offers valuable lessons for other Asian countries pursuing hybrid learning models. In the broader Asian context, the scalability of hybrid learning is particularly relevant given the region's diverse socio-economic landscapes. Programmes like Malaysia's '1BestariNet', which aims to provide internet access to schools across the country, highlight the importance of governmental support in scaling hybrid models. Countries across Asia can benefit from similar initiatives that prioritise equitable access to digital infrastructure, enabling hybrid learning to reach both urban and rural areas effectively.

The professional development of educators is another critical area where Asian countries can learn from each other's experiences. For example, Singapore has implemented extensive training programmes for educators in technology-enhanced learning (TEL), which can serve as a model for regional adoption. In Vietnam, government-led initiatives have focused on equipping teachers with digital pedagogical skills as part of national digital transformation goals. Similarly, Thailand's decentralised approach allows universities greater flexibility to tailor professional development to local needs, though disparities between urban and rural institutions remain. In contrast, China's highly centralised and technology-intensive model promotes large-scale training through state-run platforms, yet also raises concerns about equity and autonomy. Establishing a regional framework for continuous professional development could help align such efforts and ensure that educators across Asia are equipped to facilitate both traditional and online learning environments effectively.

Curriculum adaptation also presents an opportunity for shared learning. Vietnam's blended learning programmes that incorporate digital literacy and online competencies offer useful models for other Asian universities. Thailand has similarly integrated flexible learning pathways, while China has embedded AI and big data into its national curriculum strategy. Scaling such reforms across the region can better prepare students for the digital economy by fostering critical skills in digital navigation, self-regulated learning and online collaboration.

Finally, fostering cross-border collaborations among Asian universities, such as through the ASEAN University Network (AUN), can amplify the impact of hybrid learning. Collaborative platforms allow institutions in Vietnam, Thailand, China, Malaysia and Singapore to co-develop joint hybrid courses and virtual exchange programmes. These initiatives enhance educational access and inclusion, while also promoting cross-cultural understanding and regional academic mobility, thereby contributing to the broader goals of educational integration and sustainable development in Asia.

10.7 Conclusion

The exploration of blended learning within the Vietnamese higher education context highlights the substantial potential of hybrid models to transform teaching and learning. These models offer enhanced educational access, greater flexibility and systemic innovation by integrating online and traditional pedagogical methods. By addressing diverse learner needs and increasing student engagement, blended learning contributes to the development of critical digital competencies essential for 21st-century education.

Importantly, the impact of blended learning extends well beyond Vietnam, offering valuable insights for the broader Asian educational landscape. As this chapter has shown through comparative analyses with Thailand and China, each country presents distinct pathways and challenges in adopting hybrid models. Vietnam's centrally coordinated, equity-oriented approach can inspire inclusive strategies across the region. Thailand's decentralised, flexible implementation highlights institutional innovation but also underscores the risks of uneven development. Meanwhile, China's high-tech, data-driven model demonstrates the power and complexity of state-led digital transformation on a national scale.

These diverse experiences collectively illustrate how blended learning can offer flexible and scalable solutions to pressing regional challenges, such as educational access disparities, varied learning preferences and gaps in digital literacy. Countries like India and Indonesia, alongside Vietnam, Thailand and China, can draw on these shared lessons to address the digital divide and improve educational quality.

Furthermore, the integration of digital tools and technologies into hybrid learning aligns with the broader vision of Education 4.0, which prioritises personalised learning, adaptive environments and the strategic use of big data and artificial intelligence to optimise educational outcomes. By fostering learner autonomy and promoting soft skills such as critical thinking, collaboration and digital fluency, blended learning prepares students for the dynamic demands of the digital era.

To maximise the benefits of hybrid education, policymakers across Asia must invest in robust digital infrastructure, sustained professional development for educators, inclusive curriculum adaptation and comprehensive student support systems. Such coordinated efforts will enable the creation of resilient and equitable learning ecosystems capable of addressing both current needs and future uncertainties across the region.

Highlights

- Vietnam's commitment to digital transformation, exemplified by initiatives like the National Digital Transformation Programme, underscores the importance of sustained investment in digital infrastructure to scale blended learning. However, regional disparities remain, especially in rural areas.
- Comparative insights from Thailand and China highlight different national approaches: Thailand adopts a decentralised model that allows institutional flexibility but risks widening regional gaps, while China's highly centralised, technology-intensive strategy promotes rapid innovation but raises concerns around surveillance and ethical implementation.
- Professional development for educators remains a cornerstone of successful hybrid learning across all three countries. Vietnam's centrally coordinated training, Thailand's locally adapted initiatives, and China's large-scale technological upskilling illustrate varying models that other Asian countries can learn from.
- The success of hybrid learning models in Vietnam, Thailand and China offers region-wide lessons on bridging the digital divide and expanding educational access. These models are particularly effective when tailored to local needs and supported by inclusive digital policies.
- Cross-border collaborations, such as those facilitated by the ASEAN University Network (AUN), provide valuable opportunities for Asian universities to co-develop hybrid programmes, promote virtual mobility and foster intercultural exchange, strengthening regional academic integration..

References

Al-Samarraie, H., & Saeed, N. (2018). A systematic review of cloud computing tools for collaborative learning: Opportunities and challenges to the blended-learning environment. *Computers & Education, 124*, 77–91. https://doi.org/10.1016/j.compedu.2018.05.005

Anthonysamy, L., Koo, A.-C., & Hew, S.-H. (2020). Self-regulated learning strategies and non-academic outcomes in higher education blended learning environments: A one decade review. *Education and Information Technologies, 25*(5), 3677–3704. https://doi.org/10.1007/s10639-020-10134-2

Apandi, A. M., & Raman, A. (2020). Factors affecting successful implementation of blended learning at higher education. *International Journal of Instruction, Technology, and Social Sciences, 1*(1), 13–23.

Armellini, A., & Rodriguez, B. C. P. (2021). Active blended learning: Definition, literature review, and a framework for implementation. In *Cases on active blended learning in higher education* (pp. 1–22).

Bandura, A., & Walters, R. H. (1977). *Social learning theory* (vol. 1). Prentice Hall.

Boelens, R., De Wever, B., & Voet, M. (2017). Four key challenges to the design of blended learning: A systematic literature review. *Educational Research Review, 22*, 1–18. https://doi.org/10.1016/j.edurev.2017.06.001

Bordoloi, R., Das, P., & Das, K. (2021). Perception towards online/blended learning at the time of Covid-19 pandemic: An academic analytics in the Indian context. *Asian Association of Open Universities Journal, 16*(1), 41–60. https://doi.org/10.1108/aaouj-09-2020-0079

Castro, R. (2019). Blended learning in higher education: Trends and capabilities. *Education and Information Technologies, 24*(4), 2523–2546. https://doi.org/10.1007/s10639-019-09942-1

Clarke, V., Braun, V., & Hayfield, N. (2015). Thematic analysis. In *Qualitative psychology: A practical guide to research methods* (3rd ed., pp. 222–248). Sage Publications. https://doi.org/10.4135/9781526405555.n2

De Vriendt, M. (2015). Implementing blended learning, in the form of a learning network, to enhance social learning. *The Journal of Technology Enhanced Learning, Innovation & Change, 1*(1), 1–11.

Demillo, R. A. (2019). *Blended learning in practice: A guide for practitioners and researchers.* MIT Press.

Fisher, R., Perényi, A., & Birdthistle, N. (2021). The positive relationship between flipped and blended learning and student engagement, performance, and satisfaction. *Active Learning in Higher Education, 22*(2), 97–113. https://doi.org/10.1177/1469787421993050

Garrison, D. R., & Arbaugh, J. B. (2007). Researching the community of inquiry framework: Review, issues, and future directions. *The Internet and Higher Education, 10*(3), 157–172. https://doi.org/10.1016/j.iheduc.2007.04.001

Garrison, D. R., & Vaughan, N. D. (2008). *Blended learning in higher education: Framework, principles, and guidelines.* John Wiley & Sons.

Hill, J., & Reimer, T. (2022). Crossing the digital divide and the equity expanse: Reaching and teaching all students during the pandemic. *Journal of Leadership, Equity, and Research.* https://doi.org/10.31585/jler.vol1.iss2.2022.34

Islam, M. K., Sarker, M. F. H., & Islam, M. S. (2022). Promoting student-centred blended learning in higher education: A model. *E-Learning and Digital Media, 19*(1), 36–54. https://doi.org/10.1177/20427530211027721

Kazu, I. Y., & Demirkol, M. (2014). Effect of blended learning environment model on high school students' academic achievement. *Turkish Online Journal of Educational Technology-TOJET, 13*(1), 78–87.

Laorach, C., & Tuamsuk, K. (2022). Factors influencing the digital transformation of universities in Thailand. *International Journal of Innovative Research and Scientific Studies, 5*(3), 211–219. https://doi.org/10.53894/ijirss.v5i3.646

Li, S., & Wang, W. (2022). Effect of blended learning on student performance in K-12 settings: A meta-analysis. *Journal of Computer Assisted Learning, 38*(5), 1254–1272. https://doi.org/10.1111/jcal.12650

Müller, C., & Mildenberger, T. (2021). Facilitating flexible learning by replacing classroom time with an online learning environment: A systematic review of blended learning in higher education. *Educational Research Review, 34*, 100394. https://doi.org/10.1016/j.edurev.2021.100394

Munna, A. S., & Shaikh, M. S. I. (2023). Exploring the role of digital transformation in higher education institutions: Challenges and benefits of online learning in CHINA. In *Overcoming challenges in online learning* (pp. 37–46). Routledge. https://doi.org/10.4324/9781003342335-5

Mustapha, Y., Hilmi, M. F., Mansor, S., & Saad, S. M. (2022). Prospect and challenges of blended learning in Malaysia: A systematic literature review. *Asian Journal of Research in Education and Social Sciences, 4*(3), 51–61. https://doi.org/10.55057/ajress.2022.4.3.5

Nayar, B., & Koul, S. (2020). Blended learning in higher education: A transition to experiential classrooms. *International Journal of Educational Management, 34*(9), 1357–1374. https://doi.org/10.1108/ijem-08-2019-0295

Pitt, B., Huynh, T. Q., Gregson, J., Seal, T., Tran, H. H., Nguyen, H. T., Bui, T. N. T., & Nguyen, H. M. (2022). *Readiness of digital transformation in Vietnamese universities*. https://www.britishcouncil.vn/en/education/going-globalpartnerships/success-stories/report-readiness-digital-transformation-vietnamese-universities
Prime Minister of Vietnam. (2020). *National digital transformation program until 2025 with orientations to 2030*. https://vanban.chinhphu.vn/?pageid=27160&docid=200163
Prime Minister of Vietnam. (2022). *Strengthening the application of information technology and digital transformation in the education and training domain in the period of 2022–2025, toward 2030*. https://vanban.chinhphu.vn/?pageid=27160&docid=205236&classid=0
Raes, A. (2022). Exploring student and teacher experiences in hybrid learning environments: Does presence matter? *Postdigital Science and Education, 4*(1), 138–159. https://doi.org/10.1007/s42438-021-00274-0
Rasheed, R. A., Kamsin, A., & Abdullah, N. A. (2020). Challenges in the online component of blended learning: A systematic review. *Computers & Education, 144*, 103701. https://doi.org/10.1016/j.compedu.2019.103701
Sato, S. N., Condes Moreno, E., Rubio-Zarapuz, A., Dalamitros, A. A., Yañez-Sepulveda, R., Tornero-Aguilera, J. F., & Clemente-Suárez, V. J. (2023). Navigating the new normal: Adapting online and distance learning in the post-pandemic era. *Education Sciences, 14*(1), 19. https://doi.org/10.3390/educsci14010019
Stokols, S. A. (2024). *Building digital cities and digital nations: Singapore, Thailand, China* [PhD Thesis]. Massachusetts Institute of Technology. https://hdl.handle.net/1721.1/156096
Tang, S. M., & Tien, H. N. (2020). Digital transformation trend in Vietnam higher education: Blended learning model. *International Journal of Social Science and Economics Invention, 6*(07), 304–309. https://doi.org/10.23958/ijssei/vol06-i07/218
Tarnopolsky, O. (2012). *Constructivist blended learning approach: To teaching English for specific purposes*. Walter de Gruyter.
Tri, N. M., & Hoang, P. D. (2023). The impact of digital transformation in higher education: The case study from Vietnam. *Journal of Higher Education Theory and Practice, 23*(5). https://doi.org/10.33423/jhetp.v23i5.5922
Wang, N., Chen, J., Tai, M., & Zhang, J. (2021). Blended learning for Chinese university EFL learners: Learning environment and learner perceptions. *Computer Assisted Language Learning, 34*(3), 297–323. https://doi.org/10.1080/09588221.2019.1607881
Xu, Z., Zhao, Y., Liew, J., Zhou, X., & Kogut, A. (2023). Synthesizing research evidence on self-regulated learning and academic achievement in online and blended learning environments: A scoping review. *Educational Research Review, 100510*. https://doi.org/10.1016/j.edurev.2023.100510
Yan, S., & Yang, Y. (2021). Education informatization 2.0 in China: Motivation, framework, and vision. *ECNU Review of Education, 4*(2), 410–428. https://doi.org/10.1177/2096531120944929
Zaugg, H., Graham, C. R., Lim, C. P., & Wang, T. (2021). Current and future directions of blended learning and teaching in Asia. In *Blended learning for inclusive and quality higher education in Asia* (pp. 301–327). Springer. https://doi.org/10.1007/978-981-33-4106-7_16

Mai Phuc Thinh holds an MA in International Development Studies and is an emerging researcher with a focus on education and migration studies. Thinh's research interests lie at the intersection of international education, migration, and sustainable development. They are particularly passionate about exploring how educational practices and policies impact student mobility and intercultural competence, especially in transnational contexts. Thinh aims to contribute to more inclusive and sustainable educational practices through rigorous, interdisciplinary research.

Chapter 11
Mapping the Internationalisation of Higher Education in the Philippines Within ASEAN

Cluster(s) of Future Studies: Past and Futures; Humanity at the Limen; Internationalisation (new)

Nilo Jayoma Castulo and Arlyne C. Marasigan

Abstract In the recent Philippine report, EDCOM 2, highlighted the issues troubling the country's higher education system, including challenges related to access to high-quality education, quality assurance and internationalisation. This chapter explores the internationalisation status of the Philippines in comparison with the other Association of Southeast Asian Nations (ASEAN) member countries using the World System Theory (WST). It uses multiple data sources, such as bibliometric data from SciVal under Scopus, InCites under Web of Science and other public secondary data. The findings revealed that the Philippines is at the borderline between the semi-periphery and periphery using the WST based on the six key dimensions of internationalisation: Role of Quality Assurance and Accreditation, Research Publications in Scopus and Web of Science, Research Citation, Research Collaboration, Higher Education Funding and International University Ranking. These dimensions are crucial for enhancing the current state of the Philippine internationalisation in higher education, particularly the research production, citation, collaboration and shaping the future direction of ASEAN to revisit and reformulate the ASEAN University Network objectives for regionalisation.

Keywords Internationalisation · Higher education · World system theory · Transnational education · Philippines · ASEAN

N. J. Castulo (✉)
Department of Educational Leadership and Professional Services, Mindanao State University–Tawi-Tawi College of Technology and Oceanography, Bongao, Philippines
e-mail: nilocastulo@msutawi-tawi.edu.ph

A. C. Marasigan
College of Advanced Studies, Philippine Normal University, Manila, Philippines
e-mail: marasigan.ac@pnu.edu.ph

A. Pandian et al. (eds.), *Future of Education in Asia*,
https://doi.org/10.1007/978-3-032-07285-6_11

11.1 Introduction

Internationalisation and globalisation are two distinct terminologies that have become popular in the twenty-first century—globalisation is primarily interconnected and interdependent on every country's economic development. On the other hand, internationalisation refers to the organisation or institutions that integrate activities and policies to meet the demands of globalisation (Altbach & Knight, 2007). As Bourdieu (2001) noted the dominant facets of globalisation, stating: 'Globalisation' serves as a password, a watchword while in effect, it is the legitimatory mask of a policy aiming to universalise particular interests and the particular tradition of the economically and politically dominant powers (p. 84).

Technically, internationalisation may be seen as a by-product of globalisation, allowing universities to expand their reach and influence internationally. Scholars like Gao (2019) suggested five dimensions of measuring the internationalisation of higher education: governance, student, staff, curriculum and research. Moreover, Knight (2007) proposed the dimensions of quality assurance, accreditation and cross-border education. At the same time, Pohl (2015) suggests including ranking to measure the universities' internationalisation. However, scholars worry that internationalisation will become a new tool for new forms of modern colonialism (Bamberger & Morris, 2023) and the hegemony of the global north to further push for economic-driven globalisation (Shahjahan et al., 2017). Despite the divided opinion, the Philippines complied with the regionalisation through the Association of Southeast Asian Nations (ASEAN) initiative. However, the country's education sector became hesitant to engage in internationalisation due to the colonisation trauma that might threaten national security while negotiating one's place in global geocognitive spaces (Eder, 2023; Marginson, 2022).

The latest study in the Philippines, EDCOM revealed the challenges include ensuring the availability of high-quality education, maintaining quality standards and promoting internationalisation (Second Congressional Commission on Education (EDCOM), 2024). Furthermore, the Philippines has a deficiency in faculty research involvement in internationalisation efforts, mainly owing to inadequate allocation of funds (Rolando & Ralph, 2013). Lorenzo (2022) argued that cross-border research within ASEAN has not been explored to examine regionalisation in higher education and to form a knowledge society within the ASEAN context. Khalid et al. (2019) observed that the less developed ASEAN nations are falling behind in internationalisation in higher education. They argued that the policies and strategies of internationalisation should be explored, and the reasons behind the falling status should be identified. Thus, this chapter explores the internationalisation status of the Philippines in comparison with the other Association of Southeast Asian Nations (ASEAN) member countries using the World System Theory (WST).

11.2 Literature Review

11.2.1 Association of Southeast Asian Nations (ASEAN) Context

The establishment of the ASEAN countries took place on August 8, 1967, in Bangkok, Thailand, with Indonesia, Malaysia, the Philippines, Singapore and Thailand as its five founding members. Brunei Darussalam became a member of ASEAN in 1984, followed by Vietnam in 1995, Lao PDR and Myanmar in 1997, and Cambodia in 1999 (AUN, 2024b). During the fourth meeting of the ASEAN Heads of State in 1992, the leaders considered the importance of education in shaping regional identity, fostering unity and enhancing human resources development among ASEAN countries, leading to the creation of the ASEAN University Network (AUN) (AUN, 2024a).

The objectives of the ASEAN University Network (AUN, 2022, p. 2) are as follows:

(a) Provide a platform for discussion on policy development for ASEAN higher education
(b) Strengthen the existing network of cooperation among universities in ASEAN
(c) Promote educational programmes on the priority areas identified by ASEAN
(d) Promote cooperation and solidarity scholars, academicians and researchers in the Member States

These objectives paved the way in creating the 17 thematic networks, and one of the networks is the AUN Quality Assurance (QA) (AUN, 2024c). In response, the Philippine Commission on Higher (CHED) formed the Asian International Mobility for Students (AIMS) programme based on the CHED memo 11 series 2014 (Guidelines for the Participation of Selected Higher Education Institutions (HEIs). While CHED memo 19 series 2015 (Operational Guidelines for the Implementation of Asian International Mobility for Students (AIMS) Programme) and CHED memo 11 2019 (Procedure for the Issuance of Non-Immigrant Visa under Sect. 47 (A)(2) for Exchange Students under the Asian International Mobility for Students (AIMS) Programme) (CHED, 2014, 2015, 2019b).

11.2.2 Philippine Government Policies as Catalyst for Internationalisation

When international league tables in higher education increased their acceptance and popularity, scholars thought it was an excellent motivation to promote equality and global justice to the universities in the global south (Shahjahan et al., 2017). Moreover, the transformative contribution of internationalisation of higher education creates harmony to serve its aims for social, cultural, economic and political interests (Tran et al., 2023). The United Nations Educational, Scientific and Cultural Organisation (UNESCO) and Organisation for Economic Co-operation and Development (OECD) become an international body that provides Guidelines on Quality Provision in Cross-Border Higher Education to provide reciprocity in all institutions for quality education and protect its integrity when engaging cross-border education (Knight, 2007). In 2000, the CHED issued a memorandum order that provided the first guidelines for internationalisation to all universities that have partnered with foreign universities. In this memo, the commission focused on international linkages and twinning programme initiatives (CHED, 2000). Table 11.1 summarises the development of the internationalisation policies on Philippine higher education.

In 2016, CHED issued another memo no. 55 detailing more guidelines about the internationalisation of the Philippine universities. This memo details home-based internationalisation and cross-border internationalisation. Home-based or campus-based internationalisation happens in collaborative activities, events, networking and partnerships with other foreign universities. The activities within this mode include international content and dimensions on the curriculum, adding international faculty for teaching, establishing extracurricular activities with the engagement of the local culture and research collaboration with other scholars abroad. Moreover, cross-border internationalisation includes student, faculty, programme and institutional mobilities (CHED, 2016).

In 2019, the country passed a law about transnational education (RA11448). The new law serves as a guide for Philippine universities and international universities to engage in cross-border education. This new law has eleven types of transnational education: academic franchising, articulation, branch campus, international branch campus, joint degree, double degree, online learning, blended learning, open distance learning, transnational education higher education offshore institution, twinning arrangement and validation. The law allows Philippine universities to engage with transnational education with foreign universities unless it is included in the top 500 world-class universities' internationally recognised ranking by CHED (Republic Act No. 11448, 2019).

Another support system provided to Philippine universities throughout the country is the recognition of the Centre of Excellence and the Centre of Development for

Table 11.1 Timeline of Philippine policies towards internationalisation

Policy	Name
CMO 1 series of 2000	Policies and guidelines in the implementation of international linkages and twinning programmes
CMO 55 series of 2016	Policy framework and strategies on the internationalisation of Philippine higher education
Republic Act No. 11448	The transnational higher education law
CMO 62 series of 2016	Policies, standards and guidelines (PSGs) for transnational education programmes

Note CMO means CHED Memorandum Order

promoting world-class universities. Based on the Higher Education Act law of 1994, the commission is tasked to 'identify, support and develop potential centre of excellence in programme areas needed for the development of world-class scholarship, nation-building and national development (Higher Education Act of, 1994, 1994, sec 8, paragraph f)'. Based on the May 8, 2018, CHED report, the distribution of programmes with 30 COE and 47 COD in public universities while 42 COE and 63 COD in private universities out of the total of 2 393 higher education institutions in the country (CHED, 2019a).

11.2.3 Theoretical Underpinning

This study used World Systems Theory (WST), which originated from Immanuel Wallerstein in the 1970s. This theory claims that the world is divided into systems: the core (the capitalist high-income countries) and the peripheries, the middle and lower-income countries. The peripheries are divided into two: the middle class (semi-periphery countries) and the lower class (Chirot, 2015). The most current categorisation by the World Bank based on income level indicates that the Philippines, Vietnam, Cambodia, Lao PDR and Myanmar are classified as lower-middle income economies, with a range of $1,136 to $4,465. Malaysia, Indonesia and Thailand are upper-middle-income countries, with a per capita income ranging from $4,466 to $13,845. Brunei Darussalam and Singapore are high-income economies, with a per capita income of $13,846 or above (World Bank, 2023). Miyahara (2015) noted that the ASEAN objectives cannot be achieved without realising the need to improve the economic situation and the establishment of a single market.

This theory extends to understanding the role of universities in knowledge production, in which core capitalist countries like American universities disseminate knowledge to the peripheries (Sweeney & Bonney, 2021). Due to the internationalisation in higher education, the WST theory becomes relevant to understanding the movement of the world knowledge system. A study by Zhu et al. (2021) revealed that core nations have economic dominance in graduate employment in higher education and exert significant influence over academic research. Semi-peripheral and peripheral countries are urged to establish transnational networks and engage in intellectual collaboration. Scholars like Musiał (2023) have questioned the internationalisation of higher education, arguing that the semi-peripheral status of Nordic countries made universities compete in the world knowledge economy and play by the same rules of the hegemony of core societies. According to Wu et al. (2023), China's position in the WST is on the borderline between the core and the periphery. It implies that achieving success in internationalisation in higher education depends on having the financial resources necessary to support it (Miyahara, 2015; Zhu et al., 2021).

11.3 Methodology

The surge in interest in internationalisation has been seen since 2000 (Buckner, 2019). Within the ASEAN context, several Asian academics advocate for doing cross-border research to investigate the phenomenon of regionalism (Khalid et al., 2019; Lorenzo, 2022). To fill this gap, this study utilised the comparative method using the Bray and Thomas Cube framework consists of three dimensions: geographical level, aspect of education and nonlocational demographic groupings (Bray et al., 2007). This research is conducted within the geographical scope of the ASEAN region. It examines the impact of internationalisation on education and its effects on knowledge production within the ASEAN context, highlighting the Philippine status on internationalisation. Additionally, it focuses explicitly on higher education among nonlocational demographic groups.

This research collected data from multiple sources, including bibliometric data from SciVal Data (accessible under Scopus) and InCites data (available under the Web of Science), and other secondary sources from the ASEAN University Network website, UNESCO Institute for Statistics (UIS), Philippine Department of Budget and Management publicly available data. The selected period spans from 1996 to 2024. The researchers gathered data from worldwide rankings, like the Quacquarelli Symonds (QS) ranking and Times Higher Education, specifically focusing on the Asian category. The selection was based on the inclusion of these two rankings under the implementing rules and regulations of Republic Act No. 11448 (2021).

11.4 Findings

11.4.1 Dimension 1: Role of Quality Assurance

The Association of Southeast Asian Nations (ASEAN) Quality Assurance (ASEAN UN-QA) has completed 48 assessments for 154 examined degree programmes in the Philippines (see Table 11.2) (ASEAN University Network, 2023). Vietnam and Indonesia demonstrate the most substantial number of evaluated programmes, with the Philippines and Thailand ranking next. Moreover, Malaysia has a few assessed programmes for quality assurance, similar to Cambodia, Myanmar, Brunei Darussalam and Lao PDR, which have little involvement in evaluating their programmes. In contrast, Singapore has no recorded assessment participation for quality assurance in the ASEAN University Network.

Table 11.2 Association of Southeast Asian Nations (ASEAN) Quality Assurance

Country	AUN-QA number of assessment	AUN-QA assessed programme
1. Indonesia	101	340
2. Thailand	26	83
3. Malaysia	9	22
4. Philippines	48	154
5. Viet Nam	123	388
6. Singapore	0	0
7. Cambodia	1	1
8. Myanmar	0	0
9. Brunei Darussalam	1	1
10. Lao PDR	2	5

Source ASEAN University Network (2024)

11.4.2 Dimension 2: Research Publications in Scopus and Web of Science

Table 11.3 provides data on the number of journals indexed in the ASEAN citation index, Scopus and Web of Science, which helps to understand the research output of ASEAN countries. Indonesia has many indexed journals in the ASEAN citation index (ACI), indicating a strong presence in the field. Nevertheless, Singapore boasts numerous journals associated with Scopus and Web of Science. Malaysia has 152 journals in the ASEAN Citation Index, although it has 118 journals in Scopus and 93 in Web of Science. Thailand has a total of 201 journals indexed in ACI, 76 journals indexed in Scopus and 39 journals indexed in Web of Science. Consequently, the Philippines has 52 ASEAN Citation Index journals, while only 27 journals have been accepted by Scopus and Web of Science. Similarly, Vietnam is represented by 26 journal indexes in the ACI, although it is only represented by 2 journal indexes in Scopus and 9 in Web of Science. While Cambodia. Myanmar, Brunei Darussalam and Lao PDR have the lowest number of journals indexed in the three stated indexes.

The web-based analytics tools SciVal and InCites provide insights into the academic output trends of the ASEAN countries from 1996 to 2024. However, the data for 2024 was not fully complete. The researcher opted not to integrate the data because InCites provides numerous viewing approaches. The SciVal tool within Scopus provides a visual representation of scholarly output over a specific period. The InCites tool within Web of Science displays a scatter plot representing the total number of documents from the Web of Science database over a specific timeframe.

Figure 11.1 displays the distribution of published documents in Scopus. Singapore, Thailand and Malaysia have substantially increased publications from 1996 to 2008. From 2008 to 2018, Malaysia surpassed Thailand and Singapore in research outputs, experiencing consistent growth. In 2018, Indonesia experienced a significant surge in growth when compared to Malaysia and Singapore. Thailand and Vietnam

Table 11.3 Number of indexed journals from ASEAN countries

Country	ASEAN citation Index	Scopus	Web of science
1. Indonesia	218	169	85
2. Thailand	201	76	39
3. Malaysia	152	118	93
4. Philippines	52	27	27
5. Viet Nam	26	2	9
6. Singapore	10	222	134
7. Cambodia	2	1	1
8. Myanmar	2	0	0
9. Brunei Darussalam	1	1	0
10. Lao PDR	0	0	1

Source Master of journal list under Clarivate and Scimago under Scopus (ASEAN Citation Index (ACI), 2024; Clarivate, 2024; Scimago, 2024)

have experienced consistent growth and have followed a similar trajectory; however, they have not surpassed Singapore. Regarding Scopus research output performance, the Philippines lags behind Vietnam, Thailand, Indonesia, Malaysia and Singapore. Furthermore, Brunei Darussalam, Cambodia, Myanmar and Lao PDR have low research productivity in Scopus, with less than 11,000 documents from 1996 to 2024. Malaysia, Singapore, Indonesia, Thailand and Vietnam are the top five countries with the highest research output in Scopus from 1996 to 2024, which includes all forms of publications and covers all disciplines.

The scatterplot in Fig. 11.2 displays the count of Web of Science documents across all collections from 1996 to 2024. Singapore and Malaysia have the highest

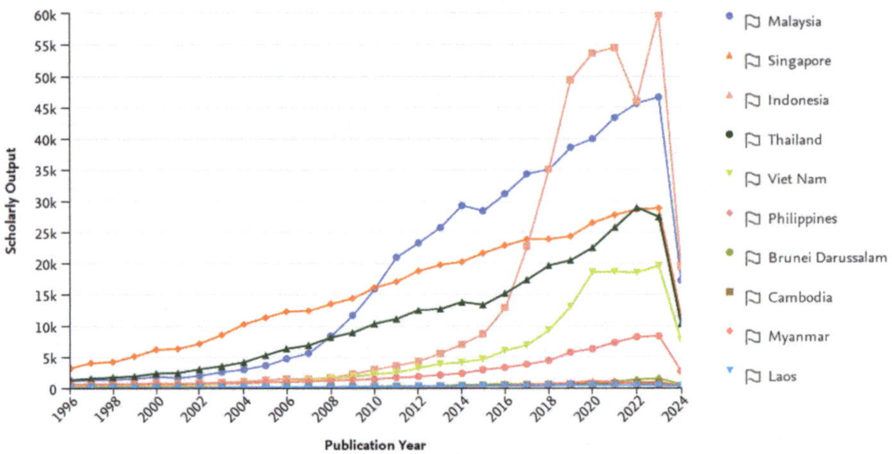

Fig. 11.1 ASEAN scholarly output in scopus between 1996 and 2024. *Source* SciVal (2024)

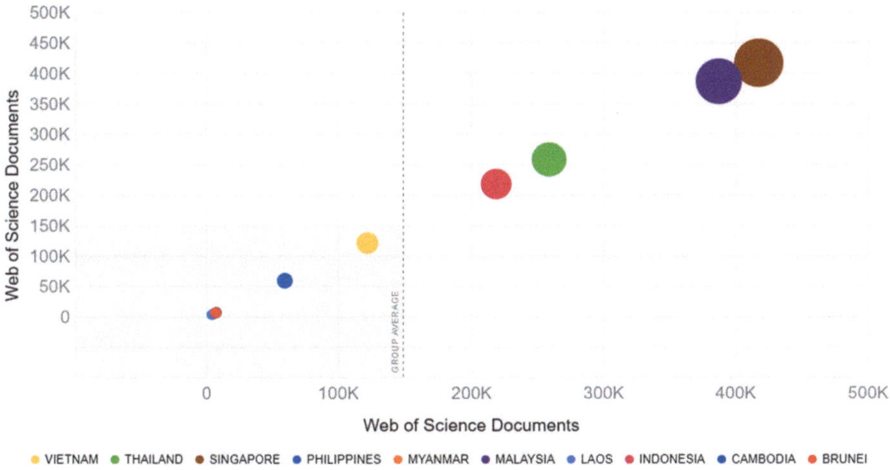

Fig. 11.2 ASEAN scholarly output in web of science between 1996 and 2024. *Notes* Includes web of science content indexed through Mar 31, 2024.Export Date: May 27, 2024. *Source* (InCites, 2024)

number of publications within the 387,000 to 418,000 document range. Thailand had around 259,000 documents. Indonesia is making progress by acquiring around 219,000 Web of Science documents. Regrettably, the remaining ASEAN countries could not achieve a research output of 150,000. Vietnam possesses 122,000 Web of Science documents, while the Philippines has 60,000 Web of Science documents. Furthermore, Cambodia, Brunei Darussalam, Myanmar and Lao PDR exhibit the lowest research production, having produced fewer than 8,000 documents in the Web of Science database between 1996 and 2024. Lao PDR has the lowest publication rate, with less than 5,000 publications.

11.4.3 Dimension 3: Research Citation

Figure 11.3 displays the cumulative number of citation counts by all types of publications, including self-citations. The data is sourced from SciVal under Scopus and covers 1996 to 2024. The data indicate that research outputs from Singapore are highly cited, followed by Malaysia and Thailand. Since 2012, research publications from Vietnam and Indonesia have shown a consistent upward trend, suggesting that other academics increasingly acknowledge their work. Nevertheless, the Philippines' citation performance has been inconsistent since 2018 and exhibits low citation counts. Brunei Darussalam, Cambodia, Lao PDR and Myanmar have citation counts below 50,000, indicating low levels of citations.

Figure 11.4 displays the times cited and the citation impact using InCites web analytics within the Web of Science. The 'times cited' refers to the total number of

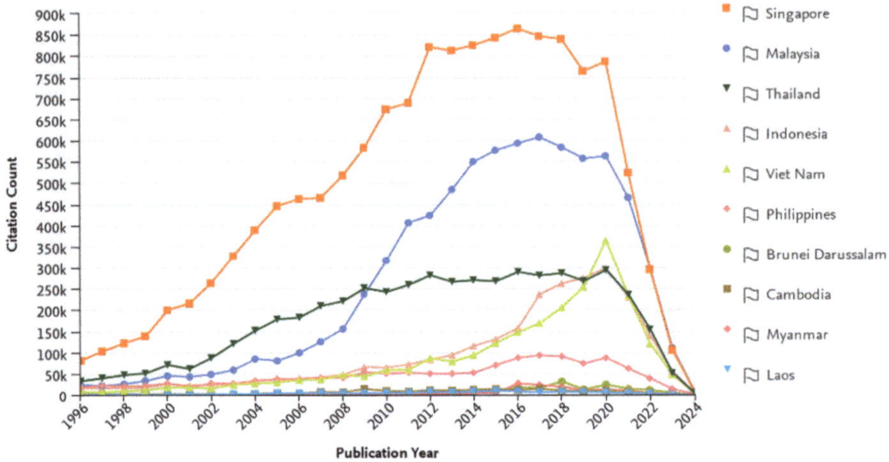

Fig. 11.3 ASEAN citation count in scopus data between 1996 and 2024. *Source* SciVal (2024)

times the publications have been referenced in other works, while 'citation impact' is the average number of citations per article. Figure 11.4 illustrates a distinct perspective, showing that research outputs from Singapore get the majority of citations with a high impact factor of 27.97. Researchers in Thailand have a higher citation impact of 15.81 than Malaysia, which has a citation impact of 14.34. Nevertheless, Malaysia outperformed Thailand in terms of the number of times cited. Brunei, Cambodia and Lao PDR have a citation impact of greater than 18 but less than 22, while the Philippines, Vietnam and Myanmar have a citation impact ranging from 15 to 17. Furthermore, Indonesia has the lowest citation impact, estimated at 8. However, it outperforms the Philippines, Myanmar, Brunei Darussalam, Cambodia and Lao PDR in terms of times cited. It illustrates that despite Indonesia having many published papers and several journals. It needed to enhance its citation performance and generate high-quality publications that attract other researchers.

11.4.4 Dimension 4: Research Collaboration

Figure 11.5 illustrates the degree of international collaboration among the countries in the Association of Southeast Asian Nations (ASEAN). Between 1996 and 2019, Singapore has shown a commendable trend of fostering research partnerships via international, national and institutional co-authorship. Since 2020, Malaysia has surpassed Singapore in terms of international collaboration. Thailand, Indonesia and Vietnam have a comparable trend of growing collaboration, with Thailand demonstrating superior performance compared to the other two. Conversely, the Philippines is behind in international collaboration and has performed poorly from 2008 to 2019. However, there has been a minor improvement in international collaboration since

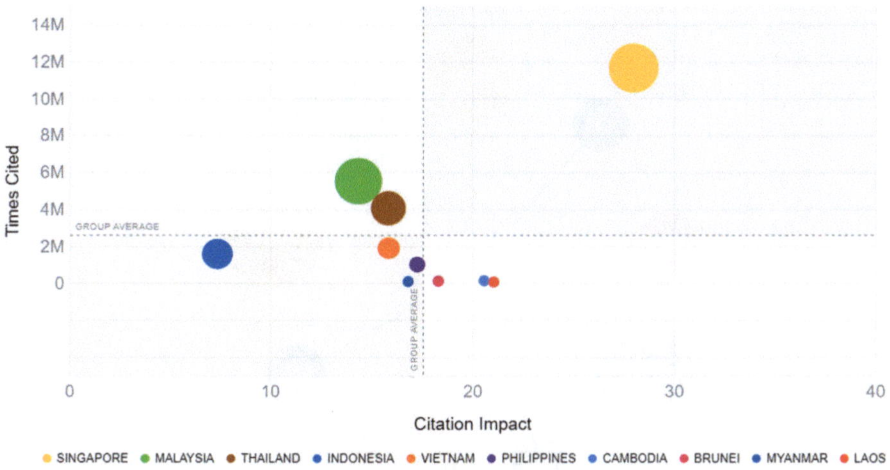

Fig. 11.4 ASEAN web of science citation impact. *Notes* Includes web of science content indexed through Mar 31, 2024.Export Date: May 28, 2024. (InCites, 2024)

2019. Nevertheless, Cambodia, Brunei Darussalam, Myanmar and Lao PDR exhibit the lowest level of international collaboration.

When analysing the collaboration between the Philippines and other ASEAN countries, Fig. 11.6 shows that the Philippines has a significant international research collaboration with Thailand, with 2782 web of science documents. It is followed by a collaboration with Malaysia, with 2745 documents; Singapore, with 2179 documents; and Indonesia, with 2126 documents. Furthermore, Vietnam has collaborated with the Philippines on research, resulting in 1702 web of science papers. However,

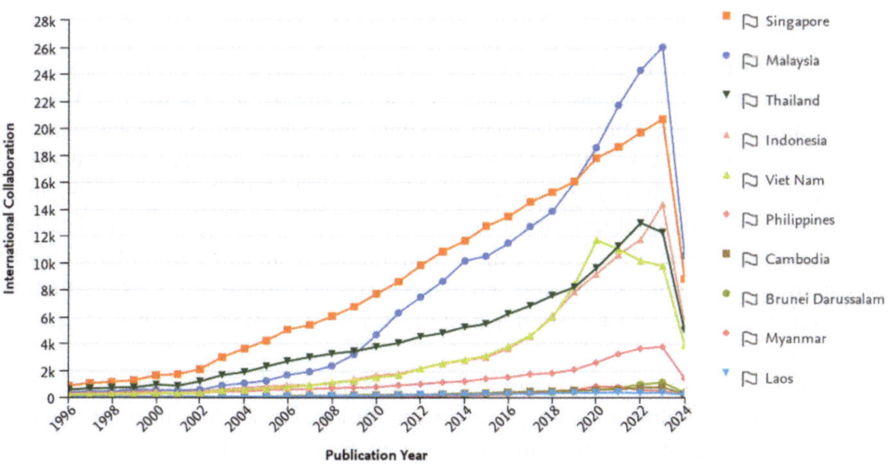

Fig. 11.5 ASEAN international research collaboration. *Source* (SciVal, 2024)

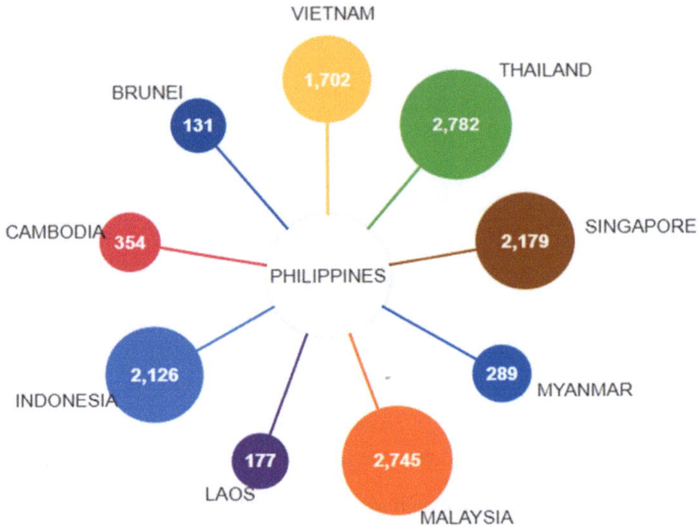

Fig. 11.6 Philippine research collaboration within ASEAN. *Notes* Includes web of science content indexed through Mar 31, 2024.Export Date: May 27, 2024. (InCites, 2024)

Cambodia and Myanmar have less international engagement with the Philippines, with fewer than 500 research collaborations. Brunei has the lowest level of international research cooperation with the Philippines, as shown by just 131 Web of Science papers.

11.4.5 Dimension 5: Higher Education Funding

Figure 11.7 displays the proportion of the Gross Domestic Product that the government spends on higher education. Figure 11.7 shows the seven countries in ASEAN, as determined by their submitted statistics to the United Nations. There was a lack of data for the years 2023 and 2024. Figure 11.7 displays the government's expenditure on tertiary education during the six years from 2017 to 2022. Malaysia has the most significant government allocation for higher education, with Singapore and Thailand following closely after. The Philippines' expenditure from 2017 to 2022 showed a consistent upward trend, surpassing Thailand's expenditures in 2022. Nevertheless, Myanmar, Lao PDR and Cambodia provide less government funding for higher education.

Figure 11.8 illustrates the distribution of government funds for Philippine state higher education institutions, categorised by institutional level. Among the 17 regions and 117 universities that receive funding from the National Expenditure Programme (NEP) of the central government, the University of the Philippines, Mindanao State

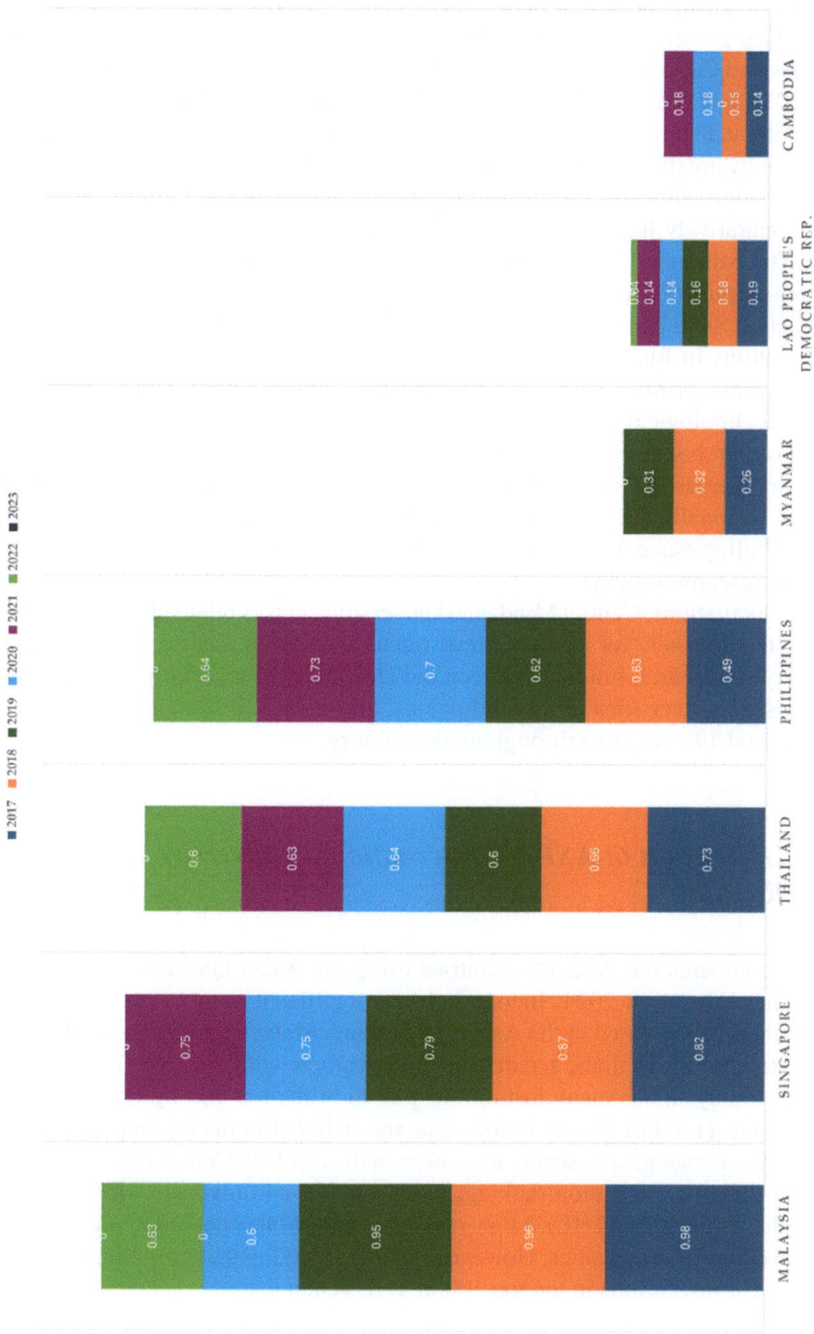

Fig. 11.7 ASEAN Government expenditure on tertiary education as a percentage of GDP (%). *Source* (UIS, 2024)

University and Polytechnic University of the Philippines have been allocated the highest shares during the period between 2019 and 2024 (Refer to Fig. 11.8).

Figure 11.9 presents the research fund allocation among the ten selected State Universities and Colleges (SUCs) from the National Expenditure Programme (NEP) between 2019 and 2024. The University of the Philippines, Mindanao State University, MSU-Iligan Institute of Technology, Visayas State University and Central Luzon State University are the five leading state institutions in the Philippines with the most substantial budget allocation for research. The University of the Philippines maintains a comparatively high level of funding, which signifies a strong emphasis on research. The Mindanao State University consistently rose in funding, reflecting a notable commitment to research.

The Visayas State University and the MSU-Iligan Institute of Technology experience fluctuations in funding but generally maintain higher levels, highlighting an ongoing research prioritisation. The funding for Central Luzon State University and Cebu Technological University fluctuates, suggesting possible difficulties in sustaining research support. In addition, the Cebu Normal University demonstrates a noteworthy increase in funding starting in 2021, indicating a growing emphasis on research. Additionally, the Philippine Normal University and the Polytechnic University of the Philippines exhibit consistent funding, indicating possible challenges in maintaining research support.

Despite fluctuations, Central Mindanao University receives relatively low funding, suggesting the necessity for increased investment in research. Figure 11.8 illustrates the disparity of research funding between the University of the Philippines and the other nine state universities in the Philippines. However, UP has eight constituent universities and 17 campuses throughout the country.

11.4.6 Dimension 6: ASEAN International University Ranking

Table 11.4 compares the ASEAN countries using the Asian QS ranking 2024 and Times Higher Education 2024. Indonesia has a significant number of institutions, with 43 of them being listed in the Asian QS ranking. Malaysia follows closely with 41 universities, while Thailand secures the third position with 25 universities. On the other hand, Malaysia has many institutions placed in the top 100 by QS Asia. The Philippines has a total of 16 institutions that are included in the ranking. However, only one of them managed to secure a position in the top 100. Conversely, according to the Times Higher Education ranking, Indonesia has 24 universities listed in the Times Higher Education (THE) Asian ranking, while Malaysia has 23 institutions and Thailand has 19 universities. However, it is worth noting that only Malaysia and Singapore have institutions in the top 100 in THE Asian list.

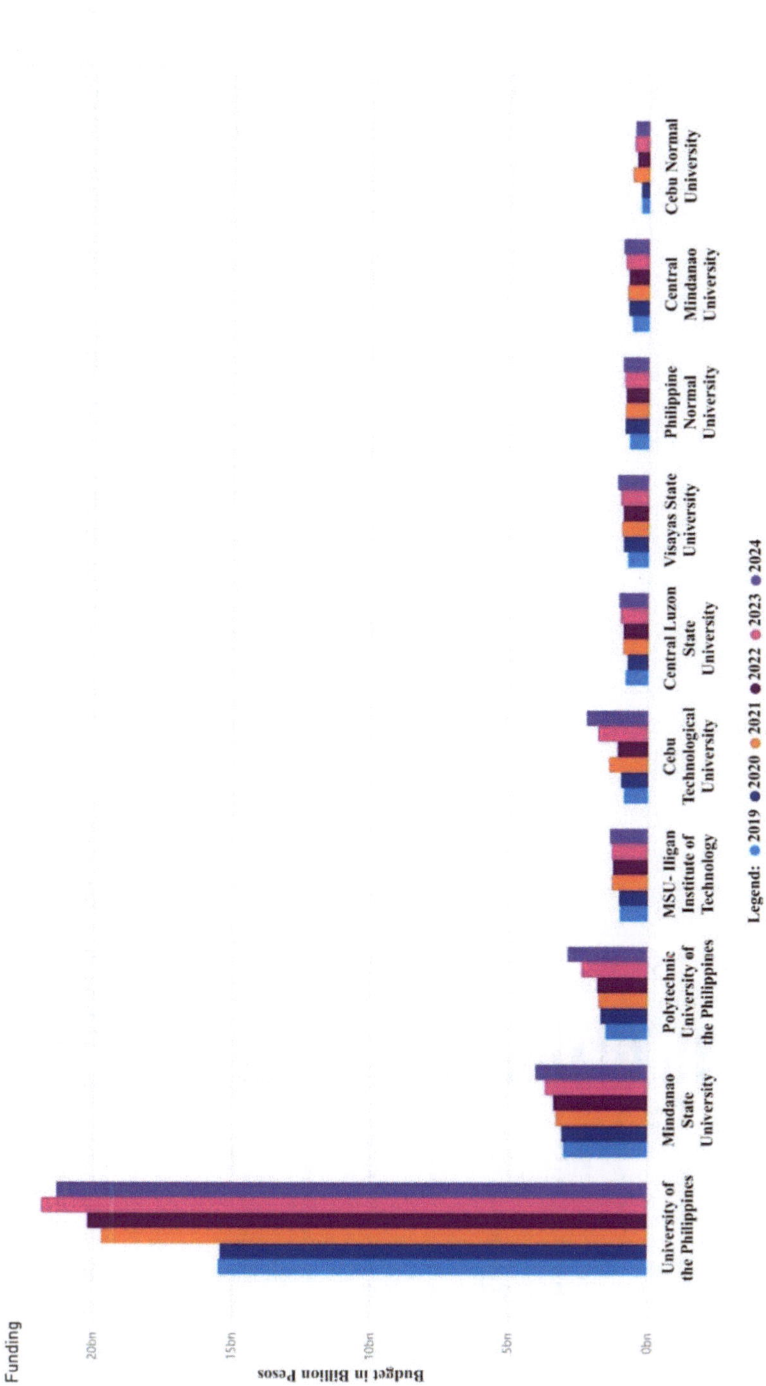

Fig. 11.8 State Universities fund allocation in the Philippines between 2019 and 2024. *Note* The selection of the ten state universities is based on the highest budget by region and international ranking inclusion

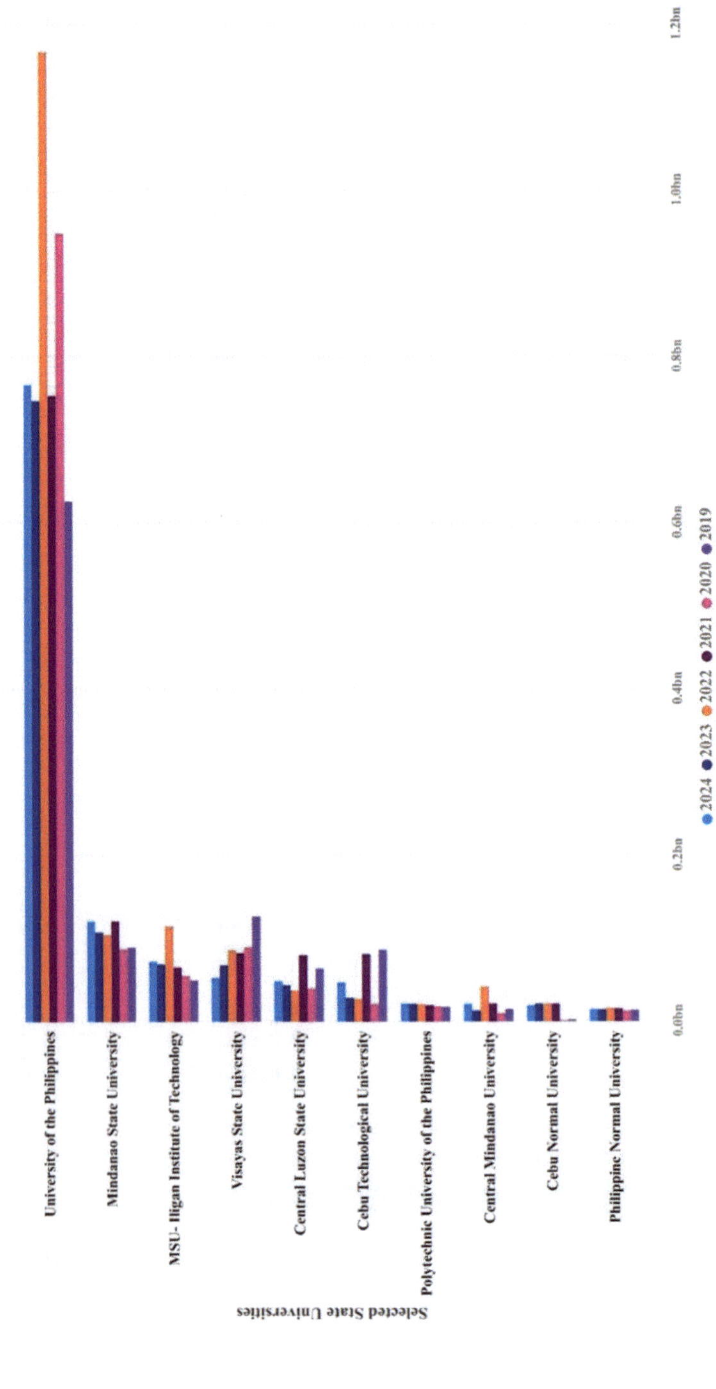

Fig. 11.9 Selected state universities research fund allocation between 2019 and 2024. *Note* The data above is based on the National Expenditure Programme (NEP) from the Philippine Department of Budget and Management

Table 11.4 Quacquarelli symonds (QS) and Times higher education (THE) 2024 Asian ranking

Country	QS ranking 2024		Times higher education 2024	
	Asian ranking	Top 100	Asian ranking	Top 100
1. Indonesia	43	4	24	0
2. Thailand	25	2	19	0
3. Malaysia	41	9	23	4
4. Philippines	16	1	5	0
5. Viet Nam	15	0	6	0
6. Singapore	5	3	2	2
7. Cambodia	1	0	0	0
8. Myanmar	2	0	0	0
9. Brunei Darussalam	3	1	1	0
10. Lao PDR	0	0	0	0

Note Quacquarelli Symonds (QS) (2024); Times Higher Education (THE) (2024a)

11.5 Discussion of Implications

11.5.1 Dimension 1: Role of Quality Assurance and Accreditation

The results suggest that the Philippines prioritised harmonisation across ASEAN by focusing on quality assurance, as the country ranks third in the number of examined programmes. Vietnam achieved the top ranking with the most evaluated programmes, while Indonesia came in second and Thailand placed fourth, respectively. However, Singapore has no assessed programmes. According to Buckner (2019), the internationalisation process might vary depending on the economic conditions of a country. In low-income countries, the main emphasis is often on implementing policies that provide quality assurance and promote capacity development within the system.

Quality assurance and accreditation are essential in cross-border internationalisation because they monitor, register and maintain the standards that universities meet and recognise that their programmes align with international standards (Knight, 2007). The Philippine Commission on Higher Education (CHED) Memo Order No. 1 series of 2005 recognised the significant contribution of quality assurance and accreditation agencies to meet the quality standards set by the prescribed requirements from the commission (Revised Policies and Guidelines on Voluntary Accreditation in Aid of Quality and Excellence in Higher Education, 2005).

11.5.2 Dimension 2: Research Publications in Scopus and Web of Science

The research productivity gap between the Philippines and the other five ASEAN member countries (Vietnam, Thailand, Indonesia, Malaysia and Singapore) is widening, prompting the need for action from the Philippines to narrow the gap. Several factors contribute to this problem. According to this dimension, most Philippine journal articles are not included in the two well-known academic databases (Scopus or Web of Science). Some early-career researchers favoured fast-track publications with low-cost article processing charges (APC) over the long-term advantages of publishing in free indexed journals that take longer to appear. Furthermore, the high cost of institutional subscriptions to indexed journals, fear of rejection and frustration with the article review and editing process all contribute to the issue (Tudy, 2023; Wa-Mbaleka, 2015).

Many scholars explored the issue of research productivity among ASEAN countries such as (Barrot, 2017; Guido & Orleans, 2020). Barrot (2017) found that approximately 2% of the published works in language and linguistics were authored by Southeast Asian countries, specifically Brunei, Singapore, Malaysia and the Philippine. While Guido & Orleans (2020) found that the Philippines ranked fifth in educational research productivity. Guerrero et al. (2022) emphasised the importance of reviewing the existing policies of the Commission on Higher Education (CHED) to better assist faculty members in their research endeavours and reduce their workload. Moreover, faculty members are expected to engage in research, community extension and instruction as part of their promotion criteria.

11.5.3 Dimension 3: Research Citation

Similarly, Singapore has higher citation counts than Malaysia. Thailand, Indonesia and Vietnam have similar performance levels in terms of citation counts. However, the Philippines experienced a downward trend in citation counts. It illustrates that other researchers do not often reference academic publications from the Philippines due to the limited research output and quality of research across all fields of study (Second Congressional Commission on Education, 2024). One way to address this issue is by fostering a culture of research (Bustamante, 2018; Guido & Orleans, 2020). Subsequently, the citation has proven valuable in assessing the quality indicator of the research productivity (Sjögårde & Didegah, 2022).

Nevertheless, utilising citations as a criterion for assessing research quality remains controversial among scholars. In addition, the emergence indicator to determine the impact of the research output depends on novelty, persistence, community and growth (Porter et al., 2019). Lee (2014) argued that this trend of using citations impacts judging scholarliness rather than focusing on the contribution to the field, which needs to be rethought. Rogayan & Corpuz (2022) propose implementing an incentive mechanism to motivate faculty members to publish their research. Additionally, they recommend allocating additional research funding to enhance research productivity and ultimately increase the number of citation counts.

11.5.4 Dimension 4: Research Collaboration

The regionalisation of the ASEAN countries changed their policies to integrate ASEAN objectives that promote collaboration. The results indicate that Singapore and Malaysia have a comparable pattern of international research collaboration. Vietnam, Thailand and Indonesia have similar patterns at the medium level. However, the degree of international research collaboration in the Philippines is critical. One of the key features of internationalisation is collaboration in research. Popular international league tables such as QS ranking include the International research network as one of the criteria for ranking (O'Callaghan, 2023), and Times Higher Education includes international co-authorship under the international outlook component in their methodology (Times Higher Education (THE), 2023).

According to Haley et al. (2024), research collaboration across the core societies (global north countries) is seen as a way to enhance competitiveness and the neoliberal market. On the other hand, the participation of the global south countries is primarily focused on enhancing research and capacity development for knowledge production. Ramos-Eclevia et al. (2018) studied the research output and collaboration among librarians in ASEAN countries. The findings indicate that Malaysian academics had the highest research output, with most of the works being authored by a single author. In addition, De Frutos-Belizón et al. (2024) emphasised the importance of social capital and team orientation in fostering international collaboration within research teams. Furthermore, it is crucial for the institution to allocate financial resources effectively, strive for academic excellence, foster motivation and encourage open and informal communication to achieve a good international research collaboration (Jeong et al., 2014).

11.5.5 Dimension 5: Higher Education Funding

Fund allocation is crucial to ASEAN higher education's harmonisation, integration and internationalisation (Adeyemo, 2019). Adeyemo (2019) believed that socioeconomic status had a significant impact on international competitiveness. Malaysia and Singapore represent the most significant percentage of ASEAN government spending on tertiary education in terms of GDP. The Philippines is inconsistent in distributing funds for research productivity, which explains the lack of outcomes in dimensions 2, 3 and 4. According to Times Higher Education's 2024 rankings, Hong Kong (China), the United Arab Emirates, the United Kingdom and Switzerland are recognised for having many internationally renowned institutions and are classified as high-income economies (Times Higher Education (THE), 2024b).

Funding plays a vital role in managing and providing support for internationalisation activities (Liu et al., 2023). Scholars noted that internationalisation in higher education could be driven by economic and educational agendas, such as the university rankings, market-oriented education and income generation from international students (Shahjahan et al., 2017; Strielkowski et al., 2021). Matei et al. (2015) introduced a framework for financing internationalisation. Within the framework, universities should carefully contemplate the funding sources (who funds?), the activities that are sponsored for internationalisation (what is funded?), the extent of these activities (where do funds go?) and the instruments and techniques used to get funding for internationalisation (how is it funded?).

11.5.6 Dimension 6: International University Ranking

The Philippines has only one institution in the top 100 of the QS Asian ranking 2024, while only universities from Singapore and Malaysia made it to the top 100 of the Times Higher Education 2024 lists. International ranking serves as the measurement of internationalisation in higher education (Gao, 2019; Gao, 2015). For instance, the City University of Hong Kong is consistently ranked as the most internationalised university in the Times Higher Education ranking 2024 (Times Higher Education (THE), 2024b). Some criticised the international ranking system, citing a lack of unanimity on the parameters used (Jeremic & Jovanovic-Milenkovic, 2014). Furthermore, it influences policy change, as seen in Europe, where European universities are urged to modernise (Hazelkorn & Ryan, 2016).

In the Philippine context, under the legislation on transnational education, the Commission on Higher Education (CHED) has identified three globally acknowledged rankings such as Times Higher Education (THE), QS Rankings and WURI as the basis for guaranteeing excellence and quality assurance on transnational education (Implementing Rules and Regulations of Republic Act No. 11448, 2021). Scholars such as Vidal and Ferreira (2020) have contended that international league tables should not be the sole basis for assessing quality, as each university has a different mission to fulfil. Nevertheless, a comparative analysis between Malaysian and Philippine universities reveals that internationalisation has become embedded in their vision-mission statements, influencing the policies and activities of these academic institutions (Dumanig & Symaco, 2022).

11.6 Future Directions

Based on the discussion of the six dimensions and the placement of the Philippines in the World System Theory reveals the borderline between the semi-periphery and the periphery (see Fig. 11.10). Singapore is classified as part of the core, but Malaysia is positioned on the boundary between the core and the semi-periphery. Indonesia and Thailand are classified as semi-peripheral countries, whereas Vietnam is considered similar to the Philippines in terms of their status in the internationalisation of higher education.

However, Brunei, Cambodia, Myanmar and Lao PDR are considered peripheral countries in internationalisation, even though Brunei is classified as a high-income economy. This classification aligns with the conclusions made by Khalid et al. (2019), whereby Singapore is classified as a high group, Indonesia, Malaysia, Brunei, Philippines and Thailand are classified as medium groups, while the other ASEAN member countries are classified as a low group in the internationalisation in higher education.

The following are future directions of the Philippine internationalisation in higher education:

- Regarding quality assurance and accreditation, adhering to the existing policy to put into practice within the ASEAN University Network and increasing research collaboration in ASEAN UN-QA is highly recommended. Moreover, combining human and financial resources is highly recommended, given the low-income status of most ASEAN member countries. This approach will facilitate research collaboration and the production of teachers and students among ASEAN universities. Berse (2018) discussed the role of policies in situating into practice to improve

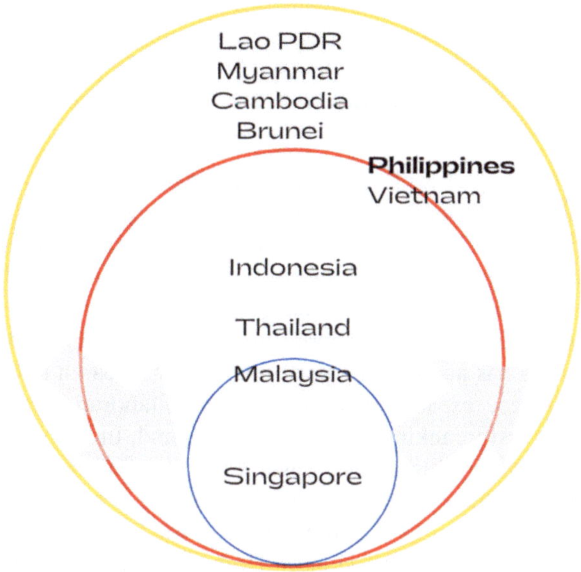

Fig. 11.10 World system theory on knowledge-economy. *Notes* Authors' own work

the Philippines' regionalisation through qualifications framework, quality assurance and credit transfer, which ranked third based on a higher number of assessed programmes (see Table 11.2 in Sect. 11.4.1).

- Data indicates that most internationally ranked Philippine universities belong to the private sector. Given the budget, allocation is not restricted by government laws and procedures for private universities. They have autonomy regarding institutional budget spending and allocations as needed for institutional competitiveness. In order to enhance the performance of state-funded universities, the government must allocate a larger budget to State Universities and Colleges (SUCs), with a specific focus on research production, allowing universities to gain more access to software and hardware (facilities and infrastructure) such as Scopus and Web of Science databases and advanced training on the utilisation of the software and hardware.
- The findings highlight the low research production in Philippine higher education. It is recommended that support systems and mechanisms be formulated to keep scholars from moving abroad and prevent brain drain. Also, to boost research production, there is a need to strengthen the inbound recruitment of scholars and academicians, such as postdoctoral fellows from other countries. However, this needs more financial investment from the government.

Given that Singapore is at par with the global north, the ASEAN plays a crucial role in improving the universities on the periphery, particularly in Brunei, Cambodia, Lao PDR and Myanmar, and in advancing Indonesia, Malaysia, Vietnam and the Philippines to catch up with the global north research and knowledge productions to narrow the gaps between the global north and global south. It is recommended to revisit the AUN charter (AUN, 2022, p. 2) to gauge the current status and review the future directions based on the existing achievement and challenges.

11.7 Conclusion

This chapter explored the internationalisation status of the Philippines in comparison with the other Association of Southeast Asian Nations (ASEAN) member countries using the World System Theory (WST). The findings indicate that the Philippines is at the borderline between the semi-periphery and periphery with respect to the six dimensions in internationalisation in higher education. Moreover, many AUN members, except Singapore, are still in the semi-periphery and periphery. The Philippines has an alarming performance for research publications in scopus and web of science, research citation, research collaboration, fund allocation and international university ranking compared to Thailand, Indonesia, Malaysia and Singapore.

Further, this chapter raises a flare for Philippine higher education to improve its performance in the knowledge economy, aiming for internationalisation. Based on the findings, it is recommended call for ASEAN University Network (AUN) and the Philippines Commission on Higher Education to revisit their existing policies and reformulate the AUN objectives and directions for a sustainable knowledge economy. This chapter does not cover all aspects of internationalisation, and future researchers may delve into different dimensions, including curriculum and engagement, to assist the government in improving its policies and strategies.

Highlights

- The Philippines is positioned at the borderline between the semi-periphery and periphery in World System Theory (WST), indicating its marginal role in the global hierarchy of higher education compared to core nations like Singapore
- To move beyond the semi-periphery in global higher education, the Philippines needs to foster stronger research collaborations, invest in state universities and engage more deeply in regional and global initiatives like the AUN to ensure sustainable internationalisation
- Transnational education is pivotal in the internationalisation of Philippine higher education, with policies such as those from the ASEAN University Network (AUN) playing a significant role in bridging gaps between the Philippines and more developed ASEAN countries
- Quality assurance and accreditation play a key role in the internationalisation process, and thus the focus must shift toward enhancing research output and citations to compete internationally, especially in high-impact academic databases like Scopus and Web of Science

References

Adeyemo, K. S. (2019). *Higher education policy in the Philippines and ASEAN integration: Demands and challenges*. Brill Sense.

Altbach, P. G., & Knight, J. (2007). The internationalization of higher education: Motivations and realities. *Journal of Studies in International Education, 11*(3–4), 290–305. https://doi.org/10.1177/1028315307303542

ASEAN Citation Index (ACI). (2024). *Number of journals by countries*. https://asean-cites.org/list_of_journal.html

ASEAN University Network. (2023). *ASEAN University Network-Quality Assurance (AUN-QA) fact check*. ASEAN. https://aunsec.org/discover-aun/thematic-networks/aun-q

ASEAN University Network. (2024). *AUN-QA assessed programmes [ASEAN]*. https://www.aunsec.org/discover-aun/thematic-networks/aun-qa/aun-qa-assessed-programmes

AUN. (2022, October 11). *Charter of the ASEAN University Network*. https://agreement.asean.org/media/download/20221103091624.pdf

AUN. (2024a). *The charter of the ASEAN University Network [ASEAN]*. https://www.aunsec.org/discover-aun/why-we-are-here

AUN. (2024b). *The founding of ASEAN [ASEAN]*. https://asean.org/the-founding-of-asean/

AUN. (2024c). *Thematic networks [ASEAN].* https://www.aunsec.org/discover-aun/thematic-networks

Bamberger, A., & Morris, P. (2023). Critical perspectives on internationalization in higher education: Commercialization, global citizenship, or postcolonial imperialism? *Critical Studies in Education*, 1–19. https://doi.org/10.1080/17508487.2023.2233572

Barrot, J. S. (2017). Research impact and productivity of Southeast Asian countries in language and linguistics. *Scientometrics, 110*(1), 1–15. https://doi.org/10.1007/s11192-016-2163-3

Berse, P. P. (2018). Harmonizing higher education at the regional level: The case of ASEAN and the Philippines. *International Journal of Higher Education, 7*(5), 44. https://doi.org/10.5430/ijhe.v7n5p44

Bourdieu, P. (2001). *Contre-feux 2: Pour un mouvement social européen.*

Bray, M., Adamson, B., & Mason, M. (2007). *Comparative education research: Approaches and methods.* The University of Hong Kong.

Buckner, E. (2019). The internationalization of higher education: National interpretations of a global model. *Comparative Education Review, 63*(3), 315–336. https://doi.org/10.1086/703794

Bustamante, C. B. S. (2018). Higher education research in the Philippines: Policies and prospects. In J. Jung, H. Horta, & A. Yonezawa (Eds.), *Researching higher education in Asia* (pp. 287–295). Springer. https://doi.org/10.1007/978-981-10-4989-7_16

CHED. (2000). *CHED Memorandum Order (CMO) No. 1 Series of 2000: Policies and guidelines in the implementation of international linkages and twinning programs.* https://drive.google.com/file/d/1aKDnURfqQ_8FZ2Shos52f5LiJsW8ndw2/view

CHED. (2014). *Guidelines for the participation of selected higher education institutions (HEIs) in Asian International Mobility for Students Program [Government]. CHED Memo 11 Series 2014.* https://drive.google.com/file/d/11pubcnpaJiosCa5-CK7i0OhkksCg67cf/view

CHED. (2015). *Operational guidelines for the implementation of Asian International Mobility for Students (AIMS) Program [Government]. CHED Memo 19 Series 2015.* https://drive.google.com/file/d/1PVTA65S8QiRQWYQkUqSNzNzevmE16ZWl/view

CHED. (2016). *Policy framework and strategies on the internationalization of Philippine higher education.* https://drive.google.com/file/d/1W9dI2sFIbTIrub2YHPIv86zwz9oV3zCz/view

CHED. (2019a). *Extension of the validity period of designated centers of excellence (COEs) and centers of development (CODs) for various disciplines [Commission on Higher Education].* https://ched.gov.ph/wp-content/uploads/Extension-of-validity-2019-List-of-COE-and-CODs_compressed.pdf

CHED. (2019b). *Procedure for the issuance of non-immigrant visa under Section 47 (A)(2) for exchange students under the Asian International Mobility for Students (AIMS) Program [Government]. CHED Memo 11 2019.* https://ieducationphl.ched.gov.ph/policies-and-guidelines/

Chirot, D. (2015). World systems theory. In *International Encyclopedia of the Social & Behavioral Sciences* (pp. 748–752). Elsevier. https://doi.org/10.1016/B978-0-08-097086-8.32172-9

Clarivate. (2024). *Master Journal list.* https://mjl.clarivate.com/search-results

De Frutos-Belizón, J., García-Carbonell, N., Guerrero-Alba, F., & Sánchez-Gardey, G. (2024). An empirical analysis of individual and collective determinants of international research collaboration. *Scientometrics, 129*(5), 2749–2770. https://doi.org/10.1007/s11192-024-04999-0

Dumanig, F. P., & Symaco, L. P. (2022). Internationalisation of higher education in Malaysia and the Philippines: A comparative analysis of mission and vision statements of selected universities. *Journal of Multilingual and Multicultural Development, 43*(2), 154–166. https://doi.org/10.1080/01434632.2020.1735401

Eder, R. (2023). *Internationalization of Philippine higher education: Between nationalism and co-optation* (Centre for Global Higher Education, Working paper no. 103).

Gao, Y. (2015). Toward a set of internationally applicable indicators for measuring university internationalization performance. *Journal of Studies in International Education, 19*(2), 182–200. https://doi.org/10.1177/1028315314559030

Gao, C. Y. (2019). *Measuring university internationalization: Indicators across national contexts.* Palgrave Macmillan.

Guerrero, J. G., Mapa-Azul, C. G., & Carmona, J. O. (2022). Creating a research culture in a Dominican university: Perspectives and productivity of the University of Santo Tomas Legazpi faculty. *Journal of Institutional Research South East Asia, 20*(2), 210–227.

Guido, R. M. D., & Orleans, A. V. O. (2020). Philippine research productivity in education research: A comparative performance in Southeast Asia. *Asia Pacific Journal of Multidisciplinary Research, 8*(4).

Haley, A., Alemu, S. K., Zerihun, Z., & Uusimäki, L. (2024). Internationalisation through research collaboration. *Educational Review, 76*(4), 675–690. https://doi.org/10.1080/00131911.2022.2054958

Hazelkorn, E., & Ryan, M. (2016). The impact of university rankings on higher education policy in Europe: A challenge to perceived wisdom and a stimulus for change. In P. Zgaga, U. Teichler, & J. Brennan (Eds.), *The globalisation challenge for European Higher Education* (pp. 37–58). Peter Lang D. https://doi.org/10.3726/978-3-653-06508-4

Higher Education Act of 1994. (1994). https://www.officialgazette.gov.ph/downloads/1994/05may/19940518-RA-07722-FVR.pdf

Implementing Rules and Regulations of Republic Act No. 11448. (2021). https://legaldex.com/laws/implementing-rules-and-regulations-of-republic-act-no-11448-transnational

InCites. (2024). Analyze by location. https://incites.clarivate.com/#/analysis/0/region

Jeong, S., Choi, J. Y., & Kim, J.-Y. (2014). On the drivers of international collaboration: The impact of informal communication, motivation, and research resources. *Science and Public Policy, 41*(4), 520–531. https://doi.org/10.1093/scipol/sct079

Jeremic, V., & Jovanovic-Milenkovic, M. (2014). Evaluation of Asian university rankings: Position and perspective of leading Indian higher education institutions. *Current Science, 106*(12), 1647–1653. Scopus.

Khalid, J., Ali, A. J., Nordin, N. M., & Shah, S. F. H. (2019). Regional cooperation in higher education: Can it lead ASEAN toward harmonization? *Center for Southeast Asian Studies, Kyoto University, 8*(1), 81–93. https://doi.org/10.20495/seas.8.1_81

Knight, J. (2007). Cross-border higher education: Quality assurance and accreditation issues and implications. In *Higher education in the world* (pp. 1–12). https://doi.org/10.1142/9789814299046_0004

Lee, I. (2014). Publish or perish: The myth and reality of academic publishing. *Language Teaching, 47*(2), 250–261. https://doi.org/10.1017/S0261444811000504

Liu, T., Hassan, A., Bin, M. A., & Anuar, M. (2023). Systemic review: Financing of higher education internationalisation. *Perspectives in Education, 41*(4).

Lorenzo, M. P. (2022). ASEAN regionalism and cross-border research of Philippine higher education: The case of the University of the Philippines Diliman campus. *Globalisation, Societies and Education, 20*(5), 590–607. https://doi.org/10.1080/14767724.2021.1973888

Marginson, S. (2022). Space and scale in higher education: The glonacal agency heuristic revisited. *Higher Education, 84*, 1365–1395. https://doi.org/10.1007/s10734-022-00955-0

Matei, L., Iwinska, J., & Crăciun, D. (2015). Patterns of funding internationalisation of higher education: A conceptual framework for the study of internationalisation. In A. Curaj, L. Matei, R. Pricopie, J. Salmi, & P. Scott (Eds.), *The European higher education area* (pp. 205–219). Springer International Publishing. https://doi.org/10.1007/978-3-319-20877-0_14

Miyahara, S. (2015). Regional quality assurance system for higher education in Southeast Asia. In *Quality assurance in LIS education: An international and comparative study* (pp. 25–38). Springer. https://doi.org/10.1007/10.1007/9781461464952_2

Musiał, K. (2023). Internationalization as myth, ceremony, and doxa in higher education: The case of the Nordic countries between centre and periphery. *Nordic Journal of Studies in Educational Policy, 9*(1), 20–36. https://doi.org/10.1080/20020317.2023.2166344

O'Callaghan, C. (2023, November 27). QS World University Rankings methodology: Using rankings to start your university search. *QS Ranking*. https://www.topuniversities.com/qs-world-university-rankings/methodology

Pohl, H. (2015). How to measure internationalization of higher education. In E. Ullberg (Ed.), *New perspectives on internationalization and competitiveness*. Springer International Publishing. https://doi.org/10.1007/978-3-319-11979-3

Porter, A. L., Garner, J., Carley, S. F., & Newman, N. C. (2019). Emergence scoring to identify frontier R&D topics and key players. *Technological Forecasting and Social Change, 146*, 628–643. https://doi.org/10.1016/j.techfore.2018.04.016

Quacquarelli Symonds (QS). (2024). *QS World University Rankings: Asia 2024*. Top University. https://www.topuniversities.com/countries=countries=countries=my

Ramos-Eclevia, M., Janio, R. V., Vinzon, M. R., Eclevia, C. L., & Apolinario, R. R. U. (2018). Researching together: Exploring research productivity and collaboration of librarians in ASEAN countries. *Journal of the Australian Library and Information Association, 67*(3), 307–320. https://doi.org/10.1080/24750158.2018.1501868

Republic Act No. 11448. (2019, September 25). An act expanding access to educational services through the establishment and administration of transnational higher education, and appropriating funds therefor. *Official Gazette*. https://www.officialgazette.gov.ph/2019/08/28/republic-act-no-11448/

Revised policies and guidelines on voluntary accreditation in aid of quality and excellence in higher education. (2005). CHED No. 1.

Rogayan, Jr. D. V., & Corpuz, L. N. (2022). Evaluating the research productivity of a state university in Central Luzon, Philippines: Basis for policy recommendations. *International Journal of Evaluation and Research in Education (IJERE), 11*(1), 128. https://doi.org/10.11591/ijere.v11i1.22099

Rolando, A. A., & Ralph, P. L. (2013). Internationalization: Investing in faculty. In R. Hanzu-Pazara & C. Stan (Eds.), *Maritime Universities, IAMU AGA 2013*. Editura Muntenia.

Scimago. (2024). *Scimago journal & country rank*. https://www.scimagojr.com/journalrank.php

SciVal. (2024). Benchmark all metrics. https://www.scival.com/benchmarking/analyse

Second Congressional Commission on Education. (2024). Miseducation: The failed system of Philippine education, EDCOM II year one report. In *Second Congressional Commission on Education (EDCOM II)*.

Shahjahan, R. A., Ramirez, G. B., & Andreotti, V. de O. (2017). Attempting to imagine the unimaginable: A decolonial reading of global university rankings. *Comparative Education Review, 61*(S1). https://doi.org/10.1086/690457

Sjögårde, P., & Didegah, F. (2022). The association between topic growth and citation impact of research publications. *Scientometrics, 127*(4), 1903–1921. https://doi.org/10.1007/s11192-022-04293-x

Strielkowski, W., Grebennikova, V., Razinkina, E., & Rudenko, E. (2021). Relationship between globalization and internationalization of higher education. *E3S Web of Conferences, 301*, 03006. https://doi.org/10.1051/e3sconf/202130103006

Sweeney, H., & Bonney, E. N. (2021). A global American university: How Midwest University exports education to the periphery. In G. Malfatti (Ed.), *Advances in higher education and professional development* (pp. 254–267). IGI Global. https://doi.org/10.4018/978-1-7998-3796-1.ch013

Times Higher Education (THE). (2023, September 20). World university rankings 2024: Methodology. *Times Higher Education*. https://www.timeshighereducation.com/world-university-rankings/world-university-rankings-2024-methodology
Times Higher Education (THE). (2024b, January 24). Most international universities in the world 2024. *Times Higher Education*. https://www.timeshighereducation.com/student/best-universities/most-international-universities-world
Times Higher Education (THE). (2024a). Asia university rankings 2024. *Times Higher Education*. https://www.timeshighereducation.com/world-university-rankings/2024/regional-ranking?fbclid=IwZXh0bgNhZW0CMTAAAR1KzF41z7pFCSY1icJDdJlHQ4_34Be0jnvyBuLilHFX0wpNnAHbztRfXk4_aem_Aby7Z7eCn4d4fYX50c1fYVZf3BNcbG_5q9W0B5eKNB_J9nexrP9NLbl3kNMBjCe_b1XvlGJ2lLh1hKQ7tdS1pbJT#!/length/100/sort_by/rank/sort_order/asc/cols/stats
Tran, L. T., Jung, J., Unangst, L., & Marshall, S. (2023). New developments in internationalisation of higher education. *Higher Education Research & Development, 42*(5), 1033–1041. https://doi.org/10.1080/07294360.2023.2216062
Tudy, R. (2023). The journey of a non-native English speaker towards Scopus publication: An autoethnography. *Recoletos Multidisciplinary Research Journal, 11*(2). https://doi.org/10.32871/rmrj2311.02.09
UIS. (2024). Government expenditure on education as a percentage of GDP. *UNESCO Institute for Statistics*. https://data.uis.unesco.org//#
Vidal, J., & Ferreira, C. (2020). Universities under pressure: The impact of international university rankings. *Journal of New Approaches in Educational Research, 9*(2), 181. https://doi.org/10.7821/naer.2020.7.475
Wa-Mbaleka. (2015). Factors leading to limited faculty publications in Philippine higher education institutions. *International Forum Journal, 18*(2), 121–141.
World Bank. (2023, June 30). World Bank Group country classifications by income level for FY24 (July 1, 2023–June 30, 2024). *World Bank*. https://datahelpdesk.worldbank.org/knowledgebase/articles/906519-world-bank-country-and-lending-groups
Wu, H., Yang, R., & Li, M. (2023). "On the centre–periphery borderline": Educational studies on/in Mainland China in the global context. *The China Quarterly*, 1–11. https://doi.org/10.1017/S0305741023001480
Zhu, M., Guo, C.-Y., Hou, A.Y.-C., & Chiu, M.-S. (2021). Graduate employment in higher education: Applying bibliometrics to world-system theory. *Journal of Education and Work, 34*(3), 356–372. https://doi.org/10.1080/13639080.2021.1922621

Nilo Jayoma Castulo is a professor in the Department of Educational Leadership and Professional Services, Mindanao State University—Tawi-Tawi College of Technology and Oceanography, Bongao, Philippines. He served as the co-chair of the Global Migration Group of the Comparative and International Education Society (CIES) in the USA and a regional editor of the Journal of Comparative and International Higher Education. He completed his PhD in Educational Leadership and Policy, specializing in Comparative Education, at the Institute of International and Comparative Education, Beijing Normal University, under the ASEAN-China Young Leaders Scholarship. He also holds a master's degree from Northeast Normal University and a bachelor's degree from the FEU Institute of Education and is a licensed professional teacher in the Philippines. Dr. Castulo actively serves as a member of the working committee of the Comparative Education and Students Critical Leadership Society at the Philippine Normal University. He has contributed scholarly works on comparative education, international higher education, and education policy, and his teaching and research interests include migration studies, comparative education, and global education leadership.

Arlyne C. Marasigan is a full professor and full-time faculty of the College of Advanced Studies at the Philippines Normal University-Manila, a founding adviser of Comparative Education and Students Critical Leadership Society (CESCLS Inc.), a member of Comparative and International Education Society (CIES)and a former Graduate Research Office director (2020–2022). She is currently engaged in national and international research collaboration. In addition, she is the co-founder of LakbayLapis2019, a lifetime member of the Women's Studies Association of the Philippines, an associate member of the National Research Council of the Philippines, a member of the Phi Lambda Theta Philippine Chapter, and an editorial board member of the Journal of International Students and Discover Education (SpringerLink). She is the current country director for the Philippines at the Star Network, United States. Additionally, she currently serves as the Country Director for the Philippines at the StarScholars Network, a U.S.-based organisation. Arlyne completed her PhD in Educational Leadership and Policy with specialisation in Comparative Education at the Beijing Normal University, China, in 2017.

Chapter 12
India and Select Asian Policies for Internationalisation of Higher Education

Cluster(s) of Future Studies: Post Normality and Complexity

D. Prachi

Abstract The majority of the Asian countries have started internationalisation of higher education (IoHE) studies to strengthen their foundational and industrial competencies. In the same direction, India has recently implemented the National Educational Policy 2020 (Ministry of Human Resource Development, Government of India, Ministry 2020) rooted in its socio-cultural belief system. Interestingly, the policy spans implementation strategies for the entire educational system of the country—primary, secondary and higher education. This chapter focuses on policy tenets related to IoHE, fulfilling the country's futuristic goals of institutional and student mobilisation. The study uses qualitative analysis and theoretical synthesis of concepts to identify overlaps between policy objectives and implementation strategies. The results highlight various transition arrangements for higher educational institutions (HEI) that guide contemporary course conceptualisation for collaborative degree programmes. Furthermore, the author has reviewed selected policies or initiatives for IoHE in China, Japan, Malaysia and Singapore. Contrasting these policies provides an insight into factors for policy initiation, implementation and collaboration options for each selected Asian country. The brief comparison suggests that IoHE practices vary from country to country, despite the analytical framework being the same. Since the implementation of IoHE relies on policy and regulatory intersections, a background and definitions of certain constructs have been included at the start.

Keywords Internationalisation of Higher Education · Higher Education Institutions (HEI) · National Educational Policy (NEP) · India · Asian

---

D. Prachi (✉)
Faculty of Media and Communication, Vishwakarma University, Pune, India
e-mail: prachi2100@gmail.com

Researcher, Mass Communication, Delhi, India

A. Pandian et al. (eds.), *Future of Education in Asia*,
https://doi.org/10.1007/978-3-032-07285-6_12

12.1 Introduction

Internationalisation of education has become a requirement of every nation willing to advance its global diplomacy, trade and innovations. The scholarly definitions of internationalisation of education consider it as the process of transmitting nationally produced knowledge to people and countries outside the nation, in order to promote cultural interactions, thereby enhancing international academic credibility (Varghese, 2020). Such a practice allows the curriculum to become cross-national and intercultural (Varghese, 2020). Countries aiming to internationalise higher education impart knowledge with universal application by integrating international, intercultural and global dimensions in the purpose, functions and delivery of educational content (Knight, 2004). Since the international dimension includes all aspects of holistic management, it can therefore improve the quality of teaching and learning, helping achieve the desired competencies (Soderqvist, 2002). These competencies, on the other hand, are a part of multiple activities, programmes and services related to international studies, international educational exchanges and technical cooperation (Arum & Van de Water, 1992).

Now, because the entire process of internationalising education relies on exchange between nations, it may take place both: (a) at home through campus-based activity and (b) abroad through cross-border educational activities (Knight, 2006). Considering the same possibilities India has panned out its educational internationalisation on both fronts, which will be discussed in the upcoming sections of the chapter. Following which, there is a discussion on selected comparable educational policy initiatives of China, Japan, Malaysia and Singapore helping contrast the approach towards internationalisation of higher education (IoHE) opted by the chosen Asian countries.

It is therefore, crucial to distinguish the concept of internationalisation of education from that of globalising education, before moving ahead. Globalising education is predominantly institutional in nature (Altbach & Knight, 2007) as it only aims to enhance the scope of 'study abroad' programmes by promoting student flows (Varghese, 2020). It does not acknowledge the intercultural, national and sectoral aspects of the curricula, its functionality and delivery patterns (Altbach & Knight, 2007), thereby ignoring the practical implementation of internationalisation aspects.

12.2 Literature Review

12.2.1 Significance of IoHE

Before proceeding to understand the implementation of IoHE, it is crucial to address the reasons a nation would attempt to internationalise its higher educational system.

Firstly, to globalise production, a nation requires higher-order skills and competencies. Such competencies can either be generated by educating citizens at home or by outsourcing talent from abroad. Given the migration patterns and the 'battle for brains', the countries are pushed to catch and train young talent for the global labour market (Varghese, 2020). Subsequently, the countries require retaining the produced or acquired talent for securing their improved competencies, which may be theoretical, technical, technological or vocational in nature and purpose. To fulfil these inter-linked requirements there begins a bi-partisan exchange between the host and foreign nations.

On the other hand, it may also be argued that the provisions for internationalised education should be implemented only at home, i.e. within a nation. Rightly so, however, this would restrict the citizens (of the implementing nation) from seeking the benefits of forex,[1] enhanced diplomatic dialogue,[2] foreign relations and several employment opportunities abroad. Cumulatively, such a nation would fail to promote its national competitiveness in a world that is becoming increasingly global and interconnected.

12.2.2 Investments in the Education Sector

While every nation considers education as one of the contributing factors to economic growth, it is one of the many industry sectors. Yet it must also be acknowledged that investing in the educational sector is equivalent to adding futuristic value to every other industry sector within a nation.

The Indian government has been significantly investing in the Education sector, with 2023 being the largest education budget allocation worth 13.67 billion USD to support Ed Tech,[3] skill development and research and development in the country (International Trade Administration, 2023). The country aims to develop three centres of excellence around IoT, AI and mechatronics, giving a push to the nation's soft skills. 280 million USD have been specifically allocated to the National Research Foundation to strengthen the country's research development (International Trade Administration, 2023).

The Chinese government on the other hand is focused towards building a skilled workforce and is spending 248.37 bn USD. The amount is directed not just for developing HEI, its infrastructure but also involves vocational training expenses (China Daily, 2024).

---

[1] Foreign exchange markets allow for various functions like investments, exchange of goods and services, financial transactions through currency conversions and global trade; making it to be an important function for global economies.

[2] Tactful discussions enabling communications for strengthening trust, reducing tensions and resolving conflicts between nations.

[3] Ed Tech or educational technologies refer to the technological implementations in the classroom or educational delivery infrastructures (e.g. in online content delivery platforms).

A combined budgetary expense of 35.55 billion USD is being allocated for higher education and sciences by the Japanese national government, which is also one of the largest budgets focusing on promotion of education, culture and sciences (statista.com, 2024).

The Malaysian government has allocated a significant 363.7 million USD for higher education in its budget for the financial year 2024 (Bernama, 2023). Where on the one hand the government is limiting the registration fee on the other, there is an extensive expenditure being incurred for implementing digitisation projects. The budgetary spending testifies that the country considers upgrading HEIs and improving student support as top national priorities.

Whereas, Singapore has a whopping 2.65 billion USD allocated for its university sector, which includes training of high-quality manpower and use-inspired[4] research efforts (FY2024 Expenditure Estimates, 2024). Furthermore, to enhance the competitive efficacy of the country, Singapore has initiated Skills Future programme that caters to adult learning and workforce upgradation requirements of the fast-aging economy.

12.2.3 Policymaking for IoHE in Selected Asian Countries (with Special Focus on India)

The following sub-sections highlight the policies of selected Asian countries, namely, China, Japan, Singapore and Malaysia, with a special focus on India. Since the policy tenets in the Indian IoHE policy evolve from ancient socio-cultural practices of the country, a detailed explanation for each such concept is also provided alongside.

(i) **Indian Policy for IoHE**

India, with the world's largest population, a median age of 28 years, and a negative net migration rate (Worldometers.info, 2024), reflects the preferences of its predominantly young population to settle abroad. While considering the contributions made to the global economy, India is set to become the world's third-largest economy in the next five years and a developed country by 2047 (World Economic Forum, 2024). Given these facts and the futuristic development goals, India has revamped its educational system by implementing the National Education Policy (Ministry of Human Resource Development, 2020); hereafter referred to as ***'NEP 2020'*** in the chapter.

NEP 2020 states guidelines and framework for internationalisation of Higher Education Institutions (HEIs) by aiming to: (a) restore the country's role as the Vishwa Guru (Ministry of Human Resource Development, 2020), (b) advance the United Nation Sustainable Development Goal 4 (SDG 4), target 4.7 by imparting Global Citizenship Education (University Grants Commission, 2021a, 2021b) and, (c) promote brand 'India for Higher Education', the country's soft power, culture

[4] Applied or experimental research done in collaboration with industry, post conduction of an extensive analysis of market needs.

and the Indian Knowledge System (IKS). In accordance with the aforementioned broader objectives, Indian HEIs are implementing new educational frameworks that foster collaboration, improve the standards of quality international education. Subsequently, establishing the country as a favourable destination for students opting to study its rich cultural heritage, literature and natural sciences[5] comprising IKS. The NEP 2020 states an educational agenda based on the core principles and constructs ingrained in the Indian socio-cultural philosophy (National Council for Educational Research and Training, 2020).

(b) the 'Vishwa Guru' Concept and Its Restoration

Following India's prehistoric 'Guru-Shishya prampara' (Master and Disciple tradition), the Indian government aims to establish the nation as 'Vishwa Guru' (world leader) in the field of educational internationalisation. The tradition did imbibe values where education was subsidised, available to those willing to educate themselves and most importantly, was multi-disciplinary, making the disciple learn both theoretical and practical implications during their training period (The Indian School of Cultural Studies, 2021). Likewise, the nation now aims to promote itself as a global study destination providing premium education at affordable costs, restoring its 'Vishwa Guru' status (Panda, 2022).

There is a two-pronged strategy devised to achieve the planned restoration, where: (a) foreign HEIs ranking in the top 100 QS[6] or THE rankings[7] will be facilitated to open up their campuses in India, and (b) top 100 NIRF[8] ranked Indian HEIs will be encouraged to set up their campus abroad. To ease and simplify the operations of such campuses within the country, the legislative framework and guidelines are designed at par with that of the autonomous Indian HEIs (Ministry of Human Resource Development, 2020).

This strategy also fulfils the need for providing access and enabling equity among students by ensuring internationalisation at home through cross-border institutional mobility. Having globally competitive foreign HEIs establish their campus within the nation will make it emerge as a global education and research hub. These changes are believed to enable at least 10 Indian HEIs to enter the Top 200 in the world QS or THE rankings by 2030 (Akhil Bharatiya Siksha Samagam - Thematic Session, 2023b).

(iii) the 'Vasudhaiva Kutumbakam' Concept and Global Citizenship

---

[5] These include sciences like Naturopathy, Yoga, knowledge of Indian scriptures, Vedas, languages like Sanskrit, Prakrut, etc.

[6] The QS World University Rankings are comparative college and university rankings compiled by a higher education analytics firm called Quacquarelli Symonds (QS World University Rankings, 2024).

[7] The Times Higher Education World University Rankings also known as the THE Rankings are an annual publication of university rankings by the magazine named Times Higher Education (Times Higher Education, 2024).

[8] National Institutional Ranking Framework is a ranking methodology adopted by the Ministry of Education, Government of India, to rank institutions of higher education in India (National Institutional Ranking Framework, 2023).

Indian socio-cultural belief system places dialogue and democratic principles at its core. It considers the world as one family (translation of 'Vasudhaiva Kutumbakam') and that is where it aims to connect every citizen with their responsibility towards one planet—the earth (Ministry of External Affairs, Government of India, 2023). This very idea of sensitising upcoming generations is also made part of the agenda for internationalisation by introducing the concept of Global Citizenship Education (GCED). NEP 2020 foresees GCED as a response to global challenges faced by an increasingly interconnected world. It believes that GCED will empower learners while also making them active promoters through learning innovative curricula designed in the areas of community engagement, environmental and value-based education (University Grants Commission, 2021a, 2021b).

Furthermore, it is an effort of the Indian government to give a push to the implementation of the United Nation's SDG 4 (Target 7) which calls for the nations to ensure that all learners are provided with the knowledge and skills to promote sustainable development (United Nations Department of Economic & Social Affairs, 2023a, 2023b). To do so, SDG 4 suggests providing education for sustainable lifestyles, human rights, gender equality, promotion of peace and non-violence, global citizenship, appreciation of cultural diversity and cultural contributions made towards sustainable development (United Nations Department of Economic & Social Affairs, 2023a, 2023b).

Considering the requirements of SDG 4 (Target 7), the democratic citizenship in India and the nation's belief in 'Unity in Diversity'; the very genesis of implementing GCED comes out to be achieving solidarity with the distant and (probably) unknown others, i.e. the foreign nationals. Fulfilling the same, NEP 2020 has mandated the embedding of GCED in the form of a spiral curriculum[9] that also conforms to the qualification or level descriptors[10] of the student (University Grants Commission, 2021a, 2021b).

The overall aim of providing GCED is to equip the learner with the required knowledge on global issues by inculcating their cognitive and non-cognitive skills for recognising, understanding, familiarising and making responsible decisions about such issues. Since doing so in a seamless manner, inevitably depended upon educational policies and systems, curricula prioritisation and availability of resources, it has been now implemented at all educational levels through NEP 2020.

(d) Brand 'India for Higher Education', Soft Power and IKS

Besides internationalising education at home through cross-border institutional mobility, India is focusing on establishing the nation as a brand recognised for its quality higher education. The two most vital and well-recognised strengths being leveraged here include the soft-skilled workforce and the Indian Knowledge System (IKS).

---

[9] A curriculum designed with key concepts repeating throughout, but the complexity of the conceptual layers or its application deepens with each repetition.

[10] The qualification or level descriptors are defined by the National Education Qualification Framework (NHEQF) in India.

Beyond setting up Indian HEIs abroad, India is highlighting the domains where students from other nationalities may opt to pursue higher studies here. The reasons are:

a. the comparative advantage of low fees and cost of living,
b. its established hold on natural sciences like naturopathy, ayurveda, yogic practices, religious scriptures, cultural ethos, art and architectural courses, textile designing, sustainable practices, etc.
c. development of Massive Open Online Courses (MOOCs) under the Study Web of Active Young Aspiring Minds (SWAYAM) initiative to attract enrolment of foreign students through the online learning platform.

The emigration of Indian students, on the other hand, is largely based on their acquired soft skills, the government has sought to extend soft power relationships with foreign countries. Furthermore, the Ministry of Human Resources Development in India plans to advance core research in the areas of Artificial Intelligence by deploying application-based research to address global challenges in the areas of healthcare, agriculture and climate change (Ministry of Human Resource Development, 2020). All these efforts require collaborative research conducted among HEIs across nations, furthering the requirement of faculty and resource exchanges, let alone student mobility.

12.2.4 Policies of Selected Asian Countries (China, Japan, Malaysia and Singapore)

All four of the selected nations, i.e. China, Japan, Malaysia and Singapore, are facing a shortage of skilled labour due to their fast-aging economies. Furthermore, these nations rely on their industrial capacities and require a workforce to sustain their economy. Doing so with an aging population also means keeping their citizens healthier for a longer span of their lives (Worldometers.info, 2024). Therefore, the majority of their budgetary expenses are also directed towards science, industrial innovations or higher education initiatives, producing self-sustaining models or an efficient workforce.

The IoHE in the given circumstance provides them with an opportunity for the inflow of skilled human resources. That is why China and Japan's efforts in IoHE policymaking are directed at developing infrastructural facilities, attracting more foreign students (Council for the Creation of Future Education, 2023; Yuan, 2024). While Malaysia and Singapore, being smaller business-oriented economies, are aiming to sustain their skilled workforce by the provision of a healthier and more supportive ecosystem (Asia-Europe Foundation, & National Educational Opportunities Network, 2021; Ministry of Education, Malaysia, 2013; Asian Development Bank Institute, 2012; Asia Europe Foundation).

The documents focused upon for writing the chapter contents comprised educational policy frameworks for the selected countries that include 'National Education

Policy, 2020' (Ministry of Human Resource Development, 2020), 'Dual Priority Agenda (DPA)' (Yuan, 2024), 'Double First Class Construction policy' (Liu et al., 2019), 'National Higher Education Equity Policy' (Asia-Europe Foundation, & National Educational Opportunities Network, 2021), 'Global 30 project' (Yonezawa, 2023) and 'Japan - Mobility and Internationalisation: Re-engaging and Accelerating Initiative (J-MIRAI)' (Council for the Creation of Future Education, 2023) and 'Malaysia Education Blueprint (MEB) 2015–2025' (Ministry of Education, Malaysia, 2013).

The policy frameworks served as a guide to search for related notifications, guidelines for HEIs (University Grants Commission, 2022a, 2022b), educational frameworks like that of Global Citizenship Education (University Grants Commission, 2021a, 2021b), Educational Internationalisation (University Grants Commission, 2021a, 2021b; Altbach & Knight, 2007) and Global trends (Fergnani, 2019; Knight, 2006). For synthesising (Jaakkola, 2020) and categorising, the author relied upon the works of Saldana J., suggesting coding strategies for qualitative content (Saldana, 2016). Studies discussing review strategies work related to mapping (Grant & Booth, 2009) and scoping reviews (Zachary et al., 2018) have been used to categorise the data related to making the NEP 2020 (Ministry of Human Resource Development, 2020).

12.3 Methodology

The conceptual chapter results from integrating multiple theoretical perspectives and policies (Jaakkola, 2020) that have been implemented in Indian and foreign contexts to define the present framework of internationalisation. The chapter includes relevant examples, theoretical constructs or policies to explain the background, structure and outcomes of the kind of international collaborative arrangements adopted by Indian HEIs. Through the synthesis of various research works and policies, the author attempts to create a diagrammatic representation interconnecting the diverse arrangements (explained in the preceding sections), leading to the maximisation of the goal of IoHE (specifically in India). To understand and compare the same with the internationalisation policies implemented in higher education at the Asian neighbouring countries, namely, China, Japan, Malaysia and Singapore, selected policies and initiatives have been analysed.

12.3.1 Theoretical Background

The findings of the chapter arise from the synthesis of Indian educational policies (Ministry of Human Resource Development, 2020), their framework (Higher & Technical Education Department Government of Maharashtra, 2022), University Grants Commission (UGC's) guidelines for implementing and regulating the framework

(University Grants Commission, 2021a, 2021b), norms and regulations implemented by HEIs (University Grants Commission, 2021a, 2021b), discussions at various educational consortiums with 'internationalisation of education' as the theme, regulatory notifications issued by statutory bodies (Akhil Bharatiya Siksha Samagam, 2023a, 2023b; The Indian School of Cultural Studies, 2021), contents of collaborative agreements between Indian and foreign HEIs (University Grants Commission, 2021a, 2021b) (University Grants Commission, 2022a, 2022b) and relevant research articles (Altbach & Knight, 2007; Knight, 2006; Soderqvist, 2002; Varghese, 2020) addressing the issue.

For understanding and comparing the Indian policy for IoHE, Dual Priority Agenda (DPA) (Yuan, 2024), Double First Class Construction policy (Liu et al., 2019) and, exchange arrangements (Cooperation & Exchanges, 2009) of China; National Higher Education Equity Policy (Asia-Europe Foundation, & National Educational Opportunities Network, 2021) and Global Schoolhouse initiative (Asian Development Bank Institute, 2012) of Singapore; Global 30 project (Yonezawa, 2023) and Japan - Mobility and Internationalisation: Re-engaging and Accelerating Initiative (J-MIRAI) (Council for the Creation of Future Education, 2023) of Japan and; Malaysia Education Blueprint (MEB) 2015–2025 (Ministry of Educaiton, Malaysia, 2013) of Malaysia have been studied.

12.3.2 Method and Theorisation

The findings result from thematic analyses of identified literature using deductive reasoning. The integration of the findings suggests that the 'internationalisation of higher education' has served as the conceptual focus, linking various theories, policymaking and practices in both Indian and global contexts. Since a majority of the examples have an Indian HEI as one of the partnering institutes in the collaborative framework (or designed course structure), the detailed perspective tends to be predominantly Indian. The author has used data attributes, namely, the title, contents, file structure, issuing agency or journal of publication, year of publication and year of effect (Saldana, 2016) for initial understanding of the collected data from various secondary sources. It was then followed by scoping and mapping reviews for categorisation and preliminary assessment of the scope of the qualitative data (Grant & Booth, 2009), respectively.

After the categorisation of data, a qualitative systematic synthesis has been carried out to arrive at the concepts addressed in the findings. The synthesis highlighted inter-connections between international course designs and the reasons for internationalisation of higher education in India, which have been proposed in a circular model (refer to Fig. 12.1). Each circle is representative of the level of implementation internationalisation aspect, leading to its consecutive level. The model specifically focuses upon the internationalisation of higher education in India, but can be used for studying layers of internationalisation of education in another country.

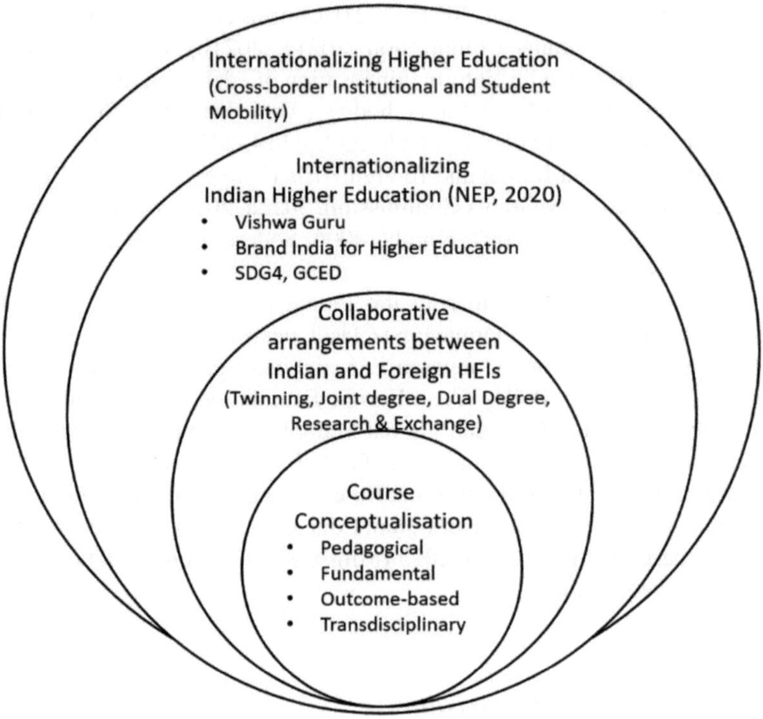

Fig. 12.1 Overlap between policy framework and course conceptualisation approach

Similarly, to draw comparisons, 'internationalisation of higher education' was used to categorise key higher educational policy initiatives for neighbouring Asian countries, namely, China, Japan, Singapore and Malaysia. The comparisons have been across four parameters (a) policy and its framework, (b) factors leading to policy adoption, (c) approach for its implementation and (d) suggested collaborative options provided by the nation (refer to Table 12.1).

12.4 Findings

Since the chapter focuses on understanding the layers of India's vision for internationalisation of higher education, followed by comparisons with selected neighbouring countries, the first set of findings relates to Indian educational policy. The subsequent section underlines the similarities and differences in the internationalisation approach in the context of India's selected neighbouring countries.

Table 12.1 Comparison of selected Asian IoHE policies based on adoption factors, implementation approach and collaborative options

Country	Policy or initiative	Factors for adoption	Implementation approach	Collaborative options suggested
India	Ministry of Human Resource Development, 2020)[a]	Indian ancient approach to Teaching Learning practices, Indian Knowledge System, soft power, SDG 4	Multi-faceted inside-out approach, brand 'INDIA' for IoHE, At home and cross-border internationalisation	Pedagogical advancements, collaborative approaches for HEIs to partner
China	Dual Priority Agenda[b]; Double First Class University[c]	National development, ideology development, need for developing skilled human resource	Educational development, at home internationalisation, outside-in approach	Infrastructural developments, curricula changes in disciplines (agriculture, industry, defence, science and technology) highlighting Chinese culture
Japan	Global 30[d] universities; J-MARAI[e]	Meritocracy, improve inbound student mobility rate (after COVID-19),	Targets for student mobility, dual degree and twinning programmes established for 2033, improved rate of employment	Improved infrastructure, in-person and online interactions, vocational training, industry tie-up for reviews and suggestions
Malaysia	MEB 2015-2025[f] (Higher Education)	Quality of education, Longevity of learning, Innovation systems, Transformation of delivery mechanisms	Ten-year value-driven plan with internationalisation as a by-product of other goals	Ministry and agency collaboration, marketing education systems, increasing post-graduate students

(continued)

Table 12.1 (continued)

Country	Policy or initiative	Factors for adoption	Implementation approach	Collaborative options suggested
Singapore	National Higher Education Equity Policy[g]; Global Schoolhouse Initiative[h]	Global reputation, educational excellence, trade, establishment as a knowledge-based economy	Mobility programmes, broadened and in-depth educational offerings	Pedagogical changes to foster life-long learning, increased vocational training courses, student & teacher mobility programmes

[a]Ministry of Human Resource Development. (2020). National Education Policy 2020. Delhi: Government of India
[b]https://www.worldscientific.com/doi/epdf/https://doi.org/10.1142/9789811246647_0001
[c]https://doi.org/10.1163/25902547-00101009
[d]https://doi.org/10.1111/aepr.12421
[e]https://www.cas.go.jp/jp/seisaku/kyouikumirai/pdf/230427outline.pdf
[f]https://www.um.edu.my/docs/um-magazine/4-executive-summary-pppm-2015-2025.pdf
[g]Asia-Europe Foundation, & National Educational Opportunities Network (2021). *National Higher Education Equity Framework.* Country Brief, Singapore
[h]https://www.adb.org/sites/default/files/publication/156243/adbi-wp388.pdf

12.4.1 Collaborative Provisions

Combining these concepts that summarise the needs and implementation strategies for internationalising higher education, India (specifically) has adopted the following five approaches to fostering collaborations between Indian and foreign HEIs. A total of 235 HEIs have been declared eligible for offering the below-mentioned provisions of collaborations by the UGC in India (University Grants Commission, 2022a, 2022b).

Twinning Programme

Under the provisions of the twinning programme, the Indian HEI provides the degree to a student enrolled primarily at the Indian HEI but has completed part of his degree (less than 30 per cent of the total credits of the programme) by enrolling at a foreign HEI. This foreign enrolment may be for completion of a specific theoretical course, an internship-based course, a research-based course or a specific semester-long or year-long programme (University Grants Commission, 2022a, 2022b).

Joint Degree Programme

Under the provisions of the joint degree programme, both the Indian and foreign HEIs will design a curriculum for a degree programme that will result in the award of a single degree jointly provided by the partnering HEIs. The enrolled student will be required to mandatorily earn 30 per cent of the total credits of the programme from each of the partnering HEIs by studying in a conventional (offline) mode (University Grants Commission, 2022a, 2022b).

Dual Degree Programme

Similar to the provisions of a joint degree programme, the dual degree programme differs in its way of execution. The enrolled student will have to fulfil the requirements of both the partnering HEIs under this arrangement. It means that the student will produce two transcripts, two theses or dissertations as per the requirements of each HEI, where each of the HEIs will also be awarding a separate degree to the student for the same level of qualification. There is no combined curriculum or examination for this arrangement (University Grants Commission, 2022a, 2022b).

Joint Research Collaborations

Where the earlier three programmes focus on student mobility, joint research collaborations also provide for faculty and research resource exchanges. These may involve working on industry, geography, language, project or discipline-specific approaches in terms of designing the initiative. There are various centres of excellence (CoE) also being established at Indian HEIs that take into consideration the funds and resources specific allocation for executing such initiatives (Akhil Bharatiya Siksha Samagam, 2023a, 2023b). Students from partnering HEIs collaborate with their mentors and principal investigators to deliver results.

Student Exchanges

There are short-duration initiatives too launched for having HEIs build stronger ties before entering degree collaborations. Eligible Indian HEIs can plan out immersion courses, research-oriented courses, semester exchange courses, internship-based courses and cultural or skill development courses in the areas of their expertise to either invite foreign nationals or send their students abroad. Such an approach prevents cultural shock and also provides students with exposure to a diverse cultural setting. These programmes are often utilised by HEIs to explore a foreign land and foresee the feasibility of entering degree-based collaborations.

12.4.2 Course Conceptualisation Approach

Since the IoHE in India stems from its philosophical ideologies being followed through generations, it is pertinent to note that the country doesn't identify education as just another industry sector. That is where India has significantly denied commoditising educational services as traded under the General Agreement on Trade in Services (GATS) framework.[11] The country has rather attempted to rightfully place the educational system on five foundational pillars of access, equity, quality, affordability and accountability while also proposing a 'light but tight' (Akhil Bharatiya Siksha Samagam, 2023a, 2023b) flexible framework for designing innovative courses under NEP 2020 (University Grants Commission, 2021a, 2021b).

After the fundamental guidelines come the propositions made for the conceptualisation of newer courses. These changes are based on the following four approaches opted for revamping the course curricula:

(a) Pedagogical approach that guides the mechanism adopted to deliver course content.
(b) Fundamental approach that includes the credit structure of the course design.
(c) Outcome-based approach is based on the expected outcomes of the course.
(d) The Transdisciplinary approach is based on integrating course contents in a manner that is discipline-agnostic and can transcend national boundaries.

Pedagogical Approach

The pedagogical approach towards internationalising courses or degree programmes focuses on changing the teaching and learning method to deliver course content effectively. Since international students have different cultural backgrounds, catering to their discipline-specific interests is paramount. In the given scenario, NEP 2020 professes to implement innovative teaching mechanisms like blended learning, flipped classrooms, experiential learning, field visits, hands-on training, case study methods, group discussions and experimental setups to allow increased interaction among students of foreign nationalities. It is also devoid of routine classroom set-up and can be easily implemented while designing courses, regardless of the collaborative arrangement opted by two partnering HEIs.

Many national universities in Malaysia, Singapore and Taiwan are leveraging their laboratory facilities by curating semester-long student exchange programmes where Indian students can have hands-on experience using advanced infrastructure, especially in the popular areas of STEM disciplines like advanced robotics, artificial intelligence, remote sensing, 3D designs and machine learning capabilities (e.g.

[11] In the context of globalisation, education is a commodity being traded under GATS framework, where the university is a provider, educator is an investor, students are the clients or customers in the market mediated framework of trade. The driving forces enabling capital flow in education are the economic rationality and commercial interests of trading nations. To serve the motivations of participating stakeholders, student mobility or institutional mobility stand relevant thereby also continuing the trade (Varghese, 2020).

National Formosa University, Taiwan; Nanyang Technological University, Singapore; University of MARA, Malaysia; just to name a few). Whereas, US and UK-based universities are experimenting with the pedagogical approach in their Master of Business Administration (MBA) programmes by inviting students to experience their labs that are partly funded or are in collaboration with an industrial corporation (e.g. American University, Stanford University, Columbia University at USA; Bournemouth University, University of Hertfordshire in UK).

Another way of internationalising education, especially for students enrolled in creative non-STEM disciplines like social sciences, is to enrich their social and cultural exposure. Several Indian universities are considering offering short-duration courses in disciplines like sociology, art & design, natural sciences, literature studies, anthropology, tourism management, international relations and policymaking, where students can travel to India to learn about the diaspora and conduct a research enquiry. Since this conceptualisation includes travel, VISA and accommodation expenses of the enrolled student, the collaborating HEIs must mutually plan and abide by a cost structure. Implementing such programmes helps foster GCED.

Fundamental Approach

The fundamental approach is directed at changing the structural framework of courses. Traditionally in India, the degree programmes have been segregated into STEM and non-STEM disciplines. To apply discipline agnosticism and do away with traditional boundaries, the NEP 2020 suggested implementing a spiral curriculum while also adding courses serving a diverse purpose. Ensuring the necessary quality and promise of premium education, the NEP 2020 mandated the requirement of skill-based, ability enhancement, field-based, value-based, co-curricular, open-elective, major and minor-track based courses for every degree programme being offered by an Indian HEI (Higher and Technical Education Department, Government of Maharashtra, 2022).

This mandate necessitated redesigning each course embedded in any degree programme, bringing gross structural changes in the degree programme (Ministry of Human Resource Development, 2020). It also enriched the programmes by adding components covering holistic education along with GCED. These changes were not just limited to the classification of courses but also affected the credit distribution among each category. Another implication of implementing the fundamental approach is that it has modernised educational delivery and outlook of various traditional courses being delivered by Indian HEIs. It will help the nation leverage and carve a niche, advancing the brand 'India for Higher Education' among foreign HEIs.

Outcome-based Approach

The outcome-based approach is about the result regardless of the course structure, discipline or pedagogy adopted to deliver the contents. This approach is best utilised for creating professional career-advancing courses like short-term certifications or diplomas where obtaining proficiency in imparted skills is the prime objective. Nations aiming to advance specific industry-oriented educational programmes may consider this approach, e.g. Taiwan may have courses in Mechatronic Engineering,

Cosmetology Studies, Bio-Industry Communication and Development; while India experiencing a boom in the IT sector has already launched courses in Data Analytics, Computer Engineering, Electronics & Communications.

Furthermore, it is not a stand-alone approach because it is linked to the changes in pedagogy and course structure. Therefore, despite utilising only a pedagogical approach to deliver a course or only the structural approach to design a course, one may experience results similar to that of applying an outcome-based approach. It is simply because pedagogy and structure inherently change the outcome of whatever course contents or design have been.

Therefore, with Indian HEIs implementing NEP 2020 at such a massive scale there is already a substantial change expected in the outcomes of its educational programmes. A cumulative effect of that is expected to help achieve its goals of internationalisation by 2047.

Transdisciplinary Approach

A curricula-designing approach that befits practical situations the most is the transdisciplinary approach. With AI invading an already globalised world, the international adaptability of academic courses is being questioned. Most courses are homogenising information in the name of introducing transdisciplinary courses. HEIs, in the present circumstance, are aiming to apply a transdisciplinary approach, integrating it with their country's national competencies.

Doing so means customising the knowledge by:

(a) adopting the fundamental approach to change course structures towards discipline agnosticism, which in turn will change course outcomes.
(b) delivering the transdisciplinary content through industrial projects or examples relevant to the collaborating nations. This approach inherently follows pragmatism. It is not about inter-relating disciplines or seeking information from specialised sources but unifying knowledge and building a holistic programme in response to global issues and challenges.

12.4.3 Contrasting India's IoHE Policy with Select Asian Countries (China, Japan, Malaysia and Singapore)

The comparisons show that internationalisation of a higher education is a function deeply embedded in (a) need for sustained economic growth, (b) migratory patterns of population, (c) sustainable long-term infrastructural developments and, often times, (d) reinforcement of a country's cultural tradition through its global presence. Upon comparison of selected ongoing educational initiatives fostering implementation of goals of internationalising higher education, it was found that every nation because of its varied demographics and psychographics, has its own blueprint for the next 10–15 years.

To begin with the selected nations, China has established a Dual Priority Agenda (Yuan, 2024) where it aims to achieve educational development through national development by reinforcing the ideology. It focuses on reducing the rate of migration to strengthen its economy by establishing Double First-Class universities (Liu et al., 2019), attracting foreign nationals to study at home. In comparison to India's approach, it is an outside-in policy where talent from outside is being invited to develop the nation.

Whereas, Japan, on the other hand, aimed to reclaim its status as the highest quality educational services by establishing Global 30 (Yonezawa, 2023) universities providing experiential learning to international students. Impacted by COVID-19, the initiative underwent changes and brought J-MIRAI (Council for the Creation of Future Education, 2023) to implementation by setting goals of inbound and outbound student mobility, initiation of dual and joint degree programmes by 2033. It also focuses on vocational studies to fulfil the futuristic requirements of the next generations.

With initiatives like SkillsFuture that also focus on Work Force Qualifications, Singapore is an exemplary nation expanding the tenure of learning to include lifelong purposes. Having autonomous and private institutes, the Singaporean government also lays additional emphasis on adult learning programmes, taking its educational system to create a Global Schoolhouse—inviting students both physically and through hybrid modes of education.

Malaysia too, reflects a similar holistic plan for revamping its higher educational sector, with internationalisation being one of the ten goals to be achieved through the implementation of its blueprint (Ministry of Education Malaysia, 2013) by 2025. It does not give additional emphasis on the mobility of students but considers it as a by-product of improved educational services and provisions.

12.5 Implications of the Study and Future Directions

As implications of the findings, the author proposes the following Venn diagram interlinking Indian philosophical values, its educational policy and resulting curriculum conceptualisation. The model represents the structural layering of a value-laden system facilitated by NEP 2020 that has led to four-course conceptualisation approaches for internationalising higher educational content delivery.

Figure 12.1 Venn diagram showing the overlap between policy framework and course conceptualisation approach

The courses being designed by Indian HEIs to meet the objectives of the internationalisation of higher education can serve as a roadmap for countries looking to attract foreign students, and vice versa. Furthermore, the flexibility offered by NEP 2020 implementation guidelines has been effectively utilised by Indian HEIs to revamp their existing curricula.

The comparisons drawn based on approaches used by the selected countries (China, Japan, Singapore and Malaysia) are mentioned in the table below.

Indian HEIs have currently launched several programmes inviting foreign students to study globally recognised IKS courses. At the same time, Indian HEIs are being encouraged to restructure all degree programmes by introducing pedagogical, structural and outcome-based changes, which will enhance the transdisciplinary content in the courses offered under the proposed degrees. Meanwhile, Singapore and Malaysia's emphasis on adult learning and vocational training, respectively, could lead to the development of post-university (or HEI) study programmes under IoHE initiatives. Additionally, the proactive efforts of Indian HEIs to secure positions among the Top 100 in QS or THE rankings are helping to establish the 'Brand India' perception globally, particularly in areas such as soft power.

Japan is developing its IoHE initiatives around a well-established scientific community, fostering innovations in cutting-edge sciences. With IITs and IIMs expanding their presence abroad—such as IIT Madras opening a campus in Zanzibar, Tanzania and IIT Delhi launching a campus in Abu Dhabi, UAE (Akhil Bharatiya Siksha Samagam, 2023a, 2023b)—India is poised to position itself as an educational hub focused on internationalised education and global citizenship. Meanwhile, China, with its well-developed infrastructure, is emerging as a competitive market attracting international students. IoHE in Asia is expected to reduce the perceived distance between nations, providing accessible and affordable foreign education to agile and astute student populations.

12.6 Conclusion

The study examines the varied approaches to the internationalisation of higher education (IoHE) across several Asian countries, with a particular emphasis on India's recent National Educational Policy (NEP 2020). By comparing the IoHE policies of India with those of China, Japan, Malaysia and Singapore, the research highlights both shared frameworks and distinctive strategies for implementation in each country. The findings underscore the crucial role of policy and regulatory frameworks in influencing the future direction of higher education, particularly in fostering collaborative degree programmes and promoting institutional mobility.

The study also stresses the importance of understanding the unique socio-political and educational contexts within each country, as it reveals that applying IoHE strategies without considering these local factors could limit their effectiveness. Ultimately, the research advocates for a more nuanced and adaptable approach to IoHE, taking into account the specific needs and challenges of each national setting to ensure the successful global integration of higher education systems.

Highlights

- India has adopted various collaborative programmes such as Twinning, Joint Degree and Dual Degree programmes, along with Joint Research Collaborations and Student Exchanges, to strengthen partnerships with foreign

higher education institutions (HEIs) and enhance the internationalisation of its education system.
- The National Educational Policy (NEP) 2020 introduces innovative course conceptualisation approaches in India, including pedagogical, fundamental, outcome-based and transdisciplinary methods, to cater to both national and international student populations while fostering global citizenship and knowledge integration.
- A comparison of India's internationalisation strategies with neighbouring Asian countries like China, Japan, Singapore and Malaysia reveals both shared and distinctive approaches to enhancing higher education, with each nation focusing on leveraging its unique strengths for global educational integration.
- India's higher education institutions, through initiatives like joint campuses abroad and the expansion of globally recognised programmes, aim to strengthen the 'Brand India' perception and position the country as a key player in the internationalisation of higher education, with a focus on innovation and global collaboration.

References

Akhil Bharatiya Siksha Samagam. (2023b). *Thematic session—Internationalization of education.* New Delhi, Delhi, India.

Akhil Bharatiya Siksha Samagam. (2023a). *Internationalization of education* [Thematic session]. New Delhi, Delhi, India.

Altbach, P. G., & Knight, J. (2007). The internationalization of higher education: Motivations and realities. *Journal of Studies in International Education, 11*(3–4), 290–305. https://doi.org/10.1177/1028315307303542

Arum, S., & Van de Water, J. (1992). The need for a definition of international education in US universities. In C. Klasek (Ed.), *Bridges to the future: Strategies for internationalizing higher education* (pp. 191–203). Association of International Education Administrators.

Asia-Europe Foundation, & National Educational Opportunities Network. (2021). *National higher education equity framework: Country brief, Singapore.* https://asef.org/wp-content/uploads/2023/04/SINGAPORE_HE-Equity-Policy_2021.pdf

Asian Development Bank Institute. (2012). *Internationalization of tertiary education services in Singapore* (ADBI Working Paper No. 388). https://www.adb.org/sites/default/files/publication/156243/adbi-wp388.pdf

Bernama. (2023, October 13). Budget 2024: Higher education receives RM 16.3b for comprehensive benefit, say educators, students. *MalayMail.* https://malaymail.com/news/2023/10/14/budget-2024-higher-education-sector-receives-rm163b-for-comprehensive-benefit-say-educators-students/96276

China Daily. (2024, May). China spending more on education, ministry says. *China Daily.* https://www.chinadailyhk.com/hk/article/588794#china-spending-more-on-education%2C-ministry-says-2024-07-23

Cooperation and exchanges. (2009, September). *Ministry of Education, People's Republic of China.* http://en.moe.gov.cn/cooperation_exchanges/201506/t20150626_191371.html

Council for the Creation of Future Education. (2023, April 27). J-MIRAI. *Cabinet Secretariat, Japan.* https://www.cas.go.jp/jp/seisaku/kyouikumirai/pdf/230427outline.pdf
Estrada, K. S. (2023). What is a Vishwaguru? Indian civilizational pedagogy as a transformative global imperative. *International Affairs, 99*(2), 433–455. https://doi.org/10.1093/ia/iiac318
Fergnani, A. (2019). Mapping futures studies scholarship from 1968 to present: A bibliometric review of thematic clusters, research trends, and research gaps. *Futures, 104*, 203–220. https://doi.org/10.1016/j.futures.2018.11.007
FY2024 expenditure estimates. (2024). *Ministry of Education, Singapore.* https://www.mof.gov.sg/docs/librariesprovider3/budget2024/downoad/pdf/27-moe-2024.pdf
Grant, M. J., & Booth, A. (2009). A typology of reviews: An analysis of 14 review types and associated methodologies. *Health Information and Libraries Journal, 26*(2), 91–108. https://doi.org/10.1111/j.1471-1842.2009.00848.x
Higher and Technical Education Department Government of Maharashtra. (2022, January 2). *NEP implementation.* https://htedu.maharashtra.gov.in/Main/KnowledgeSharingDownload
International Trade Administration. (2023, May 9). *India budget 2023.* https://www.trade.gov/market-intelligence/india-education-budget-2023
Jaakkola, E. (2020). Designing conceptual articles: Four approaches. *AMS Review, 10*(1), 18–26. https://doi.org/10.1007/s13162-020-00168-1
Knight, J. (2004). Internationalization remodeled: Definitions, rationales, and approaches. *Journal of Studies in International Education, 8*(1), 27–37. https://doi.org/10.1177/1028315303260832
Knight, J. (2006). *Higher education crossing borders: A guide to the implications of the GATS for cross-border education.* UNESCO/Commonwealth of Learning.
Liu, Q., Turner, D., & Jing, X. (2019). The "Double First-Class Initiative" in China: Background, implementation, and potential problems. *Beijing International Review of Education, 1*(1), 92–108. https://doi.org/10.1163/25902547-00101009
Ministry of Education, Malaysia. (2013). *Malaysia education blueprint 2015–2025 (higher education).* Ministry of Education. https://www.um.edu.my/docs/um-magazine/4-executive-summary-pppm-2015-2025.pdf
Ministry of External Affairs, Government of India. (2023, September 9). *Media center.* https://www.mea.gov.in/Speeches-Statements.htm?dtl/37083/English+translation+of++Prime+Ministers+remarks+at+the+G20+Summit+Session+1/
Ministry of Human Resource Development. (2020). *National education policy 2020.* Government of India. https://www.education.gov.in/sites/upload_files/mhrd/files/NEP_Final_English_0.pdf
National Council for Educational Research and Training. (2020). *NEP 2020 equitable and inclusion: Learning for all.* Government of India, Ministry of Education. Retrieved November 16, 2024, from https://www.education.gov.in/shikshakparv/docs/Inclusive_Education.pdf
National Institutional Ranking Framework. (2023, May). *NIRF rankings.* https://www.nirfindia.org/Home
Panda, P. K. (2022). Viswaguru Bharat: A beacon of ancient wisdom for modern corporate. *International Journal of Sanskrit Research, 8*(5), 230–234.
QS World University Rankings. (2024, May). *University rankings.* https://www.topuniversities.com/university-rankings
Saldana, J. (2016). *The coding manual for qualitative researchers* (3rd ed.). Sage Publications Ltd.
Soderqvist, M. (2002). *Internationalization and its management at higher education institutions: Applying conceptual, content, and discourse analysis.* Helsinki School of Economics, HeSE Print.
Statista.com. (2024, November). *Education and sciences expenditures of the government's general account budget in Japan from fiscal year 2015 to 2022 with forecasts until 2024.* https://www.statista.com/statistics/630230/japan-general-account-budget-education-science-expenditures/
The Indian School of Cultural Studies. (2021, July 7). *Guru-shishya parampara: An introduction to the ancient Indian education system.* Pratha: The Indian School of Cultural Studies. https://www.prathaculturalschool.com/post/the-guru-shishya-parampara

Times Higher Education. (2024, May). *World university rankings*. https://www.timeshighereducation.com/world-university-rankings

United Nations Department of Economic and Social Affairs. (2023a). *Ensure inclusive and equitable quality education and promote lifelong learning opportunities for all*. https://sdgs.un.org/goals/goal4#targets_and_indicators

United Nations Department of Economic and Social Affairs. (2023b). *The 17 goals*. https://sdgs.un.org/goals

University Grants Commission. (2021a). *Educational framework for global citizenship in higher education*. University Grants Commission.

University Grants Commission. (2021b). *Guidelines for internationalization of education*. University Grants Commission.

University Grants Commission. (2022a). *Key initiatives of UGC*. https://ugc.gov.in/KeyInitiatives

University Grants Commission. (2022b). *University Grants Commission (Academic collaboration between Indian and foreign higher educational institutions to offer twinning, joint degree, and dual degree programs) regulations, 2022*. The Gazette of India: Extraordinary.

Varghese, N. (2020). Internationalisation of higher education: Global trends and Indian initiatives. In V. N. P. Mittal, & P. SRD (Eds.), *Reimagining Indian universities* (pp. 1–22). Association of Indian Universities.

World Economic Forum. (2024, January 15). *India could become the world's 3rd largest economy in the next 5 years. Here's how*. https://www.weforum.org/agenda/2024/01/how-india-can-seize-its-moment-to-become-the-world-s-largest-economy/

Worldometers.info. (2024, May 30). *Countries in the world by population (2024)*. https://www.worldometers.info/world-population/population-by-country/

Yonezawa, A. (2023, February 19). Japan's higher education policies under global challenges. *Asian Economic Policy Review*. https://doi.org/10.1111/aepr.12421

Yuan, Z. (2024). Dual priority agenda: China's model for modernizing education. In S. C. Zhenguo Yuan (Ed.), *China's education policy review (2018–2021)* (vol. 8, pp. 1–34). World Scientific Publishing Company. https://doi.org/10.1142/9789811246647_0001

Zachary, M., Miccah, D., Stern, C., Tufanaru, C., McArthur, A., & Aromataris, E. (2018). Systematic review or scoping review? Guidance for authors when choosing between a systematic or a scoping review approach. *BMC Medical Research Methodology, 18*, 2–7. https://doi.org/10.1186/s12874-018-0581-1

Prachi D. is a communication researcher, formerly Assistant Professor with the Faculty of Media and Communication, Vishwakarma University, Pune, India. Acquiring seven years of academic experience, along with two years as Deputy Director, International Relations, she completed her Ph. D. in Mass Communication from Guru Gobind Singh Indraprastha University, Delhi. Her doctoral research was funded under the junior and later, senior research fellowship programs of the University Grants Commission, India. Besides that, she has scripted and presented numerous national broadcasts working as a program presenter with All India Radio. Known for her strategic communications, she has served the communications office of a premier law firm and also at the Indian Oil Corporation Limited. A consistent gold medal awardee throughout her studies, she is presently pursuing research in integrated communication technologies, over-the-top platforms, international education and, digital communications.

Conclusion

This back volume serves as a critical conclusion to *Future of Education in Asia: Sustainability, Technology and Internationalisation*, synthesising the key themes and providing insight into emerging educational trends that will shape the region's trajectory. It reflects on how Asia is not only responding to change but actively redefining the future of education through local innovation and global engagement.

Future of Education in Asia: Sustainability, Technology and Internationalisation offer a bold and comprehensive exploration of how Asian education systems are not merely adapting to global transformations but actively shaping the contours of the future. Drawing from diverse national contexts and cross-border comparisons, this volume critically interrogates the interplay between sustainability, technological advancement and internationalisation, three intersecting forces redefining the educational landscape across the region.

The chapters collectively illustrate how Asia is transitioning from being a consumer of global educational paradigms to becoming a producer of knowledge rooted in its own histories, cultural capital and innovation. Far from replicating Western frameworks, the region is generating indigenous models of reform, combining global best practices with local imperatives. Whether through India's NEP 2020, Japan's J-MIRAI or Singapore's lifelong learning ecosystems, the case studies reflect a nuanced repositioning of education as a tool not only for economic advancement but also for regional leadership and global citizenship. (**Chapter 12**: India and Select Asian Policies for Internationalisation of Higher Education).

A recurring theme across this volume is the expanding definition of internationalisation. Moving beyond traditional metrics like student mobility, institutional prestige and transnational partnerships, internationalisation is reframed here as a pedagogical and philosophical realignment. In India, this shift is embodied in efforts to integrate Indian Knowledge Systems (IKS), reconfigure curricula around transdisciplinary and outcome-based learning, and use policy instruments such as NEP 2020 to assert a distinctive global identity. In contrast, countries like China and Japan continue

A. Pandian et al. (eds.), *Future of Education in Asia*,
https://doi.org/10.1007/978-3-032-07285-6

to emphasise inbound student flows, infrastructural investments and global benchmarking as tools of educational soft power. Yet despite these differences, there is a shared discomfort with transactional models of global engagement and an emerging call for internationalisation rooted in equity, intercultural literacy and mutual knowledge production. (**Chapter 11**: Mapping the Internationalisation of Philippine Higher Education within ASEAN).

From a comparative education perspective, these diverse approaches illustrate the growing divergence in how nations conceptualise their place in the global education ecosystem. Rather than converging towards a single Westernised model, Asian countries are carving pluralistic, values-based paths that reflect unique sociopolitical contexts, historical legacies and developmental priorities. China's centralised push for educational self-reliance, Japan's recalibration towards vocational mobility and quality assurance, and Singapore's lifelong learning ecosystem all reflect tailored strategies of educational diplomacy. What unites them is not uniformity, but a shared move towards 'glocalisation', i.e. the blending of global aspirations with local meaning-making. This shift marks a deeper recognition that internationalisation, to be meaningful, must transcend competitive metrics and instead foster sustained, reciprocal partnerships built on trust, relevance and cultural resonance. (**Chapter 6**: AI's Impact on Higher Education and Academic Publishing in Asia, **Chapter 7**: Navigating the New Normal: Harnessing AI Innovations for Future-Ready Education in Malaysia, **Chapter 8**: Building the Tomorrow: Tech's Transformative Role in Asia's Built Environment Education).

Pedagogical transformation in Asia is emerging not merely through the adoption of new tools but through a rethinking of teaching and learning philosophies. Innovations such as flipped classrooms, experiential learning and outcome-driven micro-credentials, highlighted in India's NEP 2020, signalling a departure from passive, didactic modes towards more interactive, learner-centric frameworks. Singapore, through its SkillsFuture initiative, underscores lifelong learning and upskilling beyond formal education, creating pedagogical ecosystems that extend into adulthood. Similarly, Vietnam's hybrid delivery models demonstrate how blended learning is being localised to address institutional capacity gaps and student access issues. These developments align with Sustainable Development Goal 4 and the ethos of Global Citizenship Education (GCED), aiming to foster not only employability but also ethical awareness, intercultural empathy and environmental responsibility. (**Chapter 4**: Exploring Early Education for Sustainable Development: Insights from Japanese Kindergartens, **Chapter 10**: Exploring the Impact of Hybrid Learning on Student Outcomes in Vietnamese Higher Education).

However, the region's embrace of technology is neither uniform nor uncritical. Chapters from the Philippines and Malaysia reveal both the empowering and disorienting effects of AI and hybrid education models. In the Philippines, translanguaging in online mathematics learning showcases creative adaptation to local linguistic realities, while in Malaysia, rapid digitalisation has surfaced challenges related to teacher autonomy and equitable infrastructure. Rather than celebrating technology for its own sake, these cases point to a more nuanced, culturally situated engagement with digital learning, one that grapples with algorithmic bias, rural–urban divides and the need for human-centred pedagogy. In this, Asia offers not a singular narrative, but a mosaic of innovations grounded in local realities and responsive to global challenges. (**Chapter 9**: Future Directions in Filipino Education: Translanguaging Practices in Online Mathematics Learning Environments, **Chapter 3**: Teachers' Strategies for Promoting Sustainable Inclusive Practices in Heterogeneous Classrooms).

What binds these contributions is a strong undercurrent of future thinking. Educational reforms are no longer seen as merely reactive adjustments but as anticipatory strategies to address complexity, uncertainty and rapid change. The shift towards life-wide and lifelong learning ecosystems, evident in Singapore's adult learning pathways and Vietnam's flexible delivery models, points to an expanded educational imagination, one that embraces adaptability, resilience and holistic human development. (**Chapter 2**: Future of Sustainable Education in Asia: Equitable Learning through Digital and Hybrid Mobility Programmes, **Chapter 5**: Sustainable Cultural Transformations in Asia: Vietnamese Students' Intercultural Awareness following a Malaysian Exchange Programme).

Ultimately, *Future of Education in Asia* positions the region not just as a participant in global educational discourse, but as a shaper of it. It challenges the continued dominance of Western benchmarks and calls for a re-rooted cosmopolitanism, an approach to education that is both globally connected and deeply anchored in local knowledge, culture and priorities. As the geopolitical and epistemic centres of gravity shift, this volume signals that Asia's diverse educational futures are central to how the world will learn, teach and thrive in the decades ahead.

While the focus of this volume is on Asia, the discussions it inspires can also serve as a springboard for further research into educational reforms beyond the region. Countries in other parts of the Global South, such as Sub-Saharan Africa and Latin America, are experiencing similar transformations in education, often grappling with questions of sustainability, technology and internationalisation in ways that reflect their own cultural and developmental contexts. For example, Latin American nations are increasingly integrating Indigenous Knowledge Systems into education policy to preserve cultural heritage, while Sub-Saharan African countries are exploring digital tools to address infrastructural challenges.

These broader global trends offer a rich comparative landscape that invites future exploration. What makes Asia's approach distinctive, however, is the intersection of these global issues with region-specific priorities such as historical legacies, developmental stages and sociopolitical structures. This opens critical avenues for research that can bridge regional differences and foster deeper engagement with how countries in different parts of the world are adapting these broader trends to their own needs.

This book is essential reading for scholars, policymakers and educators seeking a critical, forward-looking understanding of how Asia is reimagining education for a just, sustainable and interconnected world. This, coupled with case studies with forward-thinking critique, makes the book a valuable tool for classroom discussions, research projects and policy debates.

Author Index

A. Pandian et al. (eds.), *Future of Education in Asia*,
https://doi.org/10.1007/978-3-032-07285-6

Author Index

Subject Index

A. Pandian et al. (eds.), *Future of Education in Asia*,
https://doi.org/10.1007/978-3-032-07285-6

MIX
Papier aus verantwortungsvollen Quellen
Paper from responsible sources
FSC® C105338

If you have any concerns about our products,
you can contact us on
ProductSafety@springernature.com

In case Publisher is established outside the EU,
the EU authorized representative is:
Springer Nature Customer Service Center GmbH
Europaplatz 3, 69115 Heidelberg, Germany

Printed by Libri Plureos GmbH
in Hamburg, Germany